Introduction to Analysis of the Infinite
Book II

Euler

Introduction to Analysis of the Infinite

Book II

Translated by John D. Blanton

With 49 Illustrations

Springer-Verlag
New York Berlin Heidelberg
London Paris Tokyo Hong Kong

Translator
John D. Blanton
Department of Mathematics
St. John Fisher College
Rochester, NY 14618
USA

Mathematical Subject Classification: 14H, 14J

Library of Congress Cataloging in-Publication Data
(Revised for Volume 2)
Euler, Leonard, 1707–1783
    Introduction to analysis of the infinite.
    Translation of: Introductio in analysin infinitorum.
    1. Series, Infinite—Early works to 1800.
    2. Products, Infinite—Early works to 1800.
    3. Fractions, Continued—Early works to 1800.   I. Title.
    QA295.E8413   1988   515′.243   88-18475
ISBN 0-387-96824-5 (v. 1 : alk. paper)
ISBN 0-387-97132-7 (v. 2 : alk. paper)

Printed on acid-free paper

Camera-ready copy supplied by the translator.
Printed and bound by R. R. Donnelley & Sons, Harrisonburg, Virginia.
Printed in the United States of America.

9 8 7 6 5 4 3 2 1

ISBN 0-387-97132-7 Springer-Verlag New York Berlin Heidelberg
ISBN 3-540-97132-7 Springer-Verlag Berlin Heidelberg New York

PREFACE

Often I have considered the fact that most of the difficulties which block the progress of students trying to learn analysis stem from this: that although they understand little of ordinary algebra, still they attempt this more subtle art. From this it follows not only that they remain on the fringes, but in addition they entertain strange ideas about the concept of the infinite, which they must try to use. Although analysis does not require an exhaustive knowledge of algebra, even of all the algebraic technique so far discovered, still there are topics whose consideration prepares a student for a deeper understanding. However, in the ordinary treatise on the elements of algebra, these topics are either completely omitted or are treated carelessly. For this reason, I am certain that the material I have gathered in this book is quite sufficient to remedy that defect.

I have striven to develop more adequately and clearly than is the usual case those things which are absolutely required for analysis. Moreover, I have also unraveled quite a few knotty problems so that the reader gradually and almost imperceptibly becomes acquainted with the idea of the infinite.

There are also many questions which are answered in this work by means of ordinary algebra, although they are usually discussed with the aid of analysis. In this way the interrelationship between the two methods becomes clear.

I have divided this work into two books; in the first of these I have confined myself to those matters concerning pure analysis. In the second book I have explained those things which must be known from geometry,

since analysis is ordinarily developed in such a way that its application
to geometry is shown. In both parts, however, I have omitted the elementary matters and developed only those things which, in other places,
are either completely omitted or only cursorily treated or, finally, follow
from new arguments.

Thus...in the second volume...I have treated those topics which are
commonly called higher geometry. Before I discuss conic sections, which
in other treatments almost always come first, I have proposed a theory of
curves with enough generality that it can advantageously be applied to
an examination of the nature of any curve whatsoever. I use only an
equation by which the nature of every curve is expressed, and I show how
to derive from this both the shape and its primary characteristics. It has
seemed to me that this is most especially advantageous in the case of
conic sections. Until this time they have ordinarily been treated only
from the geometric viewpoint, or if by analysis, in an awkward and
unnatural way. I have first explained their general properties from the
general equation for a second order curve. Then I have subdivided them
into genera and species, considering whether they have branches going to
infinity or the whole curve is included in a bounded region. In the
former case something else has to be considered, that is, how many
branches extend to infinity and of what nature each of these is; namely,
whether or not the branch has a straight line asymptote. In this way I
have obtained the customary three types of conic section. The first is the
ellipse, totally contained in a bounded region; the second is the hyperbola, which has four infinite branches asymptotic to two straight lines;
the third is the parabola with two infinite branches without asymptotes.
In a like manner I have described third order curves, which I have
divided into sixteen kinds, after considering their general properties.
Indeed, to these kinds I have reduced the seventy-two of Newton's

classification. I have described this method with such clarity that for curves of any higher order whatsoever, a classification may easily be made. I have made a test of the method in the case of fourth order curves. Having explained whatever concerns the order of a curve, I returned to uncovering the general properties of all curves. Thus I have explained a method for defining tangents to curves, their normals, and curvature, which is ordinarily measured by the radius of the osculating circle. Although all of these nowadays are ordinarily accomplished by means of differential calculus, nevertheless, I have here presented them using only ordinary algebra, in order that the transition from finite analysis to analysis of the infinite might be rendered easier. I have also investigated points of inflection of curves, cusps, double points, and multiple points. The definitions of these kinds of points follow without difficulty from the equations. At the same time I readily admit that these matters can be much more easily worked out by differential calculus. I have touched on the controversy over a second order cusp, where both arcs which come together in the cusp curve in the same direction. It seems to me that I have settled this question in such a way that there can remain no doubt. Finally, I have adjoined several chapters in which I have explained how to find curves with certain stated properties. I give the solution to a number of problems concerned with circles. Since there are some topics from geometry which seem to offer strong support for learning analysis, I have added an appendix in which I have presented the theory of solids and the surfaces of solids. I have shown, insofar as it is possible to do so through equations in three variables, the nature of these surfaces. Then having divided surfaces into orders, as was done in the case of curves, according to the power of the variables in the equation, I have shown that the only surface of the first order is the plane. I have divided surfaces of the second order, by considering their parts which extend to infinity, into six types. In a similar way a division can be made for surfaces of other orders. I have also considered the intersections of two surfaces. These, insofar as they can be understood through equations, I have shown generally to be curves which do not lie in a single plane.

For the rest, since many things here will be met which have already been treated by others I ought to ask pardon, since I shall not have given credit in every place to all who have toiled herein. I have endeavored to develop everything as briefly as possible. If the history of each problem had been discussed, that would have increased the size of this work beyond reasonable bounds. Many of these problems, whose solutions can be found elsewhere, in this work have solutions which arise from different arguments. For this reason I would seem in no little part to be exonerated. I do hope that both these things, but also especially those which are entirely new, will be acceptable to most of those who enjoy this work.

TRANSLATOR'S INTRODUCTION

Euler's preface to both volumes of this work was printed only in the first volume of the original 1748 edition. For the convenience of the reader of this second volume I have excerpted the pertinent parts of the original preface.

The welcome response to the publication of my English translation of the first volume of *Introduction to Analysis of the Infinte* only confirms my conviction of the relevance today of this 240 year old work. I am grateful to have had the opportunity to be part of the publication in English of this valuable text. It is my hope that Euler's thought will not only be more accessibe but more widely enjoyed by the mathematical community.

I would like to express thanks to Saint John Fisher College for a sabbatical leave during which I worked on this translation. I am also grateful to Rudiger Gebauer, Mathematics Editor and to Susan Gordon, Mathematics Assistant Editor of Springer-Verlag.

Here then I wish to thank my children Jack, Paul, Drew, and Anne for their understanding and patience in seeing me through the demands of this second volume. Finally gratitude to my wife Claire, who typed the camera-ready copy, and without whose help and encouragement this translation would not have been possible.

CONTENTS

Contents

INTRODUCTION
TO
ANALYSIS
OF THE
INFINITE

BOOK II

Containing

the theory of curves, and an appendix on surfaces.

CHAPTER I

On Curves in General

1. A variable quantity is a magnitude considered in general, and for this reason, it contains all determined quantities. Likewise in geometry a variable quantity is most conveniently represented by a straight line RS, of indefinite length, as in *figure 1,* (see Figures, page 465). Since in a line of indefinite length we can cut off any determined magnitude, the line can be associated in the mind with the variable quantity. First we choose a point A in the line RS, and associate with any determined quantity an interval of that magnitude which begin at A. Thus a determined portion of the line, AP, represents the determined value contained in the variable quantity.

2. Let x be a variable quantity which is represented by the line RS, then it is clear that any determined value of x which is real can be represented by an interval of the line RS. For instance, if P is identical with the point A then the interval AP vanishes and represents the value $x = 0$. The farther removed from A the point P is, the greater the definite value of x represented by the interval AP.

The interval AP is called the *abscissa*. The abscissas manifest the determined values of x.

3. Since the line RS extends indefinitely in either direction from A, both negative and positive values of x can be represented. We will represent positive values of x by cutting off intervals to the right of A, and negative values by intervals to the left of A. The farther to the right the point P is from A, the greater the value of x represented by the interval AP, while the farther to the left the point P is the more the value of x is decreased. If P is identical with A, the value of x is 0. For this reason if P is moved farther to the left the values of x are less than 0, that is, negative, and the intervals AP cut off to the left of A represent negative values of x, and intervals AP to the right of A are assigned positive values. The choice of which direction is to represent positive values of x is arbitrary, but whichever direction is chosen, the opposite direction will represent negative values of x.

4. Since the indefinite straight line represents the variable x, we would like to see how a function of x can be most conveniently represented. Let y be any function of x, so that y takes on a determined value when a determined value is assigned to x. After having taken a straight line RAS to denote the values of x, for any determined value of x we take the corresponding interval AP and then erect a perpendicular interval PM corresponding to the value of y. If the value of y is positive, then PM is above the line RS, while if y is negative, then PM is the perpendicular below the line. For positive values of y, the representation is above the line RS; as the value of y decreases, the representation falls onto the line RS and then below it.

5. The shape of the function of x is seen by letting y have the value AB when $x = 0$, when $x = AP$, then $y = PM$. If $x = AD$, then $y = 0$ and if when $x = AP$ and y takes on a negative value, then the perpendicular PM falls below the line RS. In a similar way, when the values of x are negative, the corresponding values of y are represented

by perpendiculars PM above or below the line RS, depending on whether the value of y is positive or negative; unless for some value x, with $x = AP$, $y = 0$, then the perpendicular vanishes.

6. For all determined values x of the line RS, at the point P we erect the perpendicular PM corresponding to the value of y, with different M's for different P's, and with M above the line RS when the value of y is positive, while M is below the line when the value of y is negative, or M falls on the line RS when the value of y vanishes, as happened at the points D and E. All of the extremities, M, of the perpendiculars form a line which may be straight or curved. In this way the line is determined by the function y. Thus any function of x is translated into geometry and determines a line, either straight or curved, whose nature is dependent on the nature of the function.

7. In this way the curve which results from the function y is completely known, since each of its points is determined by the function y. At each point P, the perpendicular PM is determined, and the point M lies on the curve. Indeed, each point on the curve is found in this way. In whatever way the curve may be viewed, for each point on the curve there is one point on the straight line RS, which is on the perpendicular to the line through the point on the curve. In this way we obtain the interval AP, which gives the value of x, and the length of the perpendicular PM, which represents the value of the function y. It follows that there is no point of the curve which is not defined by the function y.

8. Although many different curves can be described mechanically as a continuously moving point, and when this is done the whole curve can be seen by the eye, still we will consider these curves as having their origin in functions, since then they will be more apt for analytic treatment and more adapted to calculus. Any function of x gives a curve or

a straight line, and conversely a curve can define a function. Hence the nature of each curve is expressed by the function of x, determined by projecting M onto P which lies on the line RS. The length of the interval AP is the value of x, while the corresponding value of the function is the length of the perpendicular PM.

9. From this concept of a curve, there follows immediately a division into *continuous* and *discontinuous* or *mixed*. A *continuous* curve is one such that its nature can be expressed by a single function of x. If the curve is of such a nature that for its various parts, BM, MD, DM, etc. different functions of x are required for its expression, that is, after one part BM is defined by one function of x, then another function is required to express the part MD, then we call such a curve *discontinuous* or *mixed* and *irregular*. This is because such a curve cannot be expressed by one constant law, but is formed from several continuous parts.

10. In geometry we are especially concerned with continuous curves. Below we shall show that those curves which can be described by some regular constant mechanism can also be described by a single function, so that these are also continuous curves. Let $mEBMDM$ be a continuous curve whose nature is contained in some function of x which we call y. It is clear that when we take a determined value of x on the line RS from the fixed point A, then the value of y corresponds to the length of the perpendicular PM.

11. There is some terminology which is frequently used when discussing what is known about curves.

First of all the straight line RS, from which the values of x are cut off, is called the *axis* or the straight line *directrix*.

The point A, from which the values of x are measured, is called the *origin of the abscissas*.

The parts of the axis, AP, by which the determined values of x are indicated, are called *abscissas*.

The perpendiculars PM, reaching from the abscissa to the curve, are called the *ordinates*.

In this case it is called the *normal* or *orthogonal* ordinate, since it forms a right angle with the axis. When the ordinate PM forms an oblique angle with the axis, then the ordinate is called *oblique*. We will always explain the nature of curves with orthogonal ordinates unless we expressly indicate the contrary.

12. If any abscissa AP is symbolized by the variable x, then the function y gives the length of the ordinate PM and $PM = y$. The nature of the curve, provided it is continuous, is expressed through the quality of the function y, that is, the rule of formation whereby the value of y is obtained from the composition of constants and the variable x. On the axis RS, the AS part is the location of the positive abscissas, while the AR part is the location of the negative abscissas. Above the axis RS is the region of positive ordinates, while the negative ordinates lie below the axis.

13. Since each function of x gives rise to a continuous curve, this curve can be known and described by that function. First we assign the positive values of x from 0 to ∞, and for each of these values we assign the value of the function y to the ordinates, either above or below the axis depending on whether the values are positive or negative. In this way we obtain that part of the curve BMM. Next in a similar way we assign all of the values of x from 0 to $-\infty$ and the corresponding values of y determined that part of the curve BEm. Thus the whole curve contained in the function has been exhibited.

14. Since y is a function of X, either y is equal to an explicit function of x, or we have an equation in x and y whereby y is defined by x. In either case we have an equation which is said to express the nature of the curve. For this reason every curve is expressed by an equation in two variables x and y. Of these the first, x, denotes the abscissa taken on the axis from the given origin A, the second, y, is the ordinate, perpendicular to the axis. The abscissas and ordinates taken together are called the *orthogonal coordinates*. It follows that the nature of a curve is said to be defined by an equation in the orthogonal coordinates, provided the determined equation defines y as a function of x.

15. Since we have reduced our knowledge of curves to that of functions, there are as many different kinds of curves as there are of the functions we have seen before. For this reason it is reasonable to divide the curves into *algebraic* and *transcendental*. A curve will be algebraic if the ordinate y is an algebraic function of the abscissa x, or, since the nature of the curve is expressed by an algebraic equation in the coordinates, this kind of curve is also accustomed to be called *geometric*. A *transcendental* curve is one whose nature is expressed by a transcendental equation in x and y, or an equation in which y is equal to a transcendental function x. This is the principal division of continuous curves, whereby they are either *algebraic* or *transcendental*.

16. In order to describe the curve from the function of x by which the ordinate y is expressed, we must note whether the function is single-valued or multiple-valued. We suppose first that y is a single-valued function of x, or $y = P$, where P denotes some single-valued function of x. For any determined value given to x, the ordinate y also determines only one definite value. To each abscissa there corresponds a unique ordinate, and for this reason the curve is such that when from any point P on the axis RS the perpendicular PM is drawn, there is only one point

M, which makes the curve. It follows that for a single point on the axis, there corresponds a single point on the curve. Since the axis extends indefinitely in both directions, so also the curve also extends indefinitely in both directions. That is, the curve arising from such a function extends indefinitely with the axis, as is illustrated in *figure 2,* where the curve *mEBMDM* extends in both directions indefinitely without interruption.

17. Now let y be a two-valued function of x, or we let P and Q both be single valued *functions* of x and let $y^2 = 2Py - Q$. Then $y = P \pm \sqrt{P^2 - Q}$. Each abscissa x corresponds to two ordinates y, both of which are either real or complex, the former if $P^2 > Q$, the latter is $P^2 < Q$. Even when both values of y are real, the abscissa AP corresponds to two ordinates PM, PM, that is, perpendicular to the axis at P there are two points M and M. On the other hand for an abscissa x such that $P^2 < Q$, there is no corresponding ordinate. In this case no perpendicular to the axis gives a point on the curve, as for example at P in *figure 3.* When at first, $P^2 > Q$, it cannot also be true that $P^2 < Q$, except when passing through the case $P^2 = Q$, which is the boundary between ordinates which are real and complex. Where the real ordinates disappear, as at C or G, there $y = P \pm 0$ and both ordinates are equal, so that there also the curve bends back

18. In *figure 3* it is clear that while the abscissa is negative and between AC and AE, the ordinate is complex with $P^2 < Q$. Beyond E to the left the ordinates become real, which cannot happen unless at E we have $P^2 = Q$ and here both ordinates are equal. Then to the abscissas there again correspond two ordinates Pm, Pm until we come to G where the two ordinates become equal. Beyond G the ordinates again become complex. In this way a curve can consist of parts which are

distinct from each other, as *MBDBM* and *FmHm*. There may be two or more parts, nevertheless, these parts taken together are to be considered as one continuous or regular curve, since all the different parts come from a single function. Curves of the kind have the property that at each point of the axis there is a perpendicular *MM* which cuts the curve in either no points or in two points, with the exception that the two points coincide, which occurs at the points *D*, *F*, *H* and *I*.

19. If *y* is a three-valued function of *x*, or if *y* is defined by the equation $y^3 - Py^2 + Qy - R = 0$, where *P*, *Q*, and *R* are single-valued functions of *x*, then for any value of *x*, the ordinate *y* has three values which are all real or only one is real and the other two are complex. In this case each ordinate cuts the curve in three points or in only one except where two or even three of the points of intersection coincide. Since each abscissa has at least one real ordinate, the curve necessarily extends indefinitely in both directions with the axis. It follows that the curve consists of one continuously extended piece, as in *figure 4*, or in two distinct parts, as in *figure 5*, or in more distinct parts, but in any case, all the parts together are considered to be the one and the same continuous curve.

20. If *y* is a four-valued function, or if *y* is defined by the equation $y^4 - Py^3 + Qy^2 - Ry + S = 0$, then to each value of *x* there correspond four points or only two or even none, each of which cases are illustrated by *figure 6*. We should note the places I and 0, where the two points of intersection coalesce into one. For this reason, either to the left or to the right there may be none or two or four branches of the curve which extend indefinitely. In the first case, when on neither side there is any branch extending indefinitely, the curve will be closed on both sides, as the figure indicates, and a definite area will be included. In this way there are certain properties of curves which we can already be sure will

belong to those which are formed from any multiple-valued functions.

21. If for instance, y is a multiple-valued function, or is defined by an equation in which n is the largest power of y, then the number of real values of y will be either n, $n - 2$, $n - 4$, $n - 6$, etc., which will also be the total number of points in which the ordinate will intersect the curve. It follows that if one of the ordinates cuts a continuous curve in m points, then all the other ordinates intersect the curve in a number of points which differs from m by an even number. The curve cannot be cut by an ordinate in $m + 1$, $m - 1$, or $m \pm 3$, etc. points. That is, if the number of intersections of one ordinate is even or odd, all of the ordinates will intersect the curve in a number of points which is even or odd respectively.

22. It follows that if one ordinate cuts the curve in an odd number of points, then it cannot happen that some other ordinate has no intersection with the curve. Thus, there is at least one branch extending indefinitely in both directions, and if in either direction there is more than one branch extending indefinitely, then that number must be odd, since the intersection with any ordinate must be in an odd number of points. If we take together the number of branches extending indefinitely in both directions, then the sum will be even. This property is true also in case some ordinate intersects the curve in an even number of points, since then on either side there will be either none, two, four, etc. branches extending indefinitely, so that all together there will be an even number of branches extending indefinitely. Now that we have some familiarity with a few of the important properties of continuous and regular curves, we next turn our attention to discontinuous and irregular curves.

CHAPTER II

On the Change of Coordinates.

23. We recall that a given curve is described from an equation in the coordinates x and y of which the former denotes the abscissa and the latter the ordinate, all with respect to an axis RS and an origin of the abscissas A which can be arbitrarily chosen. Likewise, if a curve is already given, its nature can be expressed by an equation between its coordinates. In this case there are two things which can be arbitrarily chosen, namely, the position of the axis RS and the origin of the abscissas A. Since there are an infinite number of different ways of making these choices, so there are an infinite number of different equations for the same curve. For this reason we may obtain the same curve from different equations. However, different curves always give different equations.

24. Since by a change of axis or origin innumerable different equations express the nature of the same curve, all of these equations should be compared, so that when one is given, all of the others can be found. A curve is determined by an equation in the coordinates. If the curve is known and any line is chosen as the axis with any point on it chosen as origin, then an equation in these coordinates is defined. In this chapter we treat of a method by which if an equation of a curve is given, then an equation can be found in the coordinates depending on any choice of axis and origin, which expresses the nature of that same curve. In this way we find absolutely all equations which give the nature of that same curve. At the same time it will be easier to judge whether different curves come from different equations.

25. Suppose we have an equation in x and y. Given a straight line RS for an axis and a point A for origin, so that x denotes the abscissa AP and y the ordinate PM, then we obtain the curve CBM, as in *figure 7*, whose nature is expressed by the equation. First we keep the same axis RS, but choose a different point D as origin, so that now the point M on the curve corresponds to the abscissa DP, which we represent by t. The ordinate MP remains equal to the same y as before, but we look for the equation in t and y by means of which the nature of the curve CBM is expressed. We let the length of the interval AD be equal to f. In *figure 7 D* lies to the left of A, so that it is in the region of negative abscissas. Now $DP = t = f + x$, or $x = t - f$, so that if in the equation in x and y we substitute $t - f$ for x, then we obtain an equation in t and y which expresses the same curve. Since the size of $AD = f$ is arbitrarily chosen, we already have innumerable different equations which all express the same curve.

26. If the curve intersects the axis RS in some point, say C, then if we choose C as the origin, we obtain an equation which, when the abscissa $CP = 0$, has the ordinate PM also vanishing, provided there is only one ordinate corresponding to the point C. The intersection C of the curve and the axis, and any other points of intersection there may be, can be found from the original equation in x and y by letting $y = 0$ and solving for the values of x. This follows from the fact that wherever the curve crosses the axis, the value of y becomes zero, and conversely, if we let $y = 0$, then we can find all the values of x where the curve intersects

the axis.

27. We keep the same axis but change the origin if the given abscissa is increased or decreased; that is, if for x we substitute $t - f$, where f is a positive quantity when the new origin D is to the left of A, or it is a negative quantity when the new origin D is to the right of A.

Now consider a curve LBM, as in *figure 8,* from a given equation in $AP = x$ and $PM = y$. We choose a new axis RS which is parallel to the original axis, and we choose the point D to be the new origin. In the figure the new axis lies in the negative region for the ordinates at a distance equal to $AF = g$. We also let the interval $DF = AG = f$. In this new axis the abscissa corresponding to the point M on the curve is $DQ = t$ and the ordinate $QM = u$. Then $t = DF + FQ = f + x$ and $u = PM + PQ = g + y$. It follows that $x = t - f$, $y = u - g$ and that when $t - f$ is substituted for x, $u - g$ for y in the equation in x and y, we obtain an equation in t and u by which the nature of the same curve is expressed.

28. Since the quantities f and g are arbitrarily chosen there are infinitely many different ways of forming an equation for each of the infinite number of ways in the previous case, and all of these equations pertain to the same curve. It follows that if two equations, one in x and y, the other in t and u, differ from each other only in that one can be transformed into the other by increasing or decreasing the coordinates, then both equations, although different, still exhibit the same curve. Thus it is easy to find innumerable different equations which express the nature of the same curve.

29. Next we choose a new axis rs perpendicular to the old axis RS and crossing at the origin A. We suppose that we have an equation in x and y for the curve LM in *figure 9* from the axis RS, where $AP = x$

and $PM = y$. We drop the perpendicular MQ to the new axis rs, call the new abscissa $AQ = t$, and the new ordinate $QM = u$. Since $APMQ$ is a rectangle, $t = y$ and $u = x$. From the equation in x and y we obtain an equation in t and u by substituting t for y and u for x. The previous abscissa x has become the new ordinate $QM = u$, while the previous ordinate y has become the new abscissa $AQ = t$. For this new axis no other change in the equation takes place except that the coordinates x and y change places. For this reason the abscissa and ordinate are called coordinates, making no distinction which is called the abscissa or ordinate. Given an equation in x and y, the same curve emerges whether x or y is taken as the abscissa.

30. We positioned the new axis rs in such a way that the part As contains the positive abscissas and to the right of the axis rs we place the positive ordinates. This is all arbitrary, and it can be changed if desired. For instance, if the Ar part of the axis is designated the positive abscissas, then $AQ = -t$ and in the equation in x and y we substitute $-t$ for y. Next if we let the region to the right of the axis rs be for the negative ordinates, then $QM = -u$ and we substitute $-u$ for x. From this discussion we understand that the nature of the curve is not changed, even though in the equation in the coordinates one or the other or both are made negative. We hold to this in all transformations of equations.

31. Now we let the new axis rs intersect the old axis RS at A in some angle SAs, as in *figure 10*. The point A will be the origin in both axes. We suppose that we have an equation in x and y for some curve LM, where $AP = x$ is the abscissa and $PM = y$ is the ordinate with respect to the old axis RS. From this information we should be able to find an equation for the same curve with respect to the new axis rs. From the point M on the curve we drop the perpendicular MQ to the

new axis. The new equation is to be in t and u, where $AQ = t$ is the new abscissa and $MQ = u$ is the new ordinate. We let the angle $SAs = q$, $\sin q = m$, and $\cos q = n$, so that $m^2 + n^2 = 1$. From the point P we drop perpendiculars Pp and Pq to the new coordinates. Since $AP = x$ we have $Pp = x \sin q$ and $Ap = x \cos q$. Then, since the angles $PMQ = PAQ = q$ and $PM = y$ we have $Pq = Qp = y \sin q$ and $Mq = y \cos q$. From these it follows that

$$AQ = t = Ap - Qp = x \cos q - y \sin q$$

and

$$QM = u = Mq + Pp = x \sin q + y \cos q.$$

32. Since $\sin q = m$ and $\cos q = n$, we have $t = nx - my$ and $u = mx + ny$. It follows that

$$nt + mu = n^2x + m^2x = x$$

and

$$nu - mt = n^2y + m^2y = y.$$

The equation in u and t we are seeking is obtained by substituting, in the equation in x and y, $mu + nt$ for x and $nu - mt$ for y, provided the part of the new axis, As, contains positive abscissas and QM lies in the region of positive ordinates. We have supposed that the angle SAs lies in the region of negative ordinates. If As lies above AS, then the angle $SAs = q$ is negative, so that in the computation the sin m is also negative.

33. Now we consider the case when the new axis rs is places anywhere and any point D is the origin. Let RS be the original axis in relation to which $AP = x$ is the abscissa and $PM = y$ is the ordinate. Given an equation in x and y which expresses the nature of the curve

LM, as in *figure 11*, we look for an equation in t and u related to the new axes. We drop a perpendicular MQ from any point M on the curve to the new axis rs and we let the abscissa $DQ = t$ and the ordinate $QM = u$. In order to find the equation in t and u, we drop a perpendicular DG, from the new origin D to the old axis RS. We let $AG = f$, $DG = g$, then through D we draw a line DO, parallel to the old axis, and intersecting at O an extension of the original ordinate PM. It follows that $MO = y + g$ and $DO = GP = x + f$. Finally we let the angle $ODQ = q$, $\sin q = m$, and $\cos q = n$, so that $m^2 + n^2 = 1$.

34. Now we drop perpendiculars Op and Oq from O to the new axis DQ and the new ordinate MQ, respectively. Since the angles OMQ and ODQ are equal, $DO = x + f$ and $MO = y + g$, we have

$$Op = Qq = (x + f) \sin q = mx + mf$$

and

$$Dp = (x + f)\cos q = nx + nf.$$

Furthermore,

$$Oq = Qp = (y + g)\sin q = my + mg$$

and

$$Mq = (y + g)\cos q = ny + ng.$$

It follows that

$$DQ = t = nx + nf - my - mg$$

and

$$QM = u = mx + mf + ny + ng.$$

In this way we have defined the new coordinates t and u in terms of the old coordinates x and y. We also have

$$nt + mu = x + f$$

and

$$nu - mt = y + g,$$

since

$$m^2 + n^2 = 1.$$

It follows that

$$x = mu + nt - f$$

and

$$y = nu - mt - g.$$

These are the values which we substitute for x and y in the original equation to obtain a new equation in t and u which expresses the nature of the same curve LM.

35. Since no axis rs lying in the same plane with the curve can be concieved which is not included in this last determination, it follows that for any given curve LM, every equation for the curve is included in the equation in the coordinates t and u. Since the quantities f, g, and the angle q, upon which m and n depend, can be varied in an infinite number of ways, all the equations contained in the equation in t and u found in this way express the nature of the same curve. For this reason, the equation in t and u is usually called the general equation for the curve LM, since in this one equation are contained all equations which pertain to this curve.

36. We have previously intimated that it is difficult to decide whether two equations refer to the same curve or to two different curves. Now we have a way of making this decision. Suppose we have two equations, one in x and y, the other in t and u. In the first equation we

substitute $x = mu + nt - f$ and $y = nu - mt - g$, where m and n are related by $m^2 + n^2 = 1$. After this we determine whether the original equation in t and u is included in the new equation by a proper choice of values for f, g, m, and n. If this is possible, then both equations express the same curve; if it is not possible, the curves are different.

EXAMPLE

With this method we can show that the two equations $y^2 - ax = 0$ and $16u^2 - 24tu + 9t^2 - 55au + 10at = 0$ refer to the same curve although the equations themselves are very different. If we make the substitutions, in the first equation, $x = mu + nt - f$ and $y = nu - mt - g$, then we obtain

$$n^2 u^2 - 2mntu + m^2 t^2 - 2ngu + 2mgt$$

$$+ g^2 - mau - nat + af = 0.$$

In order to decide whether the previous equation is contained in this one we multiply it by n^2 and the new equation by 16 so that the first terms will agree. Then we have

$$16n^2 u^2 - 24n^2 tu + 9n^2 t^2 - 55n^2 au + 10n^2 at = 0$$

and

$$16n^2 u^2 - 32mntu + 16m^2 t^2 - 32ngu + 32mgt$$

$$+ 16g^2 - 16mau - 16nat + 16af = 0.$$

Next we consider what values the arbitrary constants f, g, m and n should have to make the terms equal. First we will have $24n^2 = 32mn$ and $9n^2 = 16m^2$. Then from either one we have $3n = 4m$. Since $m^2 = 1 - n^2$, we have $25n^2 = 16$ and so $n = \dfrac{4}{5}$ and $m = \dfrac{3}{5}$. Now we have the first three terms in agreement. The fourth and fifth terms give $55n^2 a = 32ng + 16ma$ and $10n^2 a = 32mn - 16na$. To determine whether they both give the same value for g, we note that from the first

we have $g = \dfrac{55na}{32} - \dfrac{ma}{2n} = \dfrac{11a}{8} - \dfrac{3a}{8} = a$. From the second we

have $g = \dfrac{5m^2a}{16m} + \dfrac{na}{2m} = \dfrac{a}{3} + \dfrac{2a}{3} = a$. Since both values of g are

equal, these terms agree. There is nothing left except that $g^2 + af = 0$,

but since f has not been determined, there is no difficulty, and we let

$f = -a$. We have shown that these two equations exhibit the same

curve.

37. Although it can happen that quite different equations represent

the same curve, still we can frequently with assurance conclude that from

different equations we have different curves. This is the case if the

degrees of the two equations are different, that is, if the highest power of

x and y is different from the highest power of t and u, then the curves

indicated by the two equations are certainly different. This follows from

the fact that the degree of an equation in x and y is the same as the

degree of the equation which results when we make the substitutions

$x = mu + nt - f$ and $y = nu - mt - g$. Hence, if the second equa-

tion in t and u has a different degree, then this indicates that the curves

are different.

38. Unless two equations, the one in x and y, and the other in t

and u, both have the same degree, we can immediately conclude that the

curves expressed by the equations are also different. The only time when

there is cause for doubt is when they both have the same degree, and in

this case only do we resort to the method given above, which is rather

tedious. If both equations have the same degree and the degree is rather

large, we will give below a more expeditious method for deciding immedi-

ately whether the curves are different.

39. The method for finding the general equation for any curve can be adapted to the case of a straight line. Let the straight line be LM, which is parallel to the axis RS, as in *figure 12*, then no matter what the abscissa may be, the ordinate will be the same, that is, $y = a$. This is the equation of a straight line parallel to the axis. We look for the general equation of a straight line related to any axis rs, as in the same *figure 12*. We let $DG = g$, the angle ODs have a sine equal to m and a cosine equal to n. We let the abscissa $DQ = t$ and the ordinate $MQ = u$. Since $y = nu - mt - g$, we have $nu - mt - g - a = 0$, and this is the general equation for a straight line. If we multiply this equation by k, let $nk = \alpha$, $mk = -\beta$, and $(g + a)k = -b$, then we have the equation $\alpha u + \beta t + b = 0$ for a straight line, which is a general first degree equation in t and u. It is clear that every first degree equation in two coordinates exhibits not a curve but a straight line.

40. It follows that whenever we have such an equation, $\alpha x + \beta y - a = 0$, in the coordinates x and y, then we have a straight line as in *figure 13*, whose position with respect to the axis RS is to be determined. First we let $y = 0$ in order to find the point C on the axis where this straight line crosses, and we see that $AC = \dfrac{a}{\alpha}$. Then we let $x = 0$, to find that $y = \dfrac{a}{\beta}$ which is the value of the ordinate AB at the origin. We have the two points B and C, which lie on the straight line, and so determine it. It follows that the straight line LM satisfies the equation, since if we let $AP = x$ be any abscissa and $MP = y$ be the corresponding ordinate, from the similar triangles CPM and CAB we have the proportion $CP:PM = CA:AB$. That is

$$\frac{a}{\alpha} - x : y = \frac{a}{\alpha} : \frac{a}{\beta},$$

so that

$$\frac{ay}{\alpha} = \frac{a^2}{\alpha\beta} - \frac{ax}{\beta},$$

or $\alpha x + \beta y = a$, which is the given equation.

41. If either $\alpha = 0$ or $\beta = 0$, then we cannot make the suggested construction, but these cases are very easy. If $\alpha = 0$ and $y = a$, then it is clear that the line satisfying these equations is parallel to the axis at a distance equal to a. If $a = 0$ or $y = 0$ then the line is coincident with the axis. If $\beta = 0$ and $x = a$, it is clear that the line satisfying these conditions is normal to the axis at a distance equal to a from the origin. In this case, every ordinate corresponds to the same abscissa, so that the abscissa is no longer variable. From this discussion it is perfectly clear how straight lines are designated by equations in the coordinates.

42. Up until this time we have assumed that the coordinates, by means of which the nature of the curve is expressed, have been perpendicular. We can still find the curve defined by a given equation if we suppose that all of the ordinates make a fixed angle with the axis. Conversely, the nature of the curve can be expressed by an equation in two coordinates which are oblique. Equations of this kind can vary in an infinite number of ways, as both the axis and the origin vary, while the curve remains the same. It follows that for any given obliquity of the coordinates there is a general equation for a curve. If we allow the obliquity to vary, we have still more equations for the same curve, and we will have this kind of equation, the most general equation for the curve. The nature of the curve is expressed not only for any axis and for any origin, but also for any obliquity of the coordinates. This most general equation becomes the general equation if we make the angle between the coordinates to be a right angle.

43. Let a curve LM be given with an equation in rectangular

coordinates, as in *figure 14*. That is, let $AP = x$, $PM = y$ and we seek an equation between x and y, where we keep the axis RS, the origin A, but the angle between the coordinates is Q. From the point M to the axis RS we draw the straight line MQ so that it makes the angle ϕ with the axis. Let $\sin \phi = \mu$ and $\cos \phi = \nu$. Then AQ is the new abscissa and MQ is the new ordinate. We let $AQ = t$, $QM = u$ so that in the right triangle PMQ, we have $\dfrac{y}{u} = \mu$ and

$$\frac{PQ}{u} = \frac{t - x}{u} = \nu.$$

It follows that $u = \dfrac{y}{\mu}$ and

$$t = \nu u + x = \frac{\nu y}{\mu} + x.$$

Conversely, $y = \mu u$ and $x = t - \nu u$. Hence, if we make the substitutions $x = t - \nu u$ and $y = \mu u$ in the equation in x and y, we obtain an equation in t and u, which together make the angle ϕ.

44. If we are given an equation in oblique coordinates AQ and QM, from this equation we can obtain an equation for the same curve in rectangular coordinates AP and PM. Let ϕ be the given angle which the ordinate MQ makes with the abscissa AQ, and let $\sin \phi = \mu$, $\cos \phi = \nu$, $AQ = t$ and $QM = u$. From M we drop the perpendicular to the axis to obtain the new ordinate MP. We let the abscissa $AP = x$ and the ordinate $MP = y$. Since $u = \dfrac{y}{\mu}$ and $t = \dfrac{\nu y}{\mu} + x$, these values can be substituted for u and t in the original equation to obtain the desired equation in x and y.

45. Now if we have an equation in the orthogonal coordinates $AP = x$ and $PM = y$ for the curve LM, as in *figure 15*, we will give a

method for the most general equation for the same curve. We let rs be any straight line which will be taken as the axis, and we choose any point D on rs as origin. The ordinate MT drawn to the new axis makes an angle $DTM = \phi$, and we let $\sin \phi = \mu$, $\cos \phi = \nu$. The new abscissa is DT and the new ordinate is TM. We seek an equation in these new coordinates. From D to the original axis RS we drop the perpendicular DG and let $AG = f$, $DG = g$. We draw DO parallel to the axis RS. We let the sine of the angle ODs be m and the cosine of that angle be n. As before, we drop the perpendicular MQ from M to the new axis rs, and we let $DQ = t$, $QM = u$. The oblique coordinates are $DT = r$, $TM = s$. We have $t = r - \nu s$ and $u = \mu s$ from section 43. Then

$$x = nr - (n\nu - m\mu)s - f$$

and

$$y = - mr + (\mu n + \nu m)s - g,$$

where $n\nu - m\mu$ is the cosine of the angle AVM, which is the angle which the new ordinate makes with the old axis RS. Note also that $\mu n + \nu m$ is the sine of this same angle AVM. Now if we substitute for x and y in the original equations these values for x and y, we obtain an equation in the oblique coordinates r and s, which is the most general equation for the curve LM.

46. Since the values substituted for x and y are first degree equations in r and s, it is clear that the most general equation is of the same degree as the original equation in x and y. It follows that no matter how the equation of a curve may be transformed by changing the axis and the origin and the inclination of the coordinates, still the equation keeps the same degree. Although the equation in the coordinates,

whether they be rectangular or oblique, can vary in infinitely many ways and still pertain to the same curve, nevertheless, the degree of the equation is never raised nor lowered. For this reason equations with different degrees, no matter how similar they may be in other respects, will always exhibit different curves.

CHAPTER III

On the Classification of Algebraic Curves by Orders.

47. Since there are an infinite variety of curves, just as there are of functions, there is no way we can acquire any knowledge of them unless we divide them into certain classes, which contain infinite numbers of curves. In this way the mind is aided in its examination of the curves. First we divide them into *algebraic* and *transcendental* curves, but each of these classes will be subdivided, due to the infinite variety. Now we consider only algebraic curves, and inquire how they may most conveniently be divided into different classes. First we define what the qualities are which will determine what class a curve will belong to so that if two curves have the same quality, they are in the same class while if they do not have the same quality, then they belong to different classes.

48. These qualities which will distinguish different classes must not arise from any place other that the equation or function, since there does not appear to be any other way to obtain knowledge of the curve, and also this is the only way of putting all algebraic curves to the same test. Functions and equations in two coordinates can be distinguished in several ways, as we have seen in the first book. The first quality which presents itself is being multiple-valued, and this seems the most apt for placing in different classes those curves which arise from functions.

Curves from single-valued functions we will call of the first genus; those from two-valued function are of the second genus; those from three-valued function are of the third genus, and so forth.

49. Although this division seems natural, still when we consider more carefully, we see that this quality does not fit the nature and characteristics of the curve very well since it depends on the position of the axis, which is arbitrary. That is, if for one axis the ordinate will give a single-valued function of the abscissa, still for a different axis the function could be multiple-valued. In this way the genus of the curve is ambiguous, and this is not allowable. For instance, the curve expressed by the equation $a^3y = a^2x^2 - x^4$ belongs to the first genus since the ordinate y is a single-valued function of x. If we interchange the coordinates, that is, take the new axis perpendicular to the original axis, then the same curve is expressed by the equation $y^4 - a^2y^2 + a^3x = 0$, so that the curve belongs to the fourth genus. For this reason the multiple valuation for a function cannot be admitted as the quality which places a curve in a certain class.

50. Equally unacceptable is the simplicity of the equation expressing the curve, in the sense of the number of terms. If we were to say that curves belonged to the first genus when the equation contained two terms, as $y^m = \alpha x^n$, to the second genus when it contained three terms, as $\alpha y^m + \beta y^p x^q + \gamma x^n = 0$, and so forth then it is clear that the same curve could belong to many different genera. For example in section 36 we discussed a curve with equation $y^2 - ax = 0$. This curve would belong to the first genus and also to the fourth genus, since when the axis is changed, the same curve has the equation

$$16u^2 - 24tu + 9t^2 - 55au + 10at = 0.$$

With a different choice of axis and origin we could have made the genus be the second, third, or fifth. From this it is clear that we cannot use this method of classification.

51. These inconveniences can be avoided if the degree of the equation, which expresses the relation between the coordinates, is the classifying principle. Since for the same curve, whatever axis and origin is chosen or however the inclination of the coordinates may be varied, the degree of the equation always remains the same, so that the same curve is not referred to different classes. Hence the classifying property is the degree of the equation in the coordinates, whether they are rectangular or oblique, nor does it change with a change in axis or origin, nor a variation in the inclination of the coordinates. Furthermore, whether the equation is a special equation, or a general equation, or even the most general equation, the curve is assigned to a fixed class. For this reason the degree of the equation is the most convenient principle of distinction for curves.

52. Since the different genera of equations have been determined by the degree, so the different genera of curves will also be designated by order. Hence a general equation of the first degree is $\alpha + \beta x + \gamma y = 0$. Every curve with coordinates x and y, whether rectangular or oblique, arising from this equation will be assigned the first order. We have seen above that the only curve in this class is the straight line, so that only the straight line is a curve of the first order, and indeed this is the simplest curve of all. Since the word "curve" is ordinarily not applied to a straight line, we could generalize the usage and call all curves "lines". Then we would say that lines of the first order contain no curved lines, but only straight lines.

53. Whether the coordinates are rectangular or oblique, we will suppose that the ordinate makes an angle ϕ with the axis and that

$\sin \phi = \mu$, $\cos \phi = \nu$. We let $y = \dfrac{u}{\mu}$ and $x = \dfrac{\nu u}{\mu} + t$ (see section 44), so that the new equation in the coordinates t and u is

$$\alpha + \beta t + \left(\frac{\beta \nu}{\mu} + \frac{\gamma}{\mu} \right) u = 0.$$ But this is no more general than the first equation, so that either is the general equation. It is clear that the equation is still valid when the ordinate makes a right angle with the axis. This same phenomenon occurs in general equations of higher degree, which are no less applicable when the coordinates are made rectangular. Since a general equation of any degree loses none of its power, its significance is not resticted by the determination of an angle with which the ordinates meets the axis, even if the angle is a right angle. Given any curve which is contained in a general equation of any degree with oblique coordinates, that same curve is contained in the same equation, if the coordinates are made rectangular.

54. All lines of the second order are contained in this general equation of the second degree,

$$\alpha + \beta x + \gamma y + \delta x^2 + \epsilon xy + \zeta y^2 = 0.$$

That is, all curves which have rectangular coordinates x and y and are contained in this equation, we classified as second order lines. These are the simplest curves, since in the first order there are only straight lines, and for this reason they are called by some first order curves. The curves contained in this equation are commonly known as *conic sections*, since all of them can be obtained as the intersection of a plane and a cone. The different species of these lines are the circle, the ellipse, the parabola, and the hyperbola. We will deduce all of these from the general equation in a subsequent chapter.

55. We will refer to lines of the third order all curves which satisfy

the following equation,

$$\alpha + \beta x + \gamma y + \delta x^2 + \epsilon xy + \zeta y^2$$
$$+ \eta x^3 + \theta x^2 y + \iota xy^2 + \kappa y^3 = 0.$$

to be rectangular coordinates, since obliquity would add nothing new to this equation, as we have remarked. Since this equation has many more arbitrary constants than the preceding general equation, so also it contains many more species, which Newton classified.

56. All curves which are exhibited by the following equation pertain to fourth order lines,

$$\alpha + \beta x + \gamma y + \delta x^2 + \epsilon xy + \zeta y^2 + \eta x^3 + \theta x^2 y$$
$$+ \iota xy^2 + \kappa y^3 + \lambda x^4 + \mu x^3 y + \nu x^2 y^2 + \xi xy^3 + o y^4 = 0.$$

Here x and y are rectangular coordinates, since obliquity of the ordinate brings no more generality to the equation. We have here fifteen arbitrary constants, so that there are many more species in this order than in the preceding. Lines of order four are frequently called third order curves, since lines of order two are reputed to be first order curves. Similarly lines of order three correspond to curves of order two.

57. From our discussion it is clear which curves belong to the fifth, sixth, seventh, etc., orders. A general equation of all fifth order lines, since it adds to the terms of the general equation for the fourth order, the terms x^5, $x^4 y$, $x^3 y^2$, $x^2 y^3$, xy^4, y^5, has all together 21 terms. The general equation of all lines of the sixth order has 28 terms, and so forth, according to the triangular numbers. That is, the general equation for the lines of order n has $\dfrac{(n+1)}{1}\dfrac{(n+2)}{2}$ terms, and the same number of arbitrary constants.

58. It is not true that when any of these constants changes, then we have a new curve. We saw in the previous chapter that for the same curve there are an infinite number of different equations, from a change of axis and of origin. We conclude that although the equations of the same degree may be different, the curves may be the same. For this reason when classifying the genera and species which belong to some order, we must be cautious, lest we assign the same curve to two or more different species.

59. Since we know the order of a curve from the degree of the equations in its coordinates, given any algebraic equation in the coordinates x and y, it is immediately settled to what order the curve is to be assigned. For instance, if at first the equation is irrational, the irrationality is to be removed. If there still remain fractions, these also must be removed. Finally we consider the highest powers of the variables x and y, and this will give the order of the curve. Thus, a curve with equation $y^2 - ax = 0$ is of order two, while the curve with equation $y^2 = x\sqrt{a^2 - x^2}$ is of order four, since when the irrationality is removed, the equation is of the fourth degree. A curve with equation $y = \dfrac{a^3 - ax^2}{a^2 + x^2}$ is of order three, since when the fraction is removed, the equation is of degree three.

60. In one and the same equation different curves may be contained, depending on whether the coordinates are rectangular or oblique. For example, the equation $y^2 = a^2 - x^2$ is of a circle if the coordinates are rectangular, while the curve is of an ellipse if the coordinates are oblique. All of these different curves belong to the same order, since when the coordinates are changed from oblique to rectangular the order of the curve remains the same. Although the general equation for a curve of any order has the same form whatever the angle made by the

ordinate and the axis, still any special equation does not determine a curve until the angle is specified.

61. In order that a curve may be properly referred to its order, it is necessary that the equation cannot be expressed as the product of rational factors. If an equation has two or more factors, then two or more equations are involved, and each of these generates its own curve, which together express the given equation. Equations of this kind, which can be resolved into factors, contain not one, but several continuous curves, each of which can be expressed by its own equation, and are not connected with each other except that their equations happen to be multiplied together. Since this connection depends on a free choice we cannot think of this equation as containing a single continuous curve. Equations of this kind, which we previously called complex, produce non-continuous curves. Indeed, it is composed of several continuous curves, and for this reason the curves are called complex.

62. For example, the equation $y^2 = ay + xy - ax$, which seems to be a second order curve, can be factored to give $(y - x)(y - a) = 0$. Hence it is made up from two equations $y - x = 0$ and $y - a = 0$, each of which is for a straight line. One of the straight lines crosses the axis at the origin and makes an angle of half of a right angle; the other is parallel to the axis at a distance equal to a. These two straight lines taken together are contained in the given equation $y^2 = ay + xy - ax$. In a similar way the equation

$$y^4 - xy^3 - a^2x^2 - ay^3 + ax^2y + a^2xy = 0$$

is complex, and for this reason it does not represent a continuous fourth degree curve. It is composed of three factors, $(y - x)(y - a)(y^2 - ax)$, and for this reason it contains three discrete lines, the two straight lines, and one curve which is contained in the equation $y^2 - ax = 0$.

63. We can form complex lines with complete freedom, and these are made up of two or more lines whether they be straight lines or curves. If for each line we write an equation expressing its nature in relation to the same axis and origin, if each equation has the form of an expression equal to zero, and these expressions are all multiplied and set equal to zero, then the equation is complex and contains all of the lines. For example, suppose we have a circle with center C and radius $CA = a$, and also a straight line LN passing through the center C, as in *figure 16*. We can find an equation, with respect to any axis, which represents both the circle and the straight line taken together, as if they were one line.

64. Let AB be the diameter of the circle which makes half of a right angle with the straight line LN. We let the axis contain the diameter AB, we take A to be the origin and we call the abscissa $AP = x$ and the ordinate $PM = y$. For the straight line $PM = CP = a - x$. Since M on the straight line falls below the axis, we have $y = -a + x$ or $y - x + a = 0$. For the circle, since $PM^2 = AP \cdot PB$ and $BP = 2a - x$, we have $y^2 + x^2 - 2ax = 0$. When we multiply the two equations we obtain a third degree complex equation,

$$y^3 - y^2 x + yx^2 - x^3 + ay^2 + 3ax^2 - 2a^2 x = 0,$$

which contains both the straight line and the circle. For instance, for the abscissa $AP = x$ we find three ordinates, two for the circle and one for the straight line. When $x = \dfrac{1}{2}a$, we have

$$y^3 + \frac{1}{2}ay^2 - \frac{3}{4}a^2 y - \frac{3}{8}a^3 = 0.$$

From this first we have $y + \dfrac{1}{2}a = 0$. When we divide by this root we obtain $y^2 - \dfrac{3}{4}a^2 = 0$, so that we have the three values $y = -\dfrac{1}{2}a$,

$y = \frac{1}{2} a \sqrt{3}$, $y = -\frac{1}{2} a \sqrt{3}$. Thus in a single equation the circle and the straight line LN are contained as if they were a single continuous curve.

65. Now that we are aware of the difference between simple and complex curves, it is clear that a second order line is either a continuous curve or is complex and made up of two straight lines. If the general equation has factors, these will be first degree and so they will indicate straight lines. A third order line is either simple, or complex and consisting of a straight line and a second order line, or it may be complex, and composed of three straight lines. Furthermore, fourth order lines are either continuous, that is, simple, or they are complex. If complex, they are made up of a straight line and a third order line, or made up of the two second order lines, or made up of a second order line and two straight lines, or consisting of four straight lines. Likewise, complex lines of fifth or higher order can be distinguished and counted. From this discussion it should be clear that in any order there can be found lines of all lower orders, not of course as simple lines, but as complex with straight lines, or with lines of the second, third, or higher order. If the orders of all the simple lines are added, the sum will indicate the order of the complex line.

CHAPTER IV

On the Special Properties of Lines of Any Order.

66. Among the special properties of lines of any order, the first place must be given to the multiplicity of intersections with a straight line which a line of any order can have. A line of the first order, that is, a straight line, can have only one intersection with another straight line, but curves can intersect a straight line in many points. For this reason it is worthwhile to investigate in how many points a curve of any order can intersect a straight line drawn in any direction. From this information we have a better idea of the nature of curves of various orders. We ascertain that a curve of the second order cannot cut a straight line in more than two points; a curve of the third order cannot cut a straight line in more than three points, and so forth.

67. Previously we have mentioned the ways in which to determine the number of points in which the axis cuts any given curve. Given an equation in the coordinates x and y, since at any point where the curve lies on the axis $y = 0$, if we let $y = 0$ in the equation, there results an equation in x alone. The solutions to this equation give the values of x where the curve intersects the axis. Thus, in the equation for a circle, which we found above to be $y^2 = 2ax - x^2$, when we let $y = 0$, we have $0 = 2ax - x^2$. We obtain the two values $x = 0$ and $x = 2a$. This

indicates, on the axis RS, first the origin A, then the point B, since $AB = 2a$, are the points where the circle cuts the axis. In a similar way with other curves, when we let $y = 0$ in the equation, the roots indicate the points of intersection between the axis and the curve.

68. Since in the general equation for any curve, any straight line can serve as the axis, when we let $y = 0$ in the general equation, the resulting equation indicates in how many points the curve crosses the straight line. That is, we have an equation in the abscissa alone as the unknown, and each of the roots give the intersection of the curve with the axis. The number of intersections, then, depends on the largest power of x in the equation, and this can be no larger than the exponent on the highest power of x. The number of intersections is the same as the exponent on the highest power of x as long as the roots are all real. If some of the roots are complex, then the number of intersections will be even less.

69. Since we have given the general equations for lines of any order, from these equations we can find, by the given method, in how many points a line of any order can intersect any straight line. Hence, we take the general equation for a line of the first order, that is, for a straight line, $\alpha + \beta x + \gamma y$. When we let $y = 0$, we obtain $\alpha + \beta x = 0$. This equation can have no more than one root, so that the straight line can intersect another straight line in only one point. However, if $\beta = 0$, the impossible equation $\alpha = 0$, shows that in this case the axis is never cut by the straight line, since both of these straight lines are parallel. This fact is clear from the equation $\alpha + \gamma y = 0$, which results when $\beta = 0$.

70. If in the general equation for lines of the second order

$$\alpha + \beta x + \gamma y + \delta x^2 + \epsilon xy + \zeta y^2 = 0$$

we let $y = 0$, we obtain the equation

$$\alpha + \beta x + \delta x^2$$

This equation has either two real roots, or none, or even one in case $\delta = 0$. It follows that a line of the second order intersects a straight line in either two points, or in one, or not at all. All of these cases are covered by the statement that a line of the second order cannot intersect a straight line in more than two points.

71. If in the general equation of a line of the third order we let $y = 0$, we obtain the equation

$$\alpha + \beta x + \gamma x^2 + \delta x^3.$$

Since this equation cannot have more than three roots, it is clear that a line of the third order cannot intersect a straight line in more than three points. It can happen that a third order line intersects a straight line in fewer points, namely, in two if $\delta = 0$ and the equation $\alpha + \beta x + \gamma x^2 = 0$ has two real roots. There will be one point of intersection if in the equation above two of the roots are complex, or if both $\delta = 0$ and $\gamma = 0$. There will be no intersection if $\delta = 0$ and the other two roots are complex, or if $\delta = 0$, $\gamma = 0$, and $\beta = 0$, but α is not equal to zero.

72. In a similar way we see that a fourth order line cannot intersect a straight line in more than four points. This property is true of lines of all orders, that is, a line of the nth order cannot intersect a straight line in more than n points. It does not follow that every n^{th} order line intersects every straight line in n points, but the number of intersections can be less than n, and there may even be no intersection. This we have seen in detail with regard to lines of the second and third

orders. In this proposition the force of the statement is only that the number of intersections cannot be greater than the degree of the equation to which the curve is referred.

73. From the number of intersections which an arbitrary straight line makes with some given curve, we cannot determine to which order the curve belongs. If the number of intersections is equal to n, it does not follow that the curve is a line of order n, since it might belong to some higher order. It could be that the curve is not even algebraic, but is transcendental. It is safe to say that any curve which intersects a straight line in n points cannot belong to any order less than n. For instance, if a curve is intersected by a straight line in four points, then it is certain that it does not belong to the second order or the third order. Whether it belongs to fourth order, or some higher order, or is transcendental we are unable to decide.

74. The general equations which we have given for any order, contain several arbitrary constants. As soon as these constants are determined, the curves are determined with respect to a given axis, so that all of the other curves, which were contained in the general equation, are excluded. For instance, although in the first degree equation $\alpha + \beta x + \gamma y = 0$ only straight lines are contained, still with respect to a given axis, the line can be varied in an infinite number of ways by the infinite number of different values which can be assigned to α, β, γ. However, as soon as definite values are given to these constants, the position of the line is fixed, and it can satisfy no other equation.

75. This equation $\alpha + \beta x + \gamma y = 0$ would seem to have three different determinations, from the three constants α, β, and γ. However, we know from the nature of equations that the equation is really determined by the ratio between these constants, that is, the ratio of two of

them to the third. For instance if β and γ are determined by α, as for example, $\beta = -\alpha$ and $\gamma = 2\alpha$, the equation $\alpha - \alpha x + 2\alpha y = 0$, is already determined, since α can be removed by division. For a similar reason the general equation for a second order line, which contains six arbitrary constants, has only five determinations. The general equation for a line of the third order has nine. In general, the general equation for a line of the nth order has $\dfrac{(n + 1)(n + 2)}{2} - 1$ determinations.

76. We can always define the arbitrary constants in such a way that the curve will pass through a given point. This gives rise to one of the determinations. Let the general equation for a line of some order be give, which is to be determined so that the curve will pass through a given point B, as in *figure 17*. We chose any axis with some origin A and drop a perpendicular from Bb to the axis. It is clear that if the curve is to pass through the point B, then the value of the ordinate y will be equal to the perpendicular Bb when the value of the abscissa x is equal to Ab. Hence we substitute in the general equation for x the value Ab and for y the value Bb, to obtain an equation from which one of the constants α, β, γ, δ, ϵ, etc. can be defined. Once this has been done, then every curve contained in the general equation so determined will pass through the point B.

77. If the curve should pass through the point C also, then we drop the perpendicular Cc to the axis and in the equation make this substitution Ac for x and Cc for y. From this equation one of the constants α, β, γ, δ, etc. will be defined. In the same way we understand that if the curve is to pass through three points B, C, and D, then three of the constants are determined. Four constants are determined by four points B, C, D, and E. If the same number of points are given, through which the curve must pass, as there are determinations to be

made in the general equation, then the curve is completely determined
and is unique.

78. Since the general equation for a first order line, that is, a
straight line, has only two determinations, if two points are given
through which the straight line must pass, then the line is completely
determined. Indeed, through two points only one straight line can pass,
and this we know from Euclid's *Elements*. If only one point is given,
then because the equation is not yet determined, there are an infinite
number of straight lines which can pass through the same point.

79. The general equation of a second order line admits five determinations, so that if five points are given, through which the curve
should pass, then the second order line is completely determined. For
this reason there is only one second order line which will pass through
five given points. If only four points, or fewer, are given, then the equation is not yet completely determined, so that there are innumerable
different line of the second order which pass through these points. If
three of the five points lie on a straight line, then, since a second order
line cannot intersect a straight line in three points, there is no continuous
curve which will pass through those points. Rather we will have a complex line, that is, two straight lines. We have already been warned that
this is contained in the general equation for lines of the second order.

80. Since the general equation for a line of the third order has nine
determinations, there is a unique third order line which will pass through
any nine points. If we choose any number of points less than nine, then
there are innumerable third order lines which pass through those points.
In a similar way there is a unique fourth order line which passes through
any fourteen points, and a unique fifth order line through twenty points,
and so forth. In general, a line of the nth order will be determined by

$$\frac{(n + 1)(n + 2)}{2} - 1 = \frac{n(n + 3)}{2}$$ points. If the number of points is less than this, then there will be innumerable n^{th} order lines through these points.

81. Unless more than $\dfrac{n(n + 3)}{2}$ points are proposed, there is always one or an infinite number of lines of order n which pass through these points. There will be exactly one line if the number of points is $\dfrac{n(n + 3)}{2}$, and an infinite number if there are fewer points. It will never happen, no matter how the points may be placed, that the solution cannot be found. The determination of the coefficients α, β, γ, δ, etc. never requires the solution of a quadratic or higher degree equation, but all the equations are linear. It follows that the quantities α, β, γ, etc. will never be complex nor multiple-valued. For this reason there is always a real line passing through the proposed points, and it is unique, provided that the number of points is the same as the number of determinations to be made in the general equation.

82. Since we are free to choose any axis, the determination of the coefficients will be facilitated if we choose the axis through one of the points and let the origin A be that point. In this case $x = 0$ and $y = 0$, so that in the general equation

$$\alpha + \beta x + \gamma y + \delta x^2 + \epsilon xy + \zeta y^2 + \eta x^3 + \cdots$$

we have immediately $\alpha = 0$. Then we can have the axis pass through another of the points. If this is done, we reduce the number of quantities by which the position of the point is determined. Finally, we can choose oblique coordinates rather than rectangular so that the ordinate of another point can pass through the origin. Knowledge of the curve and its construction follows easily from the equation, whether the coordinates are rectangular or oblique.

83. If we want to find the second order line which passes through the five points A, B, C, D, and E, then we draw the axis through two of them, A and B, as in *figure 18*. We take A as the origin, then join A to another point C. We take the angle CAB to be the angle for the obliquity of the ordinates. Hence from the other points D and E we draw the ordinates Dd and Ee to the axis parallel to AC. We let $AB = a$, $AC = b$, $Ad = c$, $Dd = d$, $Ae = e$, and $eE = f$. We take the general equation for a second order line

$$\alpha + \beta x + \gamma y + \delta x^2 + \epsilon xy + \zeta y^2 = 0.$$

It is clear that the coordinates for the five points will be

$$x = 0, y = 0 \qquad\qquad x = 0, y = b \qquad\qquad x = a, y = 0$$
$$x = c, y = d \qquad\qquad\qquad x = e, y = f.$$

From these coordinates we obtain the five equations

$$I. \quad 0 = \alpha$$

$$II. \quad 0 = \alpha + \gamma b + \zeta b^2$$

$$III. \quad 0 = \alpha + \beta a + \delta a^2$$

$$IV. \quad 0 = \alpha + \beta c + \gamma d + \delta c^2 + d\epsilon\, cd + \zeta d^2$$

$$V. \quad 0 = \alpha + \beta e + \gamma f + \delta e^2 + \epsilon ef + \zeta f^2.$$

It follows that $\alpha = 0$; $\gamma = -\zeta b$; $\beta = -\delta a$. We substitute these values into the remaining equations to obtain

$$0 = -\delta ac - \zeta bd + \delta c^2 + \epsilon cd + \zeta d^2$$

$$0 = -\delta ae - \zeta bf + \delta e^2 + \epsilon ef + \zeta f^2$$

If we multiply the first equation by ef and the second by cd and then subtract one from the other, we obtain

$$0 = - \delta acef - \zeta bdef + \delta c^2 ef + \zeta d^2 ef$$

$$+ \delta acde + \zeta bcdf - \delta cde^2 - \zeta edf^2$$

or

$$\frac{\delta}{\zeta} = \frac{bdef - bcdf - d^2 ef + cdf^2}{acde - acef - cde^2 + c^2 ef}.$$

It follows that

$$\delta = df (be - bc - de + cf)$$

$$\zeta = ce (ad - af - de + cf).$$

In this way all the coefficients are determined.

84. Once all of the coordinates have been determined for the general equation

$$0 = \alpha + \beta x + \gamma y + \delta x^2 + \cdots$$

with respect to a given axis and obliquity of coordinates, the curve can be found by an infinite number of points determined by the equation; the curve will pass through all of the proposed points. If the general equation admits of more determinations than the number of points proposed, then the other points are chosen arbitrarily so that the equation is completely determined and will pass through all of the given points. In order easily to see the drawn curve, we assign to the abscissa x successively the positive and negative integers $0, 1, 2, 3, 4, 5, 6, \cdots$, and $-1, -2, -3, -4, \cdots$, and from the completely determined equation we obtain the corresponding values of the ordinates. In this way we obtain many points of the curve which are sufficiently close together.

CHAPTER V

On Second Order Lines.

85. Since the first order lines are nothing but straight lines, whose properties are sufficiently well known from elementary geometry, we consider a bit more carefully the SECOND ORDER lines. These are the simplest of the curves and have the widest application throughout the whole of the more sophisticated geometry. These lines, which are also called CONIC SECTIONS, have quite a few significant properties, some of which were known in antiquity, and others which have come to light more recently. A knowledge of these properties has been judged to be so important, that a good number of authors treat this matter immediately after elementary geometry. However, not all of these properties can be derived from a single principle. Some become clear immediately from the equation, other properties arise from cutting the cone and still others depend on other arguments. We will investigate here only those properties which flow directly from the equation.

86. Let us consider the general equation for a second order line, which is

$$\alpha + \beta x + \gamma y + \delta x^2 + \epsilon xy + \zeta y^2.$$

We have already seen that no matter how the axis is chosen or what the obliquity of coordinates may be, every second order line is contained in this equation. We have already given this equation the following form

$$y^2 + \frac{(\epsilon x + \gamma)y}{\zeta} + \frac{\delta x^2 + \beta x + \alpha}{\zeta} = 0.$$

From this it is clear that, for each value of the abscissa x, there correspond either two or no values for the ordinate y, depending whether the two roots of y are real or complex. If ζ happens to be equal to zero, then there is only a single ordinate for each abscissa, since the other has receded to infinity, and for that reason this exception does not disturb our investigation.

87. Suppose that both values of y are real, as happens in *figure 19*, where the ordinate PMN meets the curve in two points M and N. In this case we have the sum of the two roots, $PM + PN = -\dfrac{\epsilon x - \gamma}{\zeta}$ $= \dfrac{-\epsilon AP - \gamma}{\zeta}$, where we have taken the straight line AEF as the axis, A as the origin, and the angle APN as the angle of obliquity for the coordinates, which was chosen arbitrarily. When some other ordinate npm is drawn at the same angle, and one of the values pm is negative, then in the same way we have $pn - pm = \dfrac{-\epsilon Ap - \gamma}{\zeta}$. When this equation is subtracted from the previous one, we have $PM + pm + PN - pn = \dfrac{\epsilon(Ap - AP)}{\zeta} = \dfrac{\epsilon Pp}{\zeta}$. We draw straight lines from the points m and n, parallel to the axis, until we meet the previous ordinate in the points μ and ν. Then we have $M\mu + N\nu = \dfrac{\epsilon Pp}{\zeta}$, that is, the sum $M\mu + N\nu$ is to $Pp = m\mu = n\nu$ as ϵ is to ζ, and this ratio is constant. We note that this ratio is the same, no matter where on the curve the straight lines MN and mn may be drawn, provided only that they meet the axis with the given angle

and that the straight lines $n\nu$ and $m\mu$ are drawn parallel to the axis.

88. If the ordinate PMN is moved so that the points M and N coincide, then the ordinate is tangent to the curve, since at this point the secant line becomes tangent. Now let KCI be the tangent, as in *figure 20,* which is drawn parallel to whatever straight lines MN, mn have been chosen. Straight lines of this kind, which meet the curve on either end, are usually called CHORD ORDINATES. Then from the points M, N, m, and n, we draw the straight lines MI, NK, mi, and nk to the tangent, parallel to the axis. Since the intervals CK and Ck lie on opposite sides of the point C, Ck is taken with a negative sign. Hence we have $\dfrac{CI - CK}{MI} = \dfrac{\epsilon}{\zeta}$ and $\dfrac{Ci - CK}{mi} = \dfrac{\epsilon}{\zeta}$. It follows that

$$\frac{CI - CK}{MI} = \frac{Ci - Ck}{mi}$$

or

$$\frac{MI}{mi} = \frac{CI - CK}{Ci - Ck}.$$

89. Since the position of the axis with respect to the curve is arbitrary, the straight lines MI, NK, mi, and nk can be drawn arbitrarily, provided only that they be mutually parallel. Hence we always have $\dfrac{MI}{mi} = \dfrac{CI - CK}{Ci - Ck}$. Suppose that the parallel straight lines MI and NK are drawn in such a way that $CI = CK$. This will happen if they are parallel to the straight line CL, drawn from the point of contact C to L which bisects the ordinate MN. Then since $CI - CK = 0$, we also have $Ci - Ck = \dfrac{mi}{MI}(CI - CK) = 0$. It follows that when the straight line CL is extended to l, since mi and nk are also parallel to CL, we have $ml = nl$. Hence if the straight line CLl drawn from the point of contact C bisects one of the ordinates MN, parallel to the tangent, then it

bisects all ordinates mn parallel to the same tangent.

90. Since the straight line CLl cuts all ordinates parallel to the tangent into two equal parts, this line CLl is usually called the DIAME-TER of the *second order line* or the *conic section*. It follows that a second order line has innumerable diameters, since at any point of the curve there is a tangent. Whenever there is a tangent ICK, we can draw any ordinate MN parallel to this tangent. We bisect MN in the point L, then the straight line CL is a diameter of the second order line and it bisects all ordinates which are parallel to the tangent IK.

91. From this it follows that if a straight line Ll bisects any two parallel ordinates MN and mn, then it also bisects all the other ordinates which are parallel to these. We can now find a straight line IK, tangent to the curve, which is parallel to the ordinates, so that Ll is a diameter. If follows that we have a new method for finding innumerable diameters for a second order line: we arbitrarily choose two parallel chord ordinates MN and mn, we bisect them in L and l, then the straight line through these two points also bisects all other ordinates parallel to these. It follows that it is a diameter. Furthermore, when the diameter is extended to intersect the curve at the point C, then the straight line IK through C parallel to the ordinates is tangent to the curve at C.

92. A consideration of the sum of the roots of the equation

$$y^2 + \frac{(\epsilon x + \gamma)}{\zeta} y + \frac{\delta x^2 + \beta x + \alpha}{\zeta} = 0$$

has led us to the property we have been discussing. Now from the same equation we see that the product of the two roots $PM \cdot PN = \frac{\delta x^2 + \beta x + \alpha}{\zeta}$. The expression $\frac{\delta x^2 + \beta x + \alpha}{\zeta}$ has either two real factors or none at all. There will be two real factors if the axis cuts the curve in two points E and F, as in *figure 19*. In this case $y = 0$ so that $\dfrac{\delta x^2 + \beta x + \alpha}{\zeta} = 0$, and the roots x will be AE and AF. The factors will be $(x - AE)(x - AF)$ so that

$$\frac{\delta x^2 + \beta x + \alpha}{\zeta} = \frac{\delta}{\zeta}(x - AE)(x - AF) = \frac{\delta}{\zeta}PE \cdot PF$$

since $x = AP$. For this reason we have $PM \cdot PN = \dfrac{\delta}{\zeta}PE \cdot PF$, that is, the rectangle $PM \cdot PN$ has a constant ratio of $\dfrac{\delta}{\zeta}$ to the rectangle $PE \cdot PF$. The ordinate PMN can be drawn anywhere, provided the angle of obliquity NPF is as it should be. In a similar way, if the ordinate mn is drawn, and we take pE and pm to be negative, then $pm \cdot pn = \dfrac{\delta}{\zeta}pE \cdot pF$.

93. Let us draw any straight line PEF which intersects a second order line in two points E and F. If we draw any two ordinates parallel to each other, NMP and npm, as in *figure 21,* then we always have $\dfrac{PM \cdot PN}{PE \cdot PF} = \dfrac{pm \cdot pn}{pE \cdot pF}$, since each of these quotients is equal to $\dfrac{\delta}{\zeta}$. In a similar way, since the axis is arbitrary, we could take the straight line PMN as the axis and take a straight line eqf parallel to PEF. In this case we have

$$\frac{PM \cdot PN}{PE \cdot PF} = \frac{qM \cdot qN}{qe \cdot qf} = \frac{pm \cdot pn}{pE \cdot pF}.$$

It follows that

$$\frac{qe \cdot qf}{pE \cdot pF} = \frac{qM \cdot qN}{pm \cdot pn}.$$

Hence, if we take two parallel ordinates ef and EF and any other two ordinates MN and mn which are parallel to each other, which intersect at the points P, p, q, r, then we have the following equal quotients,

$$\frac{PM \cdot PN}{PE \cdot PF} = \frac{pm \cdot pn}{pE \cdot pF} = \frac{qM \cdot qN}{qe \cdot qf} = \frac{rm \cdot rn}{re \cdot rf}.$$

This is our second general property of a second order line.

94. If the two points M and N on the curve coincide, then the straight line PMN becomes the tangent to the curve through that point, and the rectangle $PM \cdot PN$ becomes a square. From this there follows a new property for tangents. Let the straight line CPp, in *figure 24*, be tangent to a second order line at the point C, and we let PMN and pmn be any two parallel straight lines which make the same angle with the axis as the tangent. From the property we found before we have $\frac{PC^2}{PM \cdot PN} = \frac{pC^2}{pm \cdot pn}$. That is, if MN is any ordinate which is extended to the tangent and meets it with a given angle, then the ratio of the square of the straight line CP to the rectangle $PM \cdot PN$ is always constant.

95. For the same reason, if any diameter CD is drawn for a second order line, as in *figure 20,* all ordinate chords $MNmn$ which are parallel and bisected by the diameter, with the diameter meeting the curve in the two points C and D, then

$$\frac{CL \cdot LD}{LM \cdot LN} = \frac{Cl \cdot lD}{lm \cdot ln}.$$

Since $LM = LN$ and $lm = ln$, we have

$$\frac{LM^2}{lm^2} = \frac{CL \cdot LD}{Cl \cdot lD}.$$

That is, the square of LM, which is half of the ordinate chord, is always in a constant ratio to the rectangle $CL \cdot LD$. It follows that if the diameter CD is taken to be the axis and half of the ordinate chord, LM is taken as the ordinate, we obtain an equation for a second order line. That is, if we let $CD = a$, the abscissa $CL = x$, and the ordinate

$LM = y$. Since $LD = a - x$, we have $\dfrac{y^2}{ax - x^2}$ is always a constant,

for instance, $\dfrac{h}{k}$. It follows that we have the equation for a second order

line $y^2 = \dfrac{h}{k}(ax - x^2)$.

96. From the two properties of second order lines which we have already discovered, we can deduce some other properties. Given a second order line, let AB and CD be two parallel ordinate chords. We complete the quadrilateral $ACDB$, as in *figure 22*. Now if at any point M on the curve we construct the ordinate chord MN, parallel to AB and CD, and intersecting the straight lines AC and BD in the points P and Q respectively, then PM and QN are equal to each other. The reason for this is that the straight line which bisects the parallel ordinate chords AB and CD also bisects the ordinate chord MN. By elementary geometry, the same straight line which bisects the sides AB and CD also bisects PQ. Since the lines MN and PQ are bisected in the same point, it follows that $MP = NQ$ and $MQ = NP$. In summary, if we are given in addition to the four points A, B, C, and D on a second order line, a fifth point M on the line which determines a sixth N and $NQ = MP$.

97. Since $MQ \cdot QN$ is in a constant ratio to $BQ \cdot DQ$, and $QN = MP$, it follows that $MP \cdot MQ$ is also in the same constant ratio to $BQ \cdot KDQ$. For example, if any other point c of the curve is chosen, and through it the straight line GcH, parallel to AB and CD, is constructed and extended until it intersects the sides AC and BD in the points G and H, then $cG \cdot cH$ is in the same constant ratio to $BH \cdot DH$. It follows that

$$\frac{cG \cdot cH}{BH \cdot DH} = \frac{MP \cdot MQ}{BQ \cdot DQ}.$$

If we construct the straight line RMS through M, parallel to BD, and intersecting the parallel chords AB and CD in the points R and S respectively, then since $BQ = MR$ and $DQ = MS$, the same ratio $\dfrac{MP \cdot MQ}{MR \cdot MS}$ is constant. It follows that if through some point M on the curve we construct two straight lines, the one MPQ parallel to the sides AB and CD, the other RMS parallel to the base BD, then the points of intersection P, Q, and S will be such that $\dfrac{MP \cdot MQ}{MR \cdot MS}$ is a constant ratio.

98. If instead of the chord CD, which is parallel to AB, we substitute any chord Dc through the point D, and add the chord Ac, then the straight lines MQ and RMS, as before, pass through M, are parallel to AB and BD, and intersect the sides of the quadrilateral $ABDc$ in p, Q, R, and s. The same property holds, since $\dfrac{MP \cdot MQ}{BQ \cdot DQ} = \dfrac{cG \cdot cH}{BH \cdot DH}$. That is, $\dfrac{MP \cdot MQ}{MR \cdot MS} = \dfrac{cG \cdot cH}{BH \cdot DH}$, since the straight line RS is parallel to BD and is equal to it. The triangle APp is similar to AGc, as is DSs to cHD, so that we have $\dfrac{Pp}{AP} = \dfrac{Gc}{AG}$. Since $\dfrac{AP}{AG} = \dfrac{BQ}{BH}$, we have $\dfrac{Pp}{BQ} = \dfrac{Gc}{BH}$. From the other similar triangle we conclude that $\dfrac{DS}{Ss} = \dfrac{cH}{DH}$. Since $DS = MQ$, we also have $\dfrac{MQ}{Ss} = \dfrac{cH}{DH}$. It follows that

$$\frac{MQ \cdot Pp}{MR \cdot Ss} = \frac{cG \cdot cH}{BH \cdot DH}$$

since $BQ = MR$. This equation together with the one stated above gives

$$\frac{MP \cdot MQ}{MR \cdot MS} = \frac{Pp \cdot MQ}{MR \cdot Ss}.$$

Finally we have

$$\frac{MP \cdot MQ}{MR \cdot MS} = \frac{Mp \cdot MQ}{MR \cdot Ms}.$$

We conclude that wherever we choose the points c and M on the curve, the ratio $\dfrac{Mp \cdot MQ}{MR \cdot Ms}$ is always the same, provided the straight lines MQ and Rs through M are parallel to the chords AB and BD. From the equation above it follows that $\dfrac{MP}{MS} = \dfrac{Mp}{Ms}$. Since when c is changed, only the points P and s change with it, the ratio $\dfrac{Mp}{Ms}$ remains the same however c may vary, as long as M remains fixed.

99. Let A, B, C, and D be any four points on a second order line, and let the points be joined by straight lines to form the inscribed trapezium $ABDC$, as in *figure 23*. We will deduce from what has gone before the property of conic sections. That is, if from any point M of the curve, the straight lines MP, MQ, MR, and MS are drawn so as to meet the four sides of the trapezium with the same angle, then the products of the lengths of the two lines to opposite sides are always in a constant ratio. Indeed, $\dfrac{MP \cdot MQ}{MR \cdot MS}$ always has a fixed value no matter where the point M is chosen on the curve, provided the angle at P, Q, R, and S is kept the same. In order to see this, we construct the two straight line MQ and rs through M, the first parallel to AB and the second parallel to BD. We let p, q, r, and s be the points where these two lines intersect the four sides of the trapezium. From what we have seen before, $\dfrac{Mp \cdot Mq}{Mr \cdot Ms}$ has a fixed value. Because of the given angles, we have the given ratios $\dfrac{MP}{Mp}$, $\dfrac{MQ}{Mq}$, $\dfrac{MR}{Mr}$, and $\dfrac{MS}{Ms}$. From these it follows that $\dfrac{MP \cdot MQ}{MR \cdot MS}$ is the given ratio.

100. We saw above that if the parallel chords MN and mn are extended to meet some tangent CPp in the points P and p, as in *figure 24,* then

$$\frac{PM \cdot PN}{CP^2} = \frac{pm \cdot pn}{Cp^2}.$$

If we choose the points L and l so that PL is the geometric mean of PM and PN, and that pl is the geometric mean of pm and pn, then $\frac{PL^2}{CP^2} = \frac{pl^2}{Cp^2}$. It follows that $\frac{PL}{CP} = \frac{pl}{Cp}$, and it becomes clear that the points L and l lie on the same line through the point of tangency, C. Hence, if one ordinate PMN is divided by L in such a way that $PL^2 = PM \cdot PN$, then the straight line CLD drawn through the points C and L will intersect all the other ordinates pmn in a point l such that pl is the geometric mean of pm and pn. Put another way, if two ordinates PN and pn are divided by the points L and l in such a way that $PL^2 = PM \cdot PN$ and $pl^2 = pm \cdot pn$, then the straight line drawn through L and l will pass through C, the point of tangency, and all of the other parallel ordinates will be cut in the same ratio.

101. Now that we have discussed these properties of a second order line which follow immediately from the form of the equation, we proceed to investigate some of the more recondite properties. We recall the general equation of a second order line.

$$y^2 + \frac{\epsilon x + \gamma}{\zeta} y + \frac{\delta x^2 + \beta x + \alpha}{\zeta} = 0.$$

From this equation we see that for a given abscissa x with two ordinates y, namely PM and PN, we can define a diameter which bisects all ordinate chords MN. Let IG be such a diameter, as in *figure 25,* which bisects the chord MN in the point L. We let $PL = z$, and since $z = \frac{1}{2}PM + \frac{1}{2}PN$, we have $z = \dfrac{-\epsilon x - \gamma}{2\zeta}$, or $2\zeta z + \epsilon x + \gamma = 0$, which is the equation of the diameter.

102. From this we can calculate the length of the diameter IG and find the two points on the curve where the points M and N coincide, that is where $PM = PN$. From the equation we know that

$$PM + PN = \frac{-\epsilon x - \gamma}{\zeta}$$

and

$$PM \cdot PN = \frac{\delta x^2 + \beta x + \alpha}{\zeta},$$

so that

$$(PM - PN)^2 = (PM + PN)^2 - 4PM \cdot PN$$

$$= \frac{(\epsilon^2 - 4\delta\zeta)x^2 + 2(\epsilon\gamma - 2\beta\zeta)x + (\gamma^2 - 4\alpha\zeta)}{\zeta^2} = 0.$$

That is,

$$x^2 - \frac{2(2\beta\zeta - \epsilon\gamma)}{\epsilon^2 - 4\delta\zeta} x + \frac{\gamma^2 - 4\alpha\zeta}{\epsilon^2 - 4\delta\zeta} = 0.$$

Since the roots of this equation are AK and AH, we have $AK + AH = \dfrac{4\beta\zeta - 2\epsilon\gamma}{\epsilon^2 - 4\delta\zeta}$ and $AK \cdot AH = \dfrac{\gamma^2 - 4\alpha\zeta}{\epsilon^2 - 4\delta\zeta}$. It follows that

$$(AH - AK)^2 = KH^2$$

$$= \frac{4(2\beta\zeta - \epsilon\gamma)^2 - 4(\epsilon^2 - 4\delta\zeta)(\gamma^2 - 4\alpha\zeta)}{(\epsilon^2 - 4\delta\zeta)^2}.$$

If we let the coordinates be rectangular, then

$$IG^2 = \frac{\epsilon^2 + 4\zeta^2}{4\zeta^2}.$$

103. We have considered the case when the coordinates are rectangular, now we seek the equation when the coordinates are oblique. From any point M on the curve we draw an oblique ordinate Mp which makes an oblique angle MpH with the axis. We let the sine of this angle be μ and the cosine be ν. We let the new abscissa Ap be equal to t and the ordinate pM be equal to u. Then $\dfrac{y}{u} = \mu$ and $\dfrac{Pp}{u} = \nu$, so that $y = \mu u$ and $x = t + \nu u$. When we substitute these values into the equation in x and y, which is

$$\alpha + \beta x + \gamma y + \delta x^2 + \epsilon xy + \zeta y^2 = 0,$$

we obtain

$$\alpha + \beta t + \nu\beta u + \mu\gamma u + \delta t^2 + 2\nu\delta tu$$

$$+ \mu\epsilon tu + \nu^2\delta u^2 + \mu\nu\epsilon u^2 + \mu^2\zeta u^2 = 0$$

or

$$u^2 + \frac{((\mu\epsilon + 2\nu\delta)t + \mu\gamma + \nu\beta)u}{\mu^2\zeta + \mu\nu\epsilon + \nu^2\delta}$$

$$+ \frac{\delta t^2 + \beta t + \alpha}{\mu^2\zeta + \mu\nu\epsilon + \nu^2\delta} = 0.$$

104. In this case also, each ordinate has two values, namely pM and pn. For this reason, from the chord Mn, a diameter ilg is defined as before. That is, if the chord Mn is bisected at l, then l is a point on the diameter. We let $pl = v$, then

$$v = \frac{pM + pn}{2} = \frac{-(\mu\epsilon + 2\nu\delta)t - \mu\gamma - \nu\beta}{2(\mu^2\zeta + \mu\nu\epsilon + \nu^2\delta)}.$$

We drop a perpendicular lq from l to the axis AH and let $Aq = p$, $ql = q$, then $\mu = \dfrac{q}{v}$ and $v = \dfrac{pq}{v} = \dfrac{p - t}{v}$. It follows that $v = \dfrac{q}{\mu}$

and $t = p - vv = p - \dfrac{vq}{\mu}$. When we substitute these values into the equation in t and v which we found above, we obtain

$$\frac{q}{\mu} = \frac{- \mu\epsilon p - 2v\delta p + v\epsilon q + \dfrac{2v^2\delta q}{\mu} - \mu\gamma - v\beta}{2\mu^2\zeta + 2\mu v\epsilon + 2\,v^2\delta}$$

or

$$(2\mu^2\zeta + \mu v\epsilon)q + (\mu^2\epsilon + 2\mu v\delta)p + \mu^2\gamma + \mu v\beta = 0,$$

or

$$(2\mu\zeta + v\epsilon)q + (\mu\epsilon + 2v\delta)p + \gamma\mu + v\beta = 0,$$

which defines the position of the diameter ig.

105. When the previous diameter IG, whose position is determined by the equation $2\zeta z + \epsilon x + \gamma = 0$, is extended, it intersects the axis in the point O, so that $AO = \dfrac{-\gamma}{\epsilon}$. It follows that $PO = \dfrac{-\gamma}{\epsilon} - x$, the tangent of the angle LOP is equal to

$$\frac{z}{PO} = \frac{-\epsilon z}{\epsilon x + \gamma} = \frac{\epsilon}{2\zeta},$$

and the tangent of the angle MLG, in which the diameter IG bisects the chord MN, is equal to $\dfrac{2\zeta}{\epsilon}$. When the second diameter IG is extended, it intersects the axis in the point o, so that $Ao = \dfrac{-\mu\gamma - v\beta}{\mu\epsilon + 2v\delta}$, and the tangent of the angle Aol is equal to $\dfrac{\mu\epsilon + 2v\delta}{2\mu\zeta + v\epsilon}$. Since the tangent of the angle AOL is equal to $\dfrac{\epsilon}{2\zeta}$, both diameters intersect each other in the point C with an angle $OCo = Aol - AOL$, whose tangent is equal to $\dfrac{4v\delta\zeta - v\epsilon^2}{4\mu\zeta^2 + 2v\delta\epsilon + 2v\epsilon\zeta + \mu\epsilon^2}$. The angle in which the second diameter bisects its chord is $Mlo = 180^0 - lpo - Aol$ and its tangent is equal to

$$\frac{2\mu^2\zeta + 2\ \mu\nu\epsilon + 2\nu^2\delta}{\mu^2\epsilon + 2\mu\nu\delta - 2\mu\nu\zeta - \nu^2\epsilon}.$$

106. We now consider the point C where the two diameters intersect. From C we drop the perpendicular CD to the axis, we let $AD = g$, and $CD = h$. It follows first, since C lies on the diameter IG that $2\zeta h + \epsilon g + \gamma = 0$, and also, since C lies on the diameter ig, we have

$$(2\mu\zeta + \nu\epsilon)h + (\mu\epsilon + 2\nu\delta)g + \mu\gamma + \nu\beta = 0.$$

When the previous equation is multiplied by μ and subtracted from this equation, there remains $\nu\epsilon h + 2\nu\delta g + \nu\beta = 0$ or $\epsilon h + 2\delta\gamma + \beta = 0$. From this we have

$$h = \frac{-\ \epsilon g - \gamma}{2\zeta} = \frac{-\ 2\delta g - \beta}{\epsilon},$$

so that

$$(\epsilon^2 - 4\delta\zeta)g = 2\beta\zeta - \gamma\epsilon,$$

$$g = \frac{2\beta\zeta - \gamma\epsilon}{\epsilon^2 - 4\delta\zeta},$$

and

$$h = \frac{2\gamma\delta - \beta\epsilon}{\epsilon^2 - 4\delta\zeta}.$$

In these calculations the quantities μ and ν do not appear; the obliquity of the ordinate pMn depends on these, so that it is clear that the point C remains the same no matter how the obliquity varies.

107. It follows that all diameters IG and ig pass through the same point C. Once this point is found, all diameters pass through it, and conversely, all chords which pass through the point are diameters which

bisect all chords drawn with a certain angle. Since this point in any second order line is unique and all diameters pass through it, it is usually called the CENTER of the conic section. From the equation in x and y

$$\alpha + \beta x + \gamma y + \delta x^2 + \epsilon xy + \zeta y^2 = 0$$

if we take $AD = \dfrac{2\beta\zeta - \gamma\epsilon}{\epsilon^2 - 4\delta\zeta}$ we obtain $CD = \dfrac{2\gamma\delta - \beta\epsilon}{\epsilon^2 - 4\delta\zeta}$.

108. We have seen above that

$$AK + AH = \frac{4\beta\zeta - 2\gamma\epsilon}{\epsilon^2 - 4\delta\zeta}.$$

We recall that IK and GH are the perpendiculars from the ends of the diameter IG to the axis. From this it is clear that $AD = \dfrac{AK + AH}{2}$ and that the point D is the midpoint between K and H. For this reason we see also that C is the midpoint of the diameter IG. Since this is also true for any other diameter, it follows that not only do all diameters pass through the same point C, but they are also bisected by that point.

109. Now let us take any diameter AI as the axis with the chord MN as ordinate meeting the axis with an angle $APM = q$, whose sine is m and cosine is n, as in *figure 26*. We let the abscissa $AP = x$ and the ordinate $PM = y$. When the two values of the ordinate are equal in length, with one being the negative of the other, their sum is equal to zero, so that the general equation of a second order line takes the form $y^2 = \alpha + \beta x + \gamma x^2$. If we let $y = 0$, we obtain the points G and I on the axis where it crosses the curve. That is, the equation $x^2 + \dfrac{\beta}{\gamma}x + \dfrac{\alpha}{\gamma} = 0$ has the two roots $x = AG$ and $x = AI$. It follows that

$$AG + AI = \frac{-\beta}{\gamma}$$

and

$$AG \cdot AI = \frac{\alpha}{\gamma}.$$

Since the center C is the midpoint of the diameter GI, it is easy to find the center of the conic section, C; we have

$$AC = \frac{AG + AI}{2} = \frac{-\beta}{2\gamma}.$$

110. Once we know the center of the conic section C on the axis AI, it is very convenient to choose that point for the origin. With this choice we let $CP = t$ and we keep $PM = y$. Since

$$x = AC - C = \frac{-\beta}{2\gamma} - t,$$

we obtain the following equation in the coordinates t and y:

$$y^2 = \alpha - \frac{\beta^2}{2\gamma} + \frac{\beta^2}{4\gamma} - \beta t + \beta t + \gamma t^2$$

or

$$y^2 = \alpha - \frac{\beta^2}{4\gamma} + \gamma t^2.$$

When we substitute x for t we have the general equation for a second order line with any diameter for axis and the center for origin. If we change the constants we have $y^2 = \alpha - \beta x^2$. When we let $y = 0$, we obtain $CG = GI = \left(\frac{\alpha}{\beta}\right)^{\frac{1}{2}}$. It follows that the length of the diameter GI is equal to $2\left(\frac{\alpha}{\beta}\right)^{\frac{1}{2}}$.

111. If we let $x = 0$, we find the chord through the center, EF. That is, $CE = CF = \sqrt{\alpha}$, so that the length of the chord EF is equal to $2\sqrt{\alpha}$. Since this chord passes through the center, it is also a diameter, and it makes an angle $ECG = q$ with GI. This diameter EF bisects all of the chords which are parallel to the original diameter GI. We choose the ordinate aC equal to PM, and since it is parallel to PM, the line joining M to a is parallel to GI. It follows that this line is bisected by the diameter EF. These two diameters, GI and EF, are related to each other in such a way that one bisects all chords parallel to the other. Because of this reciprocal property, these two diameters are called CONJUGATE. If through the terminal points G and I of the diameter GI we draw straight lines parallel to the other diameter EF, these lines will be tangent to the curve; in like manner, if we draw straight lines through E and F parallel to the diameter GI, they will be tangent to the curve at the points E and F.

112. Now we let MQ be any oblique ordinate with the angle $AQM = \phi$, whose sine is equal to μ and cosine is equal to ν. We let the abscissa $CQ = t$ and the ordinate $MQ = u$. In the triangle PMQ, since the angle

$$PMQ = \phi - q, \qquad \sin PMQ = \mu n - \nu m,$$

and

$$\frac{y}{\mu} = \frac{u}{m} = \frac{PQ}{\mu n - \nu m},$$

so that

$$y = \frac{\mu n}{M}, \qquad PQ = \frac{(\mu n - \nu m)u}{m},$$

and

$$x = t - PQ = t - \frac{(\mu n - \nu m)u}{m}.$$

We substitute these values into the equation $y^2 = \alpha - \beta x^2$ or

$$y^2 + \beta x^2 - \alpha = 0$$

to obtain the equation

$$(\mu^2 + \beta(\mu n - \nu m)^2)u^2 - 2\beta m(\mu n - \nu m)tu$$

$$+ \beta m^2 t^2 - \alpha m^2 = 0.$$

From this equation we obtain the two values for u, QM and $- Qn$, so that

$$QM - Qn = \frac{2\beta m(\mu n - \nu m)t}{\mu^2 + \beta(\mu n - \nu m)^2}.$$

Let the chord Mn be bisected at p, then the straight line Cpg is a new diameter which bisects all chords which are parallel to Mn, and

$$Qp = \frac{\beta m(\mu n - \nu m)t}{\mu^2 + \beta(\mu n - \nu m)^2}.$$

113. From this we obtain the tangent of the angle GCg, which is equal to

$$\frac{\mu \cdot Qp}{CQ + \nu \cdot Qp} = \frac{\beta m(\mu n - \nu m)}{\mu + n\beta(\mu n - \nu m)}$$

and the tangent of the angle Mpg is equal to

$$\frac{\mu \cdot CQ}{pQ + \nu \cdot CQ} = \frac{\mu^2 + \beta(\mu n - \nu m)^2}{\mu\nu + \beta(\mu n - \nu m)(\nu n + \mu m)}.$$

This is the angle which the new chord mn makes with the diameter gi when it is bisected. Furthermore, we have

$$Cp^2 = CQ^2 + Qp^2 + 2v \cdot CQ \cdot Qp =$$

$$\frac{\mu^4 + 2\beta\mu^3 n(\mu n - vm) + \beta^2\mu^2(\mu n - vm)^2}{(\mu^2 + \beta(\mu n - vm)^2)^2} t^2.$$

It follows that

$$Cp = \frac{\mu t \sqrt{\mu^2 + 2\beta\mu n(\mu n - vm) + \beta^2(\mu n - vm)^2}}{\mu^2 + \beta(\mu n - vm)^2}.$$

If we let $Cp = r$ and $pM = s$, then

$$t = \frac{(\mu^2 + \beta(\mu n - vm)^2)r}{\mu(\mu^2 + 2\beta\mu n(\mu n - vm) + \beta^2(\mu n - vm)^2)^{\frac{1}{2}}}$$

and

$$u = s + Qp =$$

$$s + \frac{\beta m(\mu n - vm)r}{\mu(\mu^2 + 2\beta\mu n(\mu n - vm) + \beta^2(\mu n - vm)^2)^{\frac{1}{2}}}.$$

From these we obtain the values

$$y = \frac{\mu s}{m}$$

$$+ \frac{\beta(\mu n - vm)r}{(\mu^2 + 2\beta\mu n(\mu n - vm) + \beta^2(\mu n - vm)^2)^{\frac{1}{2}}}$$

$$x = - \frac{(\mu n - vm)s}{m}$$

$$+ \frac{\mu r}{(\mu^2 + 2\beta\mu n(\mu n - vm) + \beta^2(\mu n - vm)^2)^{\frac{1}{2}}}.$$

From the equation $y^2 + \beta x^2 - \alpha = 0$ we obtain

$$\frac{(\mu^2 + \beta(\mu n - \nu m)^2)s^2}{m^2}$$

$$+ \frac{\beta(\mu^2 + \beta(\mu n - \nu m)^2)r^2}{\mu^2 + 2\beta\mu n(\mu n - \nu m) + \beta^2(\mu n - \nu m)^2} - \alpha = 0.$$

114. We let the semidiameter $CG = f$ and we let the semiconjugate $CE = CF = g$, then $f = \left(\dfrac{\alpha}{\beta}\right)^{\frac{1}{2}}$ and $g = \sqrt{\alpha}$, or $\alpha = g^2$ and $\beta = \dfrac{g^2}{f^2}$. From this it follows that $y^2 + \dfrac{g^2 x^2}{f^2} = g^2$. If we let the angle $GCg = p$, then

$$\tan p = \frac{\beta m(\mu n - \nu m)}{\mu + n\beta(\mu n - \nu m)}.$$

Since the angle $GCE = q$, if we let the angle $ECe = \pi$, then $AQM = \phi = q + \pi$, so that $\mu = \sin(q + \pi)$, $\nu = \cos(q + \pi)$, $m = \sin q$, and $n = \cos q$. It follows that

$$\tan p = \frac{\beta \sin q \sin \pi}{\sin(q + \pi) + \beta \cos q \sin \pi} = \frac{\beta \tan q \tan \pi}{\tan q + \tan \pi + \beta \tan \pi}$$

and

$$\sin p = \frac{\beta \sin q \sin \pi}{(\mu^2 + 2\beta\mu n(\mu n - \nu m) + \beta^2(\mu n - \nu m)^2)^{\frac{1}{2}}}.$$

Furthermore

$$\mu^2 + \beta(\mu n - \nu m)^2 = (\sin(q + \pi))^2 + \beta(\sin \pi)^2.$$

When we use these values, we obtain the following equation in r and s,

$$\frac{((\sin(q + \pi))^2 + \beta(\sin \pi)^2)s^2}{(\sin q)^2}$$

$$+ \frac{\beta((\sin(q + \pi))^2 + \beta(\sin \pi)^2)r^2(\sin p)^2}{\beta^2(\sin q)^2(\sin \pi)^2} - \alpha = 0.$$

Since

$$\beta = \frac{\tan p \, \sin(q + \pi)}{(\sin q - \cos q \tan p)\sin \pi} = \frac{\tan p(\tan q + \tan \pi)}{\tan \pi(\tan q - \tan p)}$$

$$= \frac{g^2}{f^2} = \frac{\cot \pi \tan q + 1}{\cot p \tan q - 1},$$

or

$$\tan q = \frac{f^2 + g^2}{g^2\cot p - f^2\cot \pi}.$$

Several things can be deduced from this, among which is
$$\frac{g^2}{f^2} = \frac{\sin p \, \sin(q + \pi)}{\sin \pi \, \sin(q - p)}.$$

115. Let the semidiameter $Cg = a$ and let the semiconjugate $Ce = b$. From the equation found above, we have

$$a = \frac{\sin q \, \sin \pi \sqrt{\alpha\beta}}{\sin p\left((\sin(q + \pi))^2 + \beta(\sin \pi)^2\right)^{\frac{1}{2}}}$$

$$= \frac{g^2\sin q \, \sin \pi}{\sin p\left(f^2(\sin(q + \pi))^2 + g^2(\sin \pi)^2\right)^{\frac{1}{2}}}$$

and

$$b = \frac{fg \, \sin q}{\left(f^2(\sin(q + \pi))^2 + g^2(\sin \pi)^2\right)^{\frac{1}{2}}}.$$

From these two expressions we conclude that $\dfrac{a}{b} = \dfrac{g \, \sin \pi}{f \, \sin p}$. Since

$$(\sin(q + \pi))^2 + \frac{g^2}{f^2}(\sin \pi)^2$$

$$= \frac{\sin(q + \pi)}{\sin(q - p)}(\sin(q - p)\sin(q + \pi) + \sin p \sin \pi)$$

$$= \frac{\sin q \, \sin(q + \pi)\sin(q + \pi - p)}{\sin(q - p)},$$

it follows that

$$a = \frac{g^2 \sin \pi}{f \sin p}\left(\frac{\sin q \, \sin(q - p)}{\sin(q + \pi)\sin(q + \pi - p)}\right)^{\frac{1}{2}}.$$

That is, since

$$\frac{g^2}{f^2} = \frac{\sin p \, \sin(q + \pi)}{\sin \pi \, \sin(q - p)}$$

we have

$$a = f\left(\frac{\sin q \, \sin(q + \pi)}{\sin(q - p)\sin(q + \pi - p)}\right)^{\frac{1}{2}},$$

and

$$b = g\left(\frac{\sin q \, \sin(q - p)}{\sin(q + \pi)\sin(q + \pi - p)}\right)^{\frac{1}{2}}.$$

It follows that

$$\frac{a}{b} = \frac{f \, \sin(q + \pi)}{g \sin(q - p)}$$

and

$$ab = \frac{fg \, \sin q}{\sin(q + \pi - p)}.$$

116. We conclude that if we have two pair of conjugate diameters, GI, EF and gi, ef, then

$$\frac{Cg}{Ce} = \frac{CG \, \sin \, ECe}{\sin \, GCg}$$

Hence,

$$\frac{\sin \, GCg}{\sin \, ECe} = \frac{CE \cdot Ce}{CG \cdot Cg}.$$

If we draw the chords Ee and Gg, then the area of the triangle CGg is equal to the area of the triangle CEe. It follows that

$$\frac{Cg}{Ce} = \frac{CG \, \sin \, GCe}{CE \, \sin \, gCE}$$

or $Ce \cdot CG \, \sin \, GCe = CE \cdot Cg \, \sin \, gCE$, so that when the chords Ge and gE are drawn, the areas of the triangles GCe and gCE are equal, and the areas of the triangles ICf and iCF are equal. From the last equation,

$$ab \, \sin(q + \pi - p) = fg \, \sin \, q,$$

we conclude that

$$Cg \cdot Ce \, \sin \, gCe = CG \cdot CE \, \sin \, GCE.$$

If the chords EG and eg, are drawn, or from the sections FI and fi, we see that the triangles ICF and iCf are also equal. We conclude that all parallelograms which are described around the pairs of conjugate diameters are equal to each other.

117. We have three pair of equal triangles, namely triangles FCf and ICi are equal, fCI and FCi are equal, as are FCI and fCi. It follows that the quadrilaterals $FfCI$ and $iICf$ are equal, so that when the common triangle fCI is removed, the triangle FIf is equal to the triangle Ifi. Since both of these triangles have a common base in fI, it follows that the chords Fi and fI are parallel. From this it follows that the triangle FIi is equal to triangle ifF, so that when the equal triangles

FCi and fCi are added, the two quadrilaterals $FCIi$ and $iCfF$ are equal to each other

118. From this result we derive a method for constructing a tangent MT at the point M of any second order line. We take the diameter GI for the axis, to which EC is, as in *figure 27*, a semiconjugate. From the point M to the axis we construct MP parallel to CE, so that PM is a semichord with $PN = PM$. We draw the semidiameter CM and look for its semiconjugate CK, which will be parallel to the tangent. Let the angles $GCE = q$, $GCM = p$, and $ECK = \pi$. As we have seen

$$\frac{EC^2}{GC^2} = \frac{\sin p \, \sin(q + \pi)}{\sin \pi \, \sin(q - p)},$$

and

$$MC = CG\frac{\sqrt{\sin q \, \sin(q + \pi)}}{\sin(q - p)\sin(q + \pi - p)}.$$

In the triangle CMP we have

$$MC^2 = CP^2 + MP^2 + 2PM \cdot CP \, \cos q,$$

$$\frac{MP}{MC} = \frac{\sin p}{\sin q},$$

and

$$\frac{MP}{CP} = \frac{\sin p}{\sin(q - p)}.$$

Further, in the triangle CMT, because of the given angles, we have

$$\frac{CM}{\sin(q + \pi)} = \frac{CT}{\sin(q + \pi - p)} = \frac{MT}{\sin p}.$$

When we eliminate the angles we find that

$$MC = CG\left(\frac{MC\cdot CM}{CP\cdot CT}\right)^{\frac{1}{2}},$$

that is, $CG^2 = CP\cdot CT$. It follows that $\dfrac{CP}{CG} = \dfrac{CG}{CT}$ and from this we can find the position of the tangent. From this proportion, by division, be obtain $\dfrac{CP}{PG} = \dfrac{CG}{TG}$. Since $CG = CI$, by composition we have $\dfrac{CP}{IP} = \dfrac{CG}{TI}$.

     119. Since

$$\frac{CE^2}{CG^2} = \frac{\sin p \ \sin(q + \pi)}{\sin \pi \ \sin(q - p)}$$

and

$$\frac{CK^2}{CM^2} = \frac{\sin p \ \sin(q - p)}{\sin \pi \ \sin(q + \pi)},$$

likewise

$$\frac{CM^2}{CG^2} = \frac{\sin q \ \sin(q + \pi)}{\sin(q - p) \ \sin(q + \pi - p)}$$

and

$$\frac{CK^2}{CE^2} = \frac{\sin q \ \sin(q - p)}{\sin(q + \pi)\sin(q + \pi - p)},$$

we have

$$\frac{CE^2 + CG^2}{CG^2} = \frac{\sin p \ \sin(q + \pi) + \sin \pi \ \sin(q - p)}{\sin \pi \ \sin(q - p)},$$

and

$$\frac{CK^2 + CM^2}{CM^2} = \frac{\sin p \ \sin(q - p) + \sin \pi \ \sin(q + \pi)}{\sin \pi \ \sin(q + \pi)}.$$

Since

$$\sin A \, \sin B = \tfrac{1}{2} \cos(A - B) - \tfrac{1}{2} \cos(A + B),$$

and conversely,

$$\tfrac{1}{2} \cos A - \tfrac{1}{2} \cos B = \sin\left(\frac{A + B}{2}\right)\sin\left(\frac{B - A}{2}\right),$$

we have

$$\sin p \, \sin(q + \pi) + \sin \pi \, \sin(q - p)$$

$$= \tfrac{1}{2} \cos(q + \pi - p) - \tfrac{1}{2} \cos(q + \pi + p)$$

$$+ \tfrac{1}{2} \cos(q - \pi - p) - \tfrac{1}{2} \cos(q + \pi - p)$$

$$= \tfrac{1}{2} \cos(q - \pi - p) - \tfrac{1}{2} \cos(q + \pi + p)$$

$$= \sin q \, \sin(p + \pi).$$

We also have

$$\sin p \, \sin(q - p) + \sin \pi \, \sin(q + \pi)$$

$$= \tfrac{1}{2} \cos(q - 2p) - \tfrac{1}{2} \cos q + \tfrac{1}{2} \cos q - \tfrac{1}{2} \cos(q + 2\pi)$$

$$= \tfrac{1}{2} \cos(q - 2p) - \tfrac{1}{2} \cos (q + 2\pi) = \sin(q + \pi - p)\sin(p + \pi).$$

From these results we have

$$\frac{CE^2 + CG^2}{CG^2} = \frac{\sin q \, \sin(p + \pi)}{\sin \pi \, \sin(q - p)}$$

$$\frac{CK^2 + CM^2}{CM^2} = \frac{\sin(q + \pi - p)\sin(p + \pi)}{\sin \pi \, \sin(q + \pi)}$$

From this we conclude

$$\frac{CE^2 + CG^2}{CK^2 + CM^2} = \frac{CG^2}{CM^2}\frac{\sin q \, \sin(q + \pi)}{\sin(q - p)\sin(q + \pi - p)} = \frac{CG^2}{CM^2}\frac{CM^2}{CG^2}.$$

It follows that $CE^2 + CG^2 = CK^2 + CM^2$, and so in any second order line the sum of the squares of two conjugate semidiameters is always constant.

120. If we given two conjugate semidiameters CG and CE, and we choose arbitrarily a semidiameter CM, then we can quickly calculate its conjugate semidiameter CK, by letting $CK = \sqrt{CE^2 + CG^2 - CM^2}$. From the properties of conic sections we have found, it follows that

$$\frac{TG \cdot TI}{TM^2} = \frac{CG \cdot CI}{CK^2} = \frac{CG^2}{CK^2} = \frac{CG^2}{CE^2 + CG^2 - CM^2}.$$

We conclude that

$$TM = CG \left(\frac{CE^2 + CG^2 - CM^2}{TG \cdot TI} \right)^{\frac{1}{2}}.$$

In a similar way, if the chord MN, which passes through P, is drawn then the tangent NT can be drawn and both tangents MT and NT meet the axis TI in the same point T. After we draw the straight line CN we have

$$TN = CG \left(\frac{CE^2 + CG^2 - CN^2}{TG \cdot TI} \right)^{\frac{1}{2}},$$

so that

$$\frac{TM^2}{TN^2} = \frac{CE^2 + CG^2 - CM^2}{CE^2 + CG^2 - CN^2}.$$

Since MN is bisected by P, we have

$$\frac{\sin CTM}{\sin CTN} = \frac{TN}{TM} = \left(\frac{CE^2 + CG^2 - CN^2}{CE^2 + CG^2 - CM^2} \right)^{\frac{1}{2}}.$$

121. Consider the diameter AB, as in *figure 28*. From the terminal points A and B of the diameter we construct the tangents AK and BL at any points M on the curve we construct the tangent MT and

extend it until it intersects both of the other tangents in the points K and L. Let ECF be the conjugate diameter, to which the ordinate MP and the two tangents AK and BL are parallel. From the properties of tangents we know that $\dfrac{CP}{CA} = \dfrac{CA}{CT}$. Since $CB = CA$, we have $\dfrac{CP}{AP} = \dfrac{CA}{AT}$, and $\dfrac{CP}{BP} = \dfrac{CA}{BT}$. It follows that

$$\frac{CP}{CA} = \frac{CA}{CT} = \frac{AP}{AT} = \frac{BP}{BT},$$

and so

$$\frac{AT}{BT} = \frac{AP}{BP}.$$

Since

$$\frac{AT}{BT} = \frac{AK}{BL},$$

we have

$$\frac{AK}{BL} = \frac{AP}{BP}.$$

It follows that

$$AT = \frac{CA \cdot AP}{CP},$$

$$BT = \frac{CA \cdot BP}{CP},$$

and

$$PT = \frac{CA \cdot AP}{CP} + AP = \frac{AP \cdot BP}{CP}.$$

Hence

$$\frac{AT}{PT} = \frac{CA}{BP} = \frac{AK}{PM}.$$

In a similar way

$$\frac{BT}{PT} = \frac{CA}{AP} = \frac{BL}{PM},$$

so that

$$AK = \frac{CA \cdot PM}{BP},$$

$$BL = \frac{CA \cdot PM}{AP},$$

and

$$AK \cdot BL = \frac{CA^2 \cdot PM^2}{AP \cdot BP},$$

since

$$\frac{AP \cdot BP}{PM^2} = \frac{AC^2}{CE^2}.$$

From this there follows the outstanding property $AK \cdot BL = CE^2$. Furthermore

$$AK = CE\frac{\sqrt{AP}}{BP},$$

$$BL = CE\frac{\sqrt{BP}}{AP},$$

$$\frac{AP}{BP} = \frac{AK^2}{CE^2} = \frac{CE^2}{BL^2} = \frac{KM}{ML},$$

and

$$\frac{AK}{BL} = \frac{KM}{LM}.$$

122. We conclude that when a tangent is drawn at any point M of the curve and it meets two parallel tangents AK and BL in K and L, then the semidiameter CE which is parallel to the tangents AK and BL is the geometric mean of AK and BL, that is $CE^2 = AK \cdot BL$. If at any other point m on the curve we draw the tangent kml, then we also have $CE^2 = Ak \cdot Bl$, so that

$$\frac{AK}{Ak} = \frac{Bl}{BL}.$$

It follows that we also have

$$\frac{AK}{Kk} = \frac{Bl}{Ll}.$$

If the tangents KL and kl intersect each other in the point o, then

$$\frac{AK}{Bl} = \frac{Ak}{BL} = \frac{Kk}{Ll} = \frac{ko}{lo} = \frac{Ko}{Lo}.$$

These are the main properties of conic sections from which NEWTON found the solution to many important problems in his *Principia*.

123. Since

$$\frac{AK}{Bl} = \frac{Ko}{Lo},$$

if the tangent LB is extended to I so that $BI = AK$, then I is the point where the tangent on the other side, parallel to KL, intersects the tangent LB, just as K is the point on the tangent LK where this tangent intersects the tangent AK which is parallel to BL. Therefore, if any two tangents BL and ML are extended to I and K in the prescribed manner, and if they are intersected by a third tangent lmo in the points l and o, then $\dfrac{BI}{Bl} = \dfrac{Ko}{Lo}$ and by composition we have $\dfrac{IB}{Il} = \dfrac{Ko}{KL}$. It follows that whenever a third tangent lmo is drawn, we always have

$IB \cdot KL = Il \cdot Ko$. If we draw any fourth tangent $\lambda\mu\omega$ which intersects the two original tangents IL and KL in the points λ and ω, we also have $IB \cdot KL = I\lambda \cdot K\omega$, so that $Il \cdot Ko = I\lambda \cdot K\omega$, or $\dfrac{Il}{I\lambda} = \dfrac{K\omega}{Ko}$. If we draw the straight lines $l\omega$ and λo, in whatever ratio these two lines are divided, the straight line through the two division points divides the straight line IK in the same ratio. In particular, if the lines $l\omega$ and λo are bisected, the straight line through the two bisecting points will also bisect the straight line IK. For this reason it will also pass through the center C of the conic section.

124. Consider the straight line nmH in *figure 30,* which cuts the straight lines $l\omega$ and λo in a given ratio. It should cut the straight line KI in the same ratio. Indeed, if $\dfrac{Il}{I\lambda} = \dfrac{K\omega}{Ko}$, or $\dfrac{I\lambda}{\lambda l} = \dfrac{Ko}{o\omega}$, we give the following geometric proof. Let the straight line mn divide both $l\omega$ and λo in the ratio $m:n$, that is, let

$$\frac{\lambda m}{mo} = \frac{ln}{n\omega} = \frac{m}{n}.$$

We extend the tangents IL and KL to Q and R. Now

$$\frac{\sin Q}{\sin R} = \frac{\dfrac{ln}{Ql}}{\dfrac{n\omega}{R\omega}} = \frac{\dfrac{\lambda m}{Q\lambda}}{\dfrac{mo}{Ro}} = \frac{\dfrac{m}{Ql}}{\dfrac{n}{R\omega}}.$$

It follows that $\dfrac{Ql}{R\omega} = \dfrac{Q\lambda}{Ro}$. By division we obtain

$$\frac{l\lambda}{o\omega} = \frac{Q\lambda}{Ro} = \frac{Ql}{R\omega}.$$

Since $\dfrac{l\lambda}{o\omega} = \dfrac{I\lambda}{Ko}$, we also have $\dfrac{QI}{RK} = \dfrac{l\lambda}{o\omega}$, and

$$\frac{\sin Q}{\sin R} = \frac{\dfrac{m}{l\lambda}}{\dfrac{n}{o\,\omega}}.$$

We also have

$$\frac{\sin Q}{\sin R} = \frac{\dfrac{HI}{QI}}{\dfrac{HK}{KR}} = \frac{\dfrac{HI}{l\lambda}}{\dfrac{HK}{o\,\omega}},$$

so that

$$\frac{HI}{HK} = \frac{m}{n} = \frac{\lambda m}{mo} = \frac{ln}{n\,\omega}.$$

125. We return now to *figure 27* and consider the conjugate semi-diameters CG and CE, which meet in the oblique angle $GCE = q$. We can always find two other conjugate semidiameters CM and CK which meet in the right angle MCK. Let the angles $GCM = p$ and $ECK = \pi$, then $Q + \pi - p$ is a right angle. Hence $\sin \pi = \cos(q - p)$ and $\sin(q + \pi) = \cos p$. From section 119 we have

$$\frac{CE^2}{CG^2} = \frac{\sin p \cos p}{\sin(q - p)\cos(q - p)}$$

$$= \frac{\sin 2p}{\sin 2(q - p)}$$

$$= \frac{\sin 2p}{\sin 2q \cos 2p - \cos 2q \sin 2p}.$$

It follows that

$$\frac{CG^2}{CE^2} = \frac{\sin 2q}{\tan 2q} - \cos 2q.$$

From this we have

$$\cot 2GCM = \cot 2q + \frac{CG^2}{CE^2\sin 2q},$$

and this equation can always be solved. Since

$$\frac{CM^2}{CG^2} = \frac{\sin q \cos p}{\sin(q - p)},$$

and

$$\frac{CG^2}{CM^2} = 1 - \frac{\tan p}{\tan q}.$$

It follows that

$$\tan p = \tan q - \frac{CG^2}{CM^2}\tan q.$$

Since $CM^2 + CK^2 = CG^2 + CE^2$ and $CK \cdot CM = CG \cdot CE \sin q$, we have

$$CM + CK = \sqrt{CG^2 + 2CG \cdot CE \sin q + CE^2},$$

and

$$CM - CK = \sqrt{CG^2 - 2CG \cdot CE \sin q + CE^2}.$$

It follows that the two conjugate diameters are indeed orthogonal.

126. Let CA and CE be two orthogonal semidiameters of a conic section, as in *figure 29*. These are usually called the PRINCIPAL DIAMETERS, and they meet at the center in a right angle. Let $CP = x$ be the abscissa, and $PM = y$ the ordinate. As we have seen $y^2 = \alpha - \beta x^2$. We let the principal semidiameters $AC = a$ and $CE = b$, then $\alpha = b^2$ and $\beta = \frac{b^2}{a^2}$, so that $y^2 = b^2 - \frac{b^2 x^2}{a^2}$. We see from this equation, since the equation is not changed whether x and y are taken to be positive or negative, that the curve has four similar parts, and that the curve is equidistant on either side of the diameters

AC and EF. That is, the quadrant ACE is similar and equal to quadrant ACF, and these both are equal to the part on the other side of the diameter EF.

127. If from the center C, which we take as the origin, we draw the straight line CM, then its length is equal to

$$\sqrt{x^2 + y^2} = \left(b^2 - \frac{b^2 x^2}{a^2} + x^2 \right)^{\frac{1}{2}}.$$

From this we understand that if $b = a$, or $CE = Ca$, then $CM = \sqrt{b^2} = b = a$. In this case all straight lines which join the center to the curve are equal to each other. This is the property of a circle; it is clear that a conic section whose two principle conjugate diameters are equal is a circle. It's equation in rectangular coordinates, when we let $CP = x$ and $PM = y$, is $y^2 = a^2 - x^2$, and the radius of the circle is $CA = a$.

128. If $b \neq a$, then the straight line CM cannot be expressed rationally in terms of x. However, there is another point D on the axis, such that any straight line DM drawn from that point to the curve can be expressed rationally. In order to find that point, we let $CD = f$, and since $DP = f - x$, we have

$$DM^2 = f^2 - 2fx + x^2 + b^2 - \frac{b^2 x^2}{a^2}$$

$$= b^2 + f^2 - 2fx + \frac{(a^2 - b^2)x^2}{a^2}.$$

This expression will be a perfect square if

$$f^2 = \frac{(a^2 - b^2)(b^2 + f^2)}{a^2},$$

or $a^2 - b^2 - f^2 = 0$, so that $f = \pm \sqrt{a^2 - b^2}$. From this we see that there are a pair of points on the axis AC, either of which is at a distance

$$CD = \sqrt{a^2 - b^2}$$

from the center. In this case we have

$$DM^2 = a^2 - 2x\sqrt{a^2 - b^2} + \frac{(a^2 - b^2)x^2}{a^2},$$

so that

$$DM = a - \frac{x\sqrt{a^2 - b^2}}{a} = AC - \frac{CD \cdot CP}{AC}.$$

If $CP = 0$, then $DM = DE = a = AC$. However, if we take the abscissa $CP = CD$, that is

$$x = \sqrt{a^2 - b^2},$$

then the straight line DM becomes the ordinate DG, and so

$$DG = \frac{b^2}{a} = \frac{CE^2}{AC}$$

and we see that DG is the third proportional to AC and CE.

129. Due to the special properties which these points D possess, these points on the principal diameter are worthy of the most careful attention. They have many other outstanding properties and for this reason they have been given particular names. They are called either the FOCI or the NAVELS of the conic section. The diameter a, on which they lie, is distinguished from its conjugate b, and is called the principal and *transverse* axis, while the other is called the *conjugate* axis. The rectangular ordinate DG erected at either focus is called the SEMIPARAMETER. The whole PARAMETER is the chord through D, which is twice the length of DG, and it is also called the *latus rectum*. The

conjugate semiaxis CE is the geometric mean of the semiparameter DG and the transverse semiaxis AC. The terminal points of the transverse axis, where it intersects the curve, are called VERTICES, of which A is one. These points have the property that a tangent at such a point is perpendicular to the principal axis AC.

130. We let the semiparameter $DG = c$ and the distance from the focus to the vertex $AD = d$, then $CD = a - d = \sqrt{a^2 - b^2}$, and $DG = \dfrac{b^2}{a} = c$, so that $b^2 = ac$, and $a - d = \sqrt{a^2 - ac}$. It follows that

$$ac = 2ad - d^2, \quad a = \frac{d^2}{2d - c},$$

and

$$b = d\left(\frac{c}{2d - c}\right)^{\frac{1}{2}}.$$

Therefore, if we are given the distance from the focus to the vertex $AD = d$ and the semilatus rectum $DG = c$, then the conic section is determined. Now we let $CP = x$, so that

$$DM = a - \frac{(a - d)x}{a} = \frac{d^2}{2d - c} - \frac{(c - d)x}{d}.$$

We let

$$DP = t,$$

so that

$$x = CD - t = \frac{(c - d)d}{2d - c} - t,$$

and

$$DM = c + \frac{(c - d)t}{d}.$$

We let the angle $ADM = v$, then $\dfrac{t}{DM} = -\cos v$, so that

$$d \cdot DM = cd + (d - c)DM\cos v,$$

$$DM = \frac{cd}{d - (d - c)\cos v},$$

and

$$\cos v = \frac{d(DM - DG)}{(d - c)DM}.$$

CHAPTER VI

On the Subdivision of
Second Order Lines into Genera.

131. The Properties which were brought to light in the preceding chapter belong to all second order lines, nor did we mention any differences which would serve to distinguish one from another. Although all second order lines have the properties so for discussed, still, according to their appearance as curves, they can be quite different from each other. For this reason it is convenient to assign lines of this order to different genera so that we may more easily distinguish the different shapes which occur and to investigate those properties which are proper to each genus.

132. By changing only the axis and origin we can write the general equation for any second order line in the form $y^2 = \alpha + \beta x + \gamma x^2$, in which x and y denote the rectangular coordinates. Since for any abscissa x, the ordinate y has two values, one positive and one negative, the axis, which contains the abscissas, cuts the curve into two parts which are similar and equal. Since each second order line has an orthogonal diameter of the curve, we will assume that it is this diameter which is the axis on which we take our abscissas.

133. There are three constants, α, β, and γ which determine the equation. Since there are an infinite number of different ways of choosing these constants and so there is an innumerable variety in the different curves which arise. Nevertheless, they differ from one another in shape some more, some less. In the first place, the same shape can appear an infinite number of times from the proposed equation $y^2 = \alpha + \beta x + \gamma x^2$, since the choice of origin can vary, depending on whether the value of the abscissa x increases or decreases. It can also happen that the same shape can be magnified by the choice of the equation, so that an infinite variety is produced only in magnitude, as happens with circles with different radii. From these examples it is clear that every change in the constants α, β, and γ does not produce a change in the genus or the species of a second order line.

134. The greatest difference in the curves contained in the equation $y^2 = \alpha + \beta x + \gamma x^2$ is given by the nature of the coefficient γ, according to whether it is positive or negative. If γ is positive, then as the abscissa x increases without bound, so that the term γx^2 becomes greater than the remaining $\alpha + \beta x$, and the whole expression $\alpha + \beta x + \gamma x^2$ has a positive value, then the ordinate y has two values which increase without limit, one positive and the other negative. The same thing happens when x approaches $-\infty$, since in this case the expression $\alpha + \beta x + \gamma x^2$ still takes on an infinite positive value. For this reason, when γ is positive, the curve has four branches which go off to infinity, two when the abscissa $x = +\infty$ and two when the abscissa $x = -\infty$. These curves with four branches going to infinity are considered to constitute one genus of second order lines, and they are called HYPERBOLAS.

135. If, on the other hand, the coefficient γ has a negative value, then when $x = +\infty$ or $x = -\infty$, the expression $\alpha + \beta x + \gamma x^2$ has a

negative value so that the ordinate y becomes imaginary. In this case neither abscissa nor ordinate can be infinite, so that no portion of the curve goes off to infinity, and the whole curve is contained in a determined and bounded space. This species of second order curves is called an ELLIPSE, and its nature is contained in the equation $y^2 = \alpha + \beta x + \gamma x^2$, in which the constant γ is negative.

136. Since the value of γ, whether it is positive or negative, produces such different qualities in the second order line, that it is quite proper that different genera are established in this way. If we let $\gamma = 0$, which is the case which lies between the two discussed, the resulting curve also is a kind of species which lies between the hyperbolas and ellipses, namely one which is called a PARABOLA, and its nature is expressed by the equation $y^2 = \alpha + \beta x$. The value of β can be positive or negative, since the quality of the curve is not changed if the abscissa x is taken to be negative. Suppose now that β is positive. It is clear that if the abscissa x increases without bound, then the ordinate y will also increase without bound, both positively and negatively. From this we see that a parabola has two branches which go to infinity. It cannot have more than two, since if we let $x = -\infty$, the value of the ordinate y becomes imaginary.

137. It follows that we have three species of second order lines, namely, the ellipse, the parabola, and the hyperbola, each of which is so different from the others, that it is not possible to confuse one with the other. The essential difference lies in the number of branches which go to infinity. The ellipse has no part going to infinity, but is contained completely in a bounded space. The parabola has two such branches, and the hyperbola has four. Since we considered the generic properties of conic sections in the previous chapter, we now look at the properties of the individual species.

138.   We   begin   with   the   ellipse,   whose   equation   is $y^2 = \alpha + \beta x - \gamma x^2$. We choose our ordinates from the orthogonal diameter, as in *figure 31*. Since the choice of origin is free, if we change the coordinates by subtracting $\dfrac{\beta}{2\gamma}$ from x we can write the equation in the form $y^2 = \alpha - \gamma x^2$, in which the abscissas are taken from the center of the curve. Let C be the center and AB the orthogonal diameter. Let $CP = x$ be the abscissa and $PM = y$ the ordinate. If $y = 0$ then $x = \pm \sqrt{\alpha/\gamma}$. If x takes values outside the limits $+ \sqrt{\alpha/\gamma}$ and $- \sqrt{\alpha/\gamma}$, then the ordinate y becomes imaginary. This implies that the whole   curve   lies   between   those   two   limits.   It   follows   that $CA = CB = \sqrt{\alpha/\gamma}$. Now we let $x = 0$, so that $CD = CE = \sqrt{\alpha}$. We let the semidiameter, that is, the principal semiaxis $CA = CB = a$ and the conjugate semiaxis $CD = CE = b$, then $\alpha = b^2$, $\gamma = \dfrac{\beta^2}{\alpha^2}$. It follows that for this ellipse we have the equation $y^2 = \dfrac{b^2}{a^2}(a^2 - x^2)$.

139.   When the conjugate semiaxis a and b are equal, then the ellipse becomes a circle, since $y^2 = a^2 - x^2$, that is, $y^2 + x^2 = a^2$. In this case $CM = \sqrt{x^2 + y^2} = a$, so that each point of the curve is the same distance from the center C, and this is the defining property of a circle. If the semiaxes a and b are not equal, then the curve will be oblong, that is, either AB will be longer than DE or vice versa. Since the conjugate axes AB and DE can be interchanged and so the abscissas can be chosen from either, we suppose that AB is the major axis, that is a is greater than b and that the foci of the ellipse, F and G are on this axis. We have $CF = CG = \sqrt{a^2 - b^2}$ and the semiparameter or semi-latus rectum of the ellipse will be equal to $\dfrac{b^2}{a}$ which expresses the ordinate at either focus F or G.

140. To the point M on the curve we draw from both of the foci FM and GM. As we saw above,

$$FM = AC - \frac{CF \cdot CP}{AC} = a - \frac{x\sqrt{a^2 - b^2}}{a},$$

and

$$GM = a + \frac{x\sqrt{a^2 - b^2}}{a}.$$

It follows that $FM + GM = 2a$. We conclude that if straight lines FM and GM are drawn from both foci to any point M on the curve, then the sum of these lines is always equal to the length of the major axis $AB = 2a$. From this result we see both the special property of the foci and also an easy mechanical method of drawing an ellipse.

141. At the point M we draw the tangent TMt which meets the axes at the points T and t. As we have shown above, $CP/CA = CA/CT$, so that $CT = \dfrac{a^2}{x}$, and in a similar way we have after permuting the coordinates, $Ct = \dfrac{b^2}{y}$. It follows that

$$TP = \frac{a^2}{x} - x, \qquad TF = \frac{a^2}{x} - \sqrt{a^2 - b^2},$$

and

$$TA = \frac{a^2}{x} - a.$$

Hence

$$TP = \frac{a^2 - x^2}{x} = \frac{a^2 y^2}{b^2 x},$$

and

$$TM = \frac{y\sqrt{b^4x^2 + a^4y^2}}{b^2x}.$$

It follows that $\tan CTM = \dfrac{b^2x}{a^2y}$, $\sin CTM = \dfrac{b^2x}{(b^4x^2 + a^4y^2)^{\frac{1}{2}}}$, and

$\cos CTM = \dfrac{a^2y}{(b^4x^2 + a^4y^2)^{\frac{1}{2}}}$. We conclude that if the perpendicular

AV is erected at A on the axis, it will also be tangent to the curve, and

$$AV = \frac{a(a-x)}{x}\frac{b^2x}{a^2y} = \frac{b^2(a-x)}{ay} = b\frac{\sqrt{a-x}}{a+x},$$

since $ay = b\sqrt{a^2 - x^2}$.

142.   Since $FT = \dfrac{a^2 - x\sqrt{a^2 - b^2}}{x}$,   and

$FM = \dfrac{a^2 - x\sqrt{a^2 - b^2}}{a}$, we have $FT/FM = a/x$. In a similar way,

since $GT = \dfrac{a^2 + x\sqrt{a^2 - b^2}}{x}$, and $GM = \dfrac{a^2 + x\sqrt{a^2 - b^2}}{a}$, we

have $GT/GM = a/x$. From this it follows that $FT/TM = GT/GM$. It is also true that $FT/FM = \sin(FMT)/\sin CTM$, and $GT/GM = \sin GMt/\sin CTM$. For this reason $\sin FMT = \sin GMt$ and so the angle FMT is equal to the angle GMt. We conclude that when straight lines are drawn to any point M on the curve from the foci, they make equal angles with the tangent to the curve at that point M, and this is the greatest of the principal properties of the foci.

143. Since $GT/GM = a/x$, $CT = \dfrac{a^2}{x}$, and $CT/CA = a/x$, we have $GT/GM = CT/CA$, so that if the straight line CS is drawn from the center C parallel to GM and meeting the tangent at S, then $CS = CA = a$. In the same way, if a straight line from C is drawn to the tangent parallel to FM, then this also is equal to $CA = a$. Since

$$TM = \frac{y}{b^2 x}\sqrt{b^4 x^2 + a^4 y^2},\quad \text{and}\quad a^2 y^2 = a^2 b^2 - b^2 x^2,\quad \text{we}\quad \text{have}$$

$$TM = \frac{y}{bx}\sqrt{a^4 - x^2(a^2 - b^2)}.\ \text{However}\ FT \cdot GT = \frac{a^4 - x^2(a^2 - b^2)}{x^2}.$$

It follows that $TM = \frac{y}{b}\sqrt{FT \cdot GT}$. Since $TG/TC = TM/TS$, we have

$$TS = \frac{TM \cdot CT}{TG},\ \text{so that}$$

$$TS = \frac{y \cdot CT}{b}\sqrt{FT/GT} = \frac{y \cdot CT \cdot FT}{b(FT \cdot GT)^{\frac{1}{2}}} = \frac{y^2 \cdot CT \cdot FT}{b^2 \cdot TM}.$$

Then

$$PT = \frac{a^2 y^2}{b^2 x} = \frac{CT \cdot y^2}{b^2},$$

so that $TS = \frac{PT \cdot FT}{TM}$, and $TM/PT = FT/TS$. from this we see that
the triangles TMP and TFS are similar, so that the straight line FS
from the focus is perpendicular to the tangent. From these expressions
we can conclude that $SV = \frac{AF \cdot MV}{GM}$.

144. It follows that if we draw a straight line FS from either focus
F perpendicular to the tangent and then join the center C to the point
S with the straight line CS, then this CS is always equal to the semiaxis
$AC = a$. Since $TM/y = TF/FS$, we have

$$FS = \frac{y \cdot TF}{TM} = \frac{b \cdot TF}{(FT \cdot GT)^{\frac{1}{2}}} = b\sqrt{FT/GT}.$$

Hence $GT/FT = GM/FM = CD^2/FS^2$. The perpendicular to the
tangent from either focus is equal to $b\sqrt{GT/FT}$ so that the minor
semiaxis $CD = b$ is the geometric mean of the two perpendiculars. Now
we construct CQ from the center C perpendicular to the tangent, then

$FT/FS = CT/CQ$, therefore

$$CQ = \frac{b \cdot CT}{(FT \cdot GT)^{\frac{1}{2}}} = \frac{bx \cdot CT}{a(FM \cdot GM)^{\frac{1}{2}}} = \frac{ab}{(FM \cdot GM)^{\frac{1}{2}}},$$

so that $CQ - FS = \dfrac{b \cdot CF}{(FT \cdot GT)^{\frac{1}{2}}} = CX$ where FX is drawn parallel to

the tangent. It follows that $CQ - CX = \dfrac{b \cdot TF}{(FT \cdot GT)^{\frac{1}{2}}}$ and

$CQ + CX = \dfrac{b \cdot TG}{(FT \cdot GT)^{\frac{1}{2}}}$, so that $CQ^2 - CX^2 = b^2$ and

$CX = \sqrt{CQ^2 - b^2}$. Hence if we are given the minor axis, we can find the point X on the perpendicular CQ, through which the perpendicular from the focus passes.

145. Now that we have considered the properties of the foci, we consider any pair of conjugate diameters. Suppose that CM is the semi-diameter, whose conjugate is found by drawing CK from the center parallel to the tangent TM. We let $CM = p$, $CK = q$ and the angles $MCK = CMT = s$. First we recall from above that $p^2 + q^2 = a^2 + b^2$, and also that $pq \sin s = ab$. We have

$$p^2 = x^2 + y^2 = b^2 + \frac{(a^2 - b^2)x^2}{a^2},$$

and

$$q^2 = a^2 + b^2 - p^2 = a^2 - \frac{(a^2 - b^2)x^2}{a^2} = FM \cdot GM.$$

In the same way $p^2 = FK \cdot GK$. Then, since $CQ = \dfrac{ab}{(FM \cdot GM)^{\frac{1}{2}}}$, we have

$$\sin CMQ = \sin s = \frac{ab}{p\,(FM\cdot GM)^{\frac{1}{2}}}.$$

It follows that

$$TM/TP = \frac{y}{b\,(FT\cdot GT)^{\frac{1}{2}}}, \qquad \frac{a^2 y^2}{b^2 x} = \sqrt{FM\cdot GM} = \frac{ay}{b} = \frac{CK}{CR},$$

so that $CR = \frac{ay}{b}$, $KR = \frac{bx}{a}$, and $CR\cdot KR = CP\cdot PM$. Then

$$\sin FMS = \frac{b}{(GM\cdot FM)^{\frac{1}{2}}} = \frac{b}{q}.$$

Since

$$x = CP\frac{a\sqrt{p^2 - b^2}}{(a^2 - b^2)^{\frac{1}{2}}},$$

$$y = \frac{b\sqrt{a^2 - p^2}}{(a^2 - b^2)^{\frac{1}{2}}} = PM, \qquad \text{also} \qquad CR = \frac{a\sqrt{a^2 - p^2}}{(a^2 - b^2)^{\frac{1}{2}}}, \qquad \text{and}$$

$$KR = \frac{b\sqrt{p^2 - b^2}}{(a^2 - b^2)^{\frac{1}{2}}}, \text{ so that } \tan ACM = \frac{y}{x}, \text{ and}$$

$$\tan 2ACM = \frac{2y}{x^2 - y^2} = \frac{2ab\sqrt{(a^2 - p^2)(p^2 - b^2)}}{(a^2 + b^2)p^2 - 2a^2 b^2}.$$

Since $ab = pq \sin s$, and

$$\sqrt{(a^2 - p^2)(p^2 - b^2)} = -pq \cos s$$

we have

$$\tan 2ACM = \frac{-2q^2\cos s}{p^2 + q^2\cos 2s},$$

since $\cos s$ is negative. Finally we have $CK^2 = MT\cdot Mt$. From the above we conclude that $MV = q\sqrt{AP/BP}$, and $AV = b\sqrt{AP/BP}$, so that $AV/MV = b/q = CE/CK$. Hence if the straight lines AM and EK are drawn, they will be parallel.

146. Since $pq \sin s = ab$, it is clear that pq is greater than ab. It is also true that $p^2 + q^2 = a^2 + b^2$, so that the difference between a and b is greater than the difference between p and q. We conclude that among the conjugate axes, those which are orthogonal differ from each other the most. Let the two conjugate diameters be equal to each other. In order to find them we let $q = p$, so that $2p^2 = a^2 + b^2$, $p = q = \sqrt{(a^2 + b^2)/2}$, $\sin s = \dfrac{2ab}{a^2 + b^2}$, and $\cos s = \dfrac{-a^2 + b^2}{a^2 + b^2}$. It follows that $\sin \frac{1}{2} s = \sqrt{a^2/(a^2 + b^2)}$, and $\cos \frac{1}{2} s = \sqrt{b^2/(a^2 + b^2)}$, so that $\tan \frac{1}{2} s = \dfrac{a}{b} = \tan CEB$, and $MCK = 2CEB = AEB$. Furthermore, $CP = \dfrac{a}{2^{\frac{1}{2}}}$, $CM = \dfrac{b}{2^{\frac{1}{2}}}$, so that the equal conjugate semidiameters CM and CK are parallel respectively to the chords AE and BE.

147. If we now let the vertex A be the origin, we let $AP = x$, $PM = y$, and where before we had x we now have $a - x$, so that we now have the equation $y^2 = \dfrac{b^2}{a^2}(2ax - x^2) = \dfrac{2b^2}{a}x - \dfrac{b^2}{a^2}x^2$, where it is clear that $\dfrac{2b^2}{a}$ is the parameter or the latus rectum of the ellipse. We let the semilatus rectum, that is the ordinate at the focus, be equal to c, and we let the distance from the vertex to the focus $AF = d$. Then $\dfrac{b^2}{a} = c$, and $a - \sqrt{a^2 - b^2} = d = a - \sqrt{a^2 - ac}$, so that $2ad - d^2 = ac$ and $a = \dfrac{d^2}{2d - c}$. It follows that $y^2 = 2cx - \dfrac{c(2d - c)x^2}{d^2}$, and this is the equation of an ellipse in rec-

tangular coordinates x and y where the abscissas x lie on the principal axis AB and the vertex A is the origin, when the distance from the vertex to the focus $AF = d$ and the semilatus rectum c are given. We remark that $2d$ is always greater than c, since $AC = a = \dfrac{d^2}{2d - c}$, and $CD = b = d\sqrt{c/(2d - c)}$.

148. If it should happen that $2d = c$, then $y^2 = 2cx$. We have already seen that this is the equation for a parabola. The equation given previously was $y^2 = \alpha + \beta x$, but this can be expressed by the other equation by changing the origin by an interval equal to $\dfrac{\alpha}{\beta}$. Let the curve MAN be a parabola, as in *figure 32*. We let the abscissa $AP = x$ and the ordinate $PM = y$ then the relation between the coordinates is given by the equation $y^2 = 2cx$. The distance from the vertex to the focus $AF = d = \tfrac{1}{2}c$, and the semiparameter $FH = c$. Since it is always true that $PM^2 = 2FH \cdot AP$ when the abscissa AP increases without limit, likewise the ordinates PM and PN also increase without limit. Hence the curve has a part on either side of the axis which goes to infinity. If we should let the abscissa be negative, the ordinate would be imaginary, so that beyond A in the directions of T there is no part of the curve.

149. Since the equation of an ellipse becomes that of a parabola when $2d = c$, it is clear that a parabola is just an ellipse whose semiaxis $a = \dfrac{d^2}{2d - c}$ has become infinite in length. For this reason all of the properties which we have found for ellipses also apply to the parabola, supposing that the axis is infinite. In the first place, Since $AF = \tfrac{1}{2}c$, we have $FP = x - \tfrac{1}{2}c$. If we draw the straight line FM from the focus F to the point M on the curve, then

$$FM^2 = x^2 - cx + \frac{1}{4}c^2 + y^2 = x^2 + cx + \frac{1}{4}c^2,$$

so that

$$FM = x + \tfrac{1}{2}c = AP + AF.$$

This is the characteristic property of the focus of a parabola.

150. Since the parabola comes from the ellipse, when the principal axis increases to infinite length, we will consider the parabola as if it were an ellipse in which the semiaxis $AC = a$ has infinite length, that is, the distance from the vertex A to the center C is infinite. At the point M we draw the tangent MT which meets the axis at the point T. Since $CP/CA = CA/CT$, we have $CT \doteq \dfrac{a^2}{a - x}$, because $CP = a - x$. It follows that $AT = \dfrac{ax}{a - x}$. However, since a is an infinite quantity, in comparison to A, x vanishes, so that $a - x = a$, and so $AT = x = AP$. This result can also be shown in a different way. Since $AT = \dfrac{ax}{a - x}$, we have $AT = x + \dfrac{x^2}{a - x}$. Since the fraction $\dfrac{x^2}{a - x}$ has an infinite denominator and a finite numerator, the value of the fraction vanishes, so that $AT = AP = x$.

151. If we draw the straight line MC from the point M to the center C of the parabola, this infinite line will be parallel to the axis AC. It will also be a curve which bisects all chords parallel to the tangent MT. For instance, if we draw the chord mn parallel to MT, it will be bisected by the diameter Mp at p. Every straight line drawn in the parabola parallel to the axis AP is an oblique angled diameter. In order to clarify the nature of this kind of diameter, let $Mp = t$, and $pm = u$. We drop the perpendicular msr from m to the axis. Since $PT = 2x$ and $MT = \sqrt{4x^2 + 2cx}$, we have

$$\sqrt{4x^2 + 2cx}\,/pm = 2x/ps = \sqrt{2cx}\,/ms.$$

It follows that

$$ps = \frac{2xu}{(4x^2 + 2cx)^{\frac{1}{2}}} = u\sqrt{2x/(2x + c)},$$

and $ms = u\sqrt{c/(2x + c)}$. Hence

$$Ar = x + t + u\sqrt{2x/(2x + c)},$$

and

$$mr = \sqrt{2cx} + u\sqrt{c/(2x + c)}.$$

Since $mr^2 = 2c\cdot Ar$, we have

$$2cx + 2cu\sqrt{2x/(2x + c)} + \frac{cu^2}{2x + c}$$

$$= 2cx + 2ct + 2cu\sqrt{2x/(2x + c)},$$

so that $u^2 = 2t(2x + c) = 4FM\cdot t$. That is, $pm^2 = 4FM\cdot Mp$. But the sine of the oblique angle mps is equal to $\sqrt{c/(2x + c)} = \sqrt{AF/FM}$, and its cosine is equal to $\sqrt{2x/(2x + c)} = \sqrt{AP/FM}$, so that $\sin 2mps = \dfrac{2\sqrt{2cx}}{2x + c} = \dfrac{y}{FM} = \sin MFp$. It follows that the angle $mps = MTP = \tfrac{1}{2}MFr$.

152. Since $MF = AP + AF$ and $AP = AT$, we have $FM = FT$, so that the triangle MFT is isosceles and the angle $MFr = 2MTA$, as we have just seen. Then since $MT = 2\sqrt{x(x + \tfrac{1}{2}c)}$, we have $MT = 2\sqrt{AP\cdot FM}$, so that when we drop the perpendicular FS from the focus F to the tangent, $MS = TS = \sqrt{AP\cdot FM} = \sqrt{AT\cdot TF}$, and we conclude that $AT/TS = TS/TF$. From this proportion we see that the point S is such that the straight line A is perpendicular to the axis at the vertex A. Since $AS = \tfrac{1}{2}PM$, and $AS/TS = AF/FS$, we have

$FS = \sqrt{AF \cdot FM}$, so that FS is the geometric mean of AF and FM. Furthermore, $AS/MS = AS/TS = FS/FM = \sqrt{AF}/\sqrt{FM}$. If we erect the perpendicular MW to the tangent at M, which intersects the axis at W, then $PT/PM = PM/PW$, that is $2x/\sqrt{2cx} = \sqrt{2cx}/PW$, so that $PW = c$. We conclude that wherever the normal WM intersects the axis, the interval PW, that is the distance on the axis between the normal and the ordinate is always constant and equal to half of the latus rectum, that is the ordinate FH. Furthermore, $FW = FT = FM$, and $MW = 2\sqrt{AF \cdot FM}$.

153. We now come to the hyperbola, whose nature is expressed in the equation $y^2 = \alpha + \beta x + \gamma x^2$, where the abscissas are chosen on the orthogonal diameter. If the origin is translated by the interval $\dfrac{\beta}{2\gamma}$, we obtain the equation $y^2 = \alpha + \gamma x^2$, where the origin is at the center. It is necessary that γ is a positive quantity; as for α, it is either positive or negative and by interchanging the coordinates x and y, it can be changed from positive to negative and vice versa. For this reason we suppose that α is negative and that $y^2 = \gamma x^2 - \alpha$. It is clear that the ordinate y will vanish twice, namely when $x = \sqrt{\alpha/\gamma}$, and $x = -\sqrt{\alpha/\gamma}$. We let C be the center and A and B be the places where the curve crosses the axis, as in *figure 33*. We let the semiaxis $CA = CB = a$, so that $a = \sqrt{\alpha/\gamma}$, and $\alpha = \gamma a^2$, with the result that $y^2 = \gamma x^2 - \gamma a^2$. As long as x^2 is less than a^2, the ordinate will be imaginary, so that no part of the curve corresponds to points on the axis from A to B. Whenever x^2 is greater than a^2 the ordinates increase continuously and finally go to infinity, so that the hyperbola has four branches: AI, Ai, BK, Bk, which go to infinity and are equal and similar to each other. This is the principal property of a hyperbola.

154. If we let $x = 0$, then $y^2 = -\gamma a^2$, so that the hyperbola, unlike the ellipse, has no real conjugate axis, since the ordinate at the center C is imaginary. For the imaginary conjugate axis we choose bi, so as to preserve some similarity to the ellipse, with the result that $\gamma a^2 = b^2$ and $\gamma = \dfrac{b^2}{a^2}$. We let the abscissa $CP = x$ and the ordinate $PM = y$, so that $y^2 = \dfrac{b^2}{a^2}(x^2 - a^2)$. It follows that the equation for an ellipse, which we found above, is transformed into an equation for the hyperbola by substituting $-b^2$ for b^2. Because of this affinity, the properties which we have found for ellipses can easily be transformed to the hyperbola. In the first place, since the distance of the foci from the center in an ellipse is equal to $\sqrt{a^2 - b^2}$, we have for the hyperbola, $CF = CG = \sqrt{a^2 + b^2}$. From this we have $FP = x - \sqrt{a^2 + b^2}$ and $GP = x + \sqrt{a^2 + b^2}$. Since $y^2 = -b^2 + \dfrac{b^2 x^2}{b^2}$, we have

$$FM = \left[a^2 + x^2 + \frac{b^2 x^2}{a^2} - 2x\sqrt{a^2 + b^2}\right]^{\frac{1}{2}}$$

$$= \frac{x\sqrt{a^2 + b^2}}{a} - a$$

and

$$GM = \left[a^2 + x^2 + \frac{b^2 x^2}{a^2} + 2x\sqrt{a^2 + b^2}\right]^{\frac{1}{2}}$$

$$= \frac{x\sqrt{a^2 + b^2}}{a} + a.$$

From this it follows that when we draw the straight lines FM and GM from the foci to the point M on the curve, we have $FM + AC = \dfrac{CP \cdot CF}{CA}$, and $GM - AC = \dfrac{CP \cdot CF}{CA}$ so that the difference in the lengths of these two straight lines, $GM - FM = 2AC$. Insofar as the sum of these two lines in an ellipse is equal to the principal axis AB, so in a hyperbola the difference is equal to the principal axis AB.

155. The position for the tangent MT can be found, since for any second order line we have $CP/CA = CA/CT$, so that $CT = \dfrac{a^2}{x}$ and $PT = \dfrac{x^2 - a^2}{x} = \dfrac{a^2 y^2}{b^2 x}$, so that

$$MT = \frac{y}{b^2 x} \sqrt{b^4 x^2 + a^4 y^2} = \frac{y}{bx} \sqrt{a^2 x^2 + b^2 x^2 - a^4}.$$

Since $\quad FM \cdot GM = \dfrac{a^2 x^2 + b^2 x^2 - a^4}{a^2}, \quad$ it $\qquad$ follows $\qquad$ that

$MT = \dfrac{ay}{bx} \sqrt{FM \cdot GM}.\quad$ Hence $\quad FT = \sqrt{a^2 + b^2} - \dfrac{a^2}{x},\quad$ and

$GT = \sqrt{a^2 + b^2} + \dfrac{a^2}{x}.\quad$ It $\quad$ follows $\quad$ that $\quad FT/FM = a/x \quad$ and $GT/GM = a/x$ so that $FT/GT = FM/GM$. This proportion shows that the angle FMG is bisected by the tangent MT so that $FMT = GMT$. When the straight line CM is extended, it will be an oblique diameter which bisects every chord which is parallel to the tangent MT.

156. Drop the perpendicular CQ from the center C to the tangent. Then $TM/CT = PT/TQ = PM/CQ$, that is

$$\frac{\dfrac{ay}{bx}\sqrt{FM \cdot GM}}{\dfrac{a^2}{x}} = \frac{\dfrac{a^2 y^2}{b^2 x}}{TQ} = \frac{y}{CQ}.$$

It follows that $TQ = \dfrac{a^3 y}{bx\,(FM \cdot GM)^{\frac{1}{2}}}$, and $CQ = \dfrac{ab}{(FM \cdot GM)^{\frac{1}{2}}}$. Likewise we drop the perpendicular FS from the focus to the tangent, so that $TM/FT = PT/TS = PM/FS$, that is ,

$$\frac{ay}{bx}\sqrt{FM \cdot GM} \Big/ \frac{aFM}{x} = \frac{a^2y^2}{b^2x}/TS = y/FS.$$

It follows that $TS = \dfrac{a^2 yFM}{bx(FM \cdot GM)^{\frac{1}{2}}}$ and $FS = \dfrac{bFM}{(FM \cdot GM)^{\frac{1}{2}}}$. Furthermore, if the perpendicular Gs is dropped from the other focus G to the tangent, then $Ts = \dfrac{a^2 yGM}{bx(FM \cdot GM)^{\frac{1}{2}}}$, and $Gs = \dfrac{bGM}{(FM \cdot GM)^{\frac{1}{2}}}$. From this we have

$$TS \cdot Ts = \frac{a^4 y^2}{b^2 x^2} = \frac{a^2(x^2 - a^2)}{x^2} = CT \cdot PT,$$

and $TS/CT = PT/Ts$. Then we have $FS \cdot Gs = b^2$. Since $QS = Qs$, we have

$$QS = \frac{TS + Ts}{2} = \frac{a^2 y(FM + GM)}{2bx(FM \cdot GM)^{\frac{1}{2}}} = \frac{ay\sqrt{a^2 + b^2}}{b(FM \cdot GM)^{\frac{1}{2}}} = Qs.$$

From this it follows that

$$CS^2 = CQ^2 + QS^2 = \frac{a^2b^4 + a^4y^2 + a^2b^2y^2}{b^2FM \cdot GM}$$

$$= \frac{a^2b^4 + (a^2 + b^2)(b^2x^2 - a^2b^2)}{b^2FM \cdot GM} = \frac{(a^2 + b^2)x^2 - a^4}{FM \cdot GM} = a^2.$$

As in the ellipse, so also here we have the straight line $CS = a = CA$. Then we have

$$CQ + FS = \frac{bx\sqrt{a^2 + b^2}}{a(FM \cdot GM)^{\frac{1}{2}}},$$

so that

$$(CQ + FS)^2 - CQ^2 = \frac{b^2 x^2 (a^2 + b^2) - a^4 b^2}{a^2 FM \cdot GM} = b^2.$$

It follows that if the straight line FX is drawn from the focus F parallel to the tangent, meeting the perpendicular CQ at X, then $CX = \sqrt{b^2 + CQ^2}$. This is similar to a property we found for the ellipse.

157. If we erect perpendiculars to the axis at the vertices A and B which meet the tangent in V and v, since $AT = \dfrac{a(x - a)}{x}$, $BT = \dfrac{a(x + a)}{x}$, and $PT/PM = AT/AV = BT/Bv$, we have $AV = \dfrac{b^2(x - a)}{ay}$, and $Bv = \dfrac{b^2(x + a)}{ay}$. It follows that $AV \cdot Bv = \dfrac{b^4(x^2 - a^2)}{a^2 y^2} = b^2$, that is $AV \cdot Bv = FS \cdot Gs$. Then

$$PT/TM = AT/TV = BT/Tv,$$

so that

$$TV = \frac{b(x - a)\sqrt{FM \cdot GM}}{xy}$$

and

$$Tv = \frac{b(x + a)\sqrt{FM \cdot GM}}{xy}.$$

It follows that

$$TV \cdot Tv = \frac{a^2}{x^2} FM \cdot GM = FT \cdot GT.$$

In a similar way many other consequences can be deduced.

158. Since $CT = \dfrac{a^2}{x}$, it is clear that the larger we take the abscissa $CP = x$ the smaller the interval CT becomes, and so when the

curve is extended to infinity, the tangent will pass through the center C and $CT = 0$. Since $\tan PTM = \dfrac{PM}{PT} = \dfrac{b^2 x}{a^2 y}$, when the point M goes to infinity, that is when we let $x = \infty$, then $y = \dfrac{b}{a}\sqrt{x^2 - a^2} = \dfrac{bx}{a}$. Hence the tangent to the curve when it is extended to infinity, passes through the center C, it makes an angle ACD with the axis, and the tangent of this angle is equal to $\dfrac{b}{a}$. Hence, if we erect the perpendicular AD to the axis at the vertex A with length equal to b, then the straight line CD extended in both directions to infinity never touches the curve, but the curve approaches the line more and more closely, until at infinity the curve and the line coincide. The same thing happens with the part of the line Ck and the branch Bk which finally coincide. Further, if we draw the straight line KCi on the other side, then the branches of the curve BK and Bi also coincide when extended to infinity. Straight lines of this kind, to which curves approach more and more closely, but meeting only at infinity, are called ASYMPTOTES, wherefore the straight lines ICk and KCi are two asymptotes of the hyperbola.

159. The asymptotes cross at the center C of the hyperbola and make an angle $ACD = ACd$, whose tangent is equal to $\dfrac{b}{a}$, and the double angle CDd has a tangent equal to $\dfrac{2ab}{a^2 - b^2}$. From this it is clear that if $b = a$, then the angle at which the asymptotes intersect, DCd, is a right angle. In this case the hyperbola is called equilateral. Since $AC = a$ and $AD = b$ we have $CD = Cd = \sqrt{a^2 + b^2}$, so that if from the focus G perpendiculars GH are dropped to both asymptotes, since $CG = \sqrt{a^2 + b^2} = CD$, we have $CH = AC = BC = a$ and $GH = b$.

160. Let the chord $MPN = 2y$ be extended in both directions until it meets the asymptotes in m and n. Then $Pm = Pn = \dfrac{bx}{a}$ and

$$Cm = Cn = \frac{x\sqrt{a^2 + b^2}}{a} = FM + AC = GM - AC.$$

Furthermore, $Mm = Nn = \dfrac{bx - ay}{a}$ and $Nm = Mn = \dfrac{bx + ay}{a}$, so

that $\qquad Mm \cdot Nm = Nn \cdot Mn = \dfrac{b^2x^2 - a^2y^2}{a^2} = b^2,$ $\qquad$ since

$a^2y^2 = b^2x^2 - a^2b^2$. Hence everywhere

$$Mm \cdot Nm = Mm \cdot Mn = Nn \cdot Nm = Nn \cdot Mn = b^2 = AD^2.$$

From M we draw Mr parallel to the asymptote Cd, then $2b\sqrt{a^2 + b^2} = Mm/mr$, so that

$$mr = Mr = \frac{(bx - ay)\sqrt{a^2 + b^2}}{2ab}$$

and

$$Cm - mr = Cr = \frac{(bx + ay)\sqrt{a^2 + b^2}}{2ab}.$$

From this it follows that

$$Mr \cdot Cr = \frac{(b^2x^2 - a^2y^2)(a^2 + b^2)}{4a^2b^2} = \frac{a^2 + b^2}{4}.$$

That is, when we draw from A the straight line AE parallel to the asymptote Cd, we have $AE = CE = \frac{1}{2}\sqrt{a^2 + b^2}$, so that $Mr \cdot Cr = AE \cdot CE$. This is the principal property of the relationship between a hyperbola and its asymptotes.

161. We now take one of the asymptotes as the axis, with the center C as the origin, as in *figure 34*. Let the abscissa $CP = x$, and we take ordinates $PM = y$ parallel to the other asymptote, so that $yx = \dfrac{a^2 + b^2}{4}$, since $AC = BC = a$ and $AD = Ad = b$. That is, if $AE = CE = h$, then $yx = h^2$ and $y = \dfrac{h^2}{x}$. It follows that when

$x = 0$, we have $y = \infty$, and conversely, if $x = \infty$, then $y = 0$. Now we draw the straight line $QMNR$ which is parallel to an arbitrary straight line GH and which passes through the point M on the curve, and we let $CQ = t$ and $QM = u$, so that $GH/u + CH/PQ = CG/PM$. It follows that $PQ = \dfrac{CH}{GH} u$ and $PM = \dfrac{CG}{GH} u$, so that $y = \dfrac{CG}{GH} u$ and $x = t - \dfrac{CH}{GH} u$. When we substitute the given values we obtain $\dfrac{CG}{GH} tu - \dfrac{CH \cdot CG}{CG^2} u^2 = h^2$, that is, $u^2 - \dfrac{GH}{CH} tu + \dfrac{GH^2}{CH \cdot CG} h^2 = 0$. From this equation we see that the ordinate u has two values, namely QM and QN, their sum is equal to $\dfrac{GH}{CH} t = QR$, and the rectangle $QM \cdot QN = \dfrac{GH^2}{CH \cdot CG} h^2$.

162. Since $QM + QN = QR$, it follows that $QM = RN$ and $QN = RM$. Therefore, if the points M and N coincide, which occurs when the straight line QR is tangent to the curve, then the line is bisected in that point of contact. For instance, if the straight line XY is tangent to the hyperbola, then the point of contact Z is the midpoint. It follows that if from the point Z we draw ZV parallel to the other asymptote, then $CV = VY$. From this we have an expeditious method for drawing the tangent at any point Z of the hyperbola. That is, we take $VY = CV$, and then the straight line through Y and the point Z on the curve is tangent to the hyperbola at that point Z. Since $CV \cdot ZV = h^2 = \dfrac{a^2 + b^2}{4}$, we have

$$CX \cdot CY = a^2 + b^2 = CD^2 = CD \cdot Cd.$$

It follows that if the straight lines DX and dY are drawn, they will be parallel to each other. From this we have a very easy method for drawing any tangents to the curve.

163. Since the rectangle $QM \cdot QN = \dfrac{GH^2}{CH \cdot CG} h^2$, it is clear that whenever we draw the straight line QR, parallel to HG, the rectangle $QM \cdot QN$ always has the same area. We also have $QM \cdot QN = QM \cdot MR = QN \cdot NR = \dfrac{GH^2}{CH \cdot CG} h^2$. Now if we consider the tangent line parallel to QR, since the part of it lying between the asymptotes is bisected by the point of contact, we let half of it be called q, then we always have $QM \cdot QN = QM \cdot MR = RN \cdot RM = RN \cdot NQ = q^2$. This is a significant property of hyperbolas drawn within asymptotes.

164. Since the hyperbola consists of two diametrically opposed parts, IAi and KBk, these properties apply not only to the straight lines which are drawn within the asymptotes and intersect that part of the curve in two points. They also apply to those which reach the opposite part. For instance if the straight line $Mqrn$ is drawn from M parallel to Gh, and we let $Cq = t$, $qM = u$, then because of the similarity of the triangles CGh and PMq, we have $PM = y = \dfrac{CG}{Gh} u$ and $qP = x - t = \dfrac{Ch}{Gh} u$, so that $x = t + \dfrac{Ch}{Gh} u$. Since $xy = h^2$, we have

$$\frac{CG}{Gh} tu + \frac{CG \cdot Ch}{Gh^2} u^2 = h^2.$$

That is,

$$u^2 + \frac{Gh}{Ch} tu - \frac{Gh^2}{CG \cdot Ch} h^2 = 0.$$

165. Since the ordinate u has two values, qM and $-qn$, where the qn is negative since it is directed towards the other direction of the asymptote CP, which was chosen as the axis. The sum of these two roots qM and $-qn$ is equal to $-\dfrac{Gh}{Ch} t = -qr$, so that $qM = rn$

and $qn = rM$. We also know from the equation that the product of the roots, $-qM \cdot qn = \dfrac{Gh^2}{CG \cdot Ch} h^2$, that is

$$qM \cdot qn = qM \cdot rM = rn \cdot qn = rn \cdot rM = \frac{Gh^2}{CG \cdot Ch} h^2.$$

It follows that this rectangle always has the same area no matter what straight line Mn, parallel to Gh, is chosen. These are the principal properties of the different kinds of second order lines. These along with the general properties make up a great multitude of remarkable properties.

CHAPTER VII

On the Investigation of Branches
Which Go to Infinity.

166. If any curve has a branch or part which goes to infinity and from a point on the curve a rectangular ordinate is dropped an infinite distance to any axis, then either the abscissa x or the ordinate y, or both will be infinite; for unless one or the other or both were infinite, then the distance of the point on the curve from the origin would be finite, that is equal to $\sqrt{x^2 + y^2}$, contrary to the hypothesis. For this reason, if the curve has a branch which goes to infinity, then either corresponding to some finite abscissa there is an infinite real ordinate, or an infinite abscissa corresponds to a real ordinate, either finite or infinite. On this basis we can investigate the branches which go to infinity.

167. Suppose that we are given an algebraic equation of any degree in the coordinates x and y, for example of degree n. We consider separately those terms in x and y with degree exactly equal to n, which are $\alpha y^n + \beta y^{n-1} x + \gamma y^{n-2} x^2 + \delta y^{n-3} x^3 + \cdots + \xi x^n$. This expression can be represented as the product of linear factors of the form $Ay + Bx$, which are either real or complex. If there are complex factors, then they are even in number and taken two at a time they give real quadratic factors of the form $A^2 y^2 - 2ABxy \cos \phi + B^2 x^2$. Any factor of this kind always has a value equal to ∞^2 if x or y or both are set equal to ∞, since the term $2AByx \cos \phi$ is always less than the sum of the other two, $A^2y^2 + B^2x^2$; for neither A nor B can be equal to zero. It follows that the factor $A^2y^2 - 2ABxy \cos \phi + B^2x^2$ cannot be equal to zero, or a finite number, or even an infinite quantity, since when x or y or both are set equal to infinity, the factor is equal to ∞^2, which is infinitely greater than ∞.

168. If that part of the equation which consists of the sum $\alpha y^n + \beta y^{n-1}x + \gamma y^{n-2}x^2 + \cdots + \xi x^n$ has no real linear factor, which can occur only if n is even, then it will consist of a product of quadratic factors of the form $A^2y^2 - 2ABxy \cos \phi + B^2x^2$. For this reason if x or y becomes infinite, then this expression takes on the value ∞^n and it cannot be equal to a finite quantity nor even an infinite quantity ∞^m where m is less than n. The other members of the equation in which the variables x and y have a lower degree cannot equal the first expression. It follows that the equation could not hold if x or y or both were to become infinite.

169. We conclude that any curve which is expressed by an equation in the coordinates x and y whose highest degree terms have no real linear factor will have no branches which go to infinity. It follows that the whole curve is contained in a bounded region, for instance in an ellipse or a circle. For this reason, if in the general equation of a second order line, $\alpha y^2 + \beta xy + \gamma x^2 + \delta y + \epsilon x + \zeta = 0$, the highest member, $\alpha y^2 + \beta xy + \gamma x^2$, in which the variables x and y have the second degree has no real linear factors. This happens if β^2 is greater than $4\alpha\gamma$, then the curve has no branches going to infinity, so that it must be an ellipse.

170. In order that we may develop these things more clearly, we

will distinguish the members of an equation in the coordinates x and y. To the first, or highest, member we assign all the terms of the equation with degree n. To the second member we assign all terms of degree $n - 1$; to the third member are assigned all $n - 2$ degree terms, and so forth, until we arrive at the last member, which is of degree zero, that is, it is the constant term. Let P be the first or highest member, Q the second member, R the third member, S the fourth, and so forth.

171. If the highest member P has no real linear factor, then the curve corresponding to the equation $P + Q + R + S + \cdots = 0$ has no branch going to infinity. For this reason we suppose that the highest member P has exactly one real linear factor, $ay - bx$, so that $P = (ay - bx)M$, where M is a function in x and y of degree $n - 1$, which has no real linear factors. If we let x or y or both become infinite, then $M = \infty^{n-1}$. However, Q can also become infinite, while $R, S, \cdots$ become infinite in a lower degree. We conclude that the equation $P + Q + R + \cdots = 0$ can hold, provided $ay - bx$ is equal to a finite quantity or it vanishes, so that the curve can go to infinity.

172. Let $ay - bx = p$ where p is a finite quantity which is such that when the curve goes to infinity, we have

$$pM + Q + R + S + \cdots = 0,\qquad\text{that}\qquad\text{is}$$

$$p = \frac{-Q - R - S - \cdots}{M}.$$ Since M becomes infinite to a higher degree than do R, S, etc., the rational functions R/M, S/M, $\cdots$ all vanish, so that $p = -Q/M$. For this reason the rational function $-Q/M$ gives the value p when the variables x and y become infinite. Since $ay - bx = p$, we have $y = \dfrac{bx + p}{a}$ and $\dfrac{y}{x} = \dfrac{b}{a} + \dfrac{p}{ax} = \dfrac{b}{a}$ because $\dfrac{p}{ax} = 0$ when $x = \infty$. It follows that when the curve goes to infinity it becomes $y = \dfrac{bx}{a}$.

173. Since Q and M are homogeneous functions of degree $n - 1$, we have $-Q/M$ is a function of zero degree. If we let $y = \dfrac{bx}{a}$, $-Q/M$ gives the constant value for p. That is, because the function $-Q/M$ is determined, if we know only the ratio of y to x, which is b/a, then the value of p can be found by substituting b for y and a for x in the expression $-Q/M$. Once we have found p in this way, we have $ay - bx = p$, which is contained in the equation $P + Q + R + S + \cdots = 0$ if the curve goes to infinity.

174. The part of the curve which goes to infinity is expressed by the equation $ay - bx = p$. This is the equation for a straight line, which when extended to infinity, finally coincides with the curve. Hence this straight line is an asymptote for the curve, since when the curve is extended to infinity is coincides with the line, so that the curve approaches the line more and more closely. Furthermore since when the given equation $P + Q + R + S + \cdots = 0$ becomes the equation $ay - bx = p$ when x or y become infinite, we understand that this straight line extended in either direction to infinity coincides with the curve. For this reason the curve has two branches which go to infinity in opposite directions, one with the straight line in one direction, the other with the line in the opposite direction.

175. The curve has two branches which go to infinity and each branch approaches the same straight line which is called its asymptote, provided the highest member P of the equation $P + Q + R + S + \cdots = 0$ has exactly one real linear factor. Now we suppose that the highest member P has two real linear factors $ay - bx$ and $cy - dx$, so that $P = (ay - bx)(cy - dx)M$ where M is a homogeneous function of degree $n - 2$. We have two cases to consider, depending on whether the two factors are equal to each other or

are different.

176. Suppose the two factors are different. It is clear that the equation $(ay - bx)(cy - dx)M + Q + R + S + \cdots = 0$ can hold in two different ways for infinite abscissas or ordinates; either $ay - bx$ or $cy - dx$ is equal to a finite quantity. Suppose that $ay - bx = p$ and, since p is finite, at infinity we have $y/x = b/a$ and as before $p = \dfrac{-Q - R - S - \cdots}{(cy - dx)M} = \dfrac{-Q}{(cy - dx)M}$ which is a function of degree zero in x and y. Hence if we let $y/x = b/a$, or which comes to the same thing, we substitute b for y and a for x, then we obtain the value of the constant p. Then we have $p = \dfrac{-Q}{(bc - ad)M}$ and since the factors are different, $bc - ad$ does not vanish, nor does M since it contains no real linear factor. It follows that p is finite or even equal to zero, which occurs if either there is no member Q, or if it contains the factor $ay - bx$.

177. Since the highest member P has a real linear factor $ay - bx$, the curve has an asymptote whose position is given by $ay - bx = p$, as in the previous case. In a similar way, because of the other factor $cy - dx$, the curve has another asymptote given by the equation $cy - dx = q$, where $q = \dfrac{-Q}{(ay - bx)M}$ after d is substituted for y and c for x. Hence the curve has two different asymptotes, with four branches which go to infinity and finally coincide with the straight lines. This is the case which we discovered above when we discussed the hyperbola. We conclude that if in the equation for a second order line $\alpha y^2 + \beta xy + \gamma x^2 + \delta y + \epsilon x + \zeta = 0$, the highest member $\alpha y^2 + \beta xy + \gamma x^2$ has two real linear factors which are different, which occurs if β^2 is greater than $4\alpha\gamma$, then the curve is a hyperbola.

178. Suppose that both factors, $ay - bx$ and $cy - dx$, are equal to each other, so that $P = (ay - bx)^2 M$. Since $P + Q + R + S + \cdots = 0$, we have

$$(ay - bx)^2 = \frac{-Q - R - S - \cdots}{M}.$$

Since A is of degree $n - 1$, R of degree $n - 2$, S of degree $n - 3$ and because M is of degree $n - 2$, in the infinite case we have $\dfrac{S}{M} = 0$, so that

$$(ay - bx)^2 = -\frac{Q}{M} - \frac{R}{M} = \frac{-q}{M(\mu y + \nu x)}(\mu y + \nu x) - \frac{R}{M}.$$

We note that $\dfrac{Q}{M(\mu y + \nu x)}$ and $\dfrac{R}{M}$ are both zero degree functions in x and y. Since at infinity we have $y/x = b/a$, is we substitute b/a for y/x, that is b for y and a for x, then both of these functions become constant.

179. When we made these substitutions we obtain $\dfrac{Q}{M(\mu y + \nu x)} = A$ and $\dfrac{R}{M} = B$, so that

$$(ay - bx)^2 = -A(\mu y + \nu x) - B.$$

This is an equation for a curve with which the curve with equation $P + Q + R + S + \cdots = 0$ becomes identified at infinity. Indeed, since μ and ν are arbitrary, when we let $\mu = b$ and $\nu = a$ and change the coordinates we have $ay - bx = u\sqrt{a^2 + b^2}$ and $by + ax = t\sqrt{a^2 + b^2}$. Now for this curve we have the equation $u^2 + \dfrac{At}{(a^2 + b^2)^{\frac{1}{2}}} + \dfrac{B}{a^2 + b^2} = 0$, which is clearly the equation for a parabola. The curve we are seeking is therefore like a hyperbola as it goes to infinity. It follows that it has only two branches which go to infinity and its asymptotes are not straight lines, but the parabola expressed above.

180. This is what happens provided A is not equal to zero. However, if $A = 0$, which happens if the second member Q vanishes or is divisible by $ay - bx$, then the equation is no longer a parabola. The equation becomes $u^2 + \dfrac{B}{a^2 + b^2} = 0$, the there are three cases which we will consider. In the first case B is a negative quantity, so that if we let $\dfrac{B}{a^2 + b^2} = -f^2$, then the equation $u^2 - f^2 = 0$ is made up of two equations $u - f = 0$ and $u + f = 0$. These are two parallel straight lines each of which is an asymptote of the curve as in the first case above so that the curve has four branches going to infinity which coincide with these lines at infinity.

181. In the second case B is a positive quantity, for example $+ f^2$. Since in this case the equation $u^2 + f^2 = 0$ is impossible, the curve has no branches going to infinity, and it is contained in a bounded region. Not only does the curve with equation $P + Q + R + S + \cdots = 0$ have no branch going to infinity when the highest member P has no real linear factor, but also the same phenomenon can occur when P has such factors, as we have just seen. There are still other cases when this occurs.

182. The third case is that in which $B = 0$, and this lies between the two previous cases, so it is not clear how it will go. For this reason we consider the other terms in the defining equation. That is, since $P + Q + R + S + \cdots = 0$ and $P = (ay - bx)^2 M$, at infinity we have $\dfrac{y}{x} = \dfrac{b}{a}$ and $(ay - bx)^2 + \dfrac{Q}{M} + \dfrac{R}{M} + \dfrac{S}{M} + \dfrac{T}{M} + \cdots = 0$. As before we make the substitution $\dfrac{y}{x} = \dfrac{b}{a}$ so that $\dfrac{Q}{M} = A(by + ax)$

and $\dfrac{R}{M} = B.$ Since S, T, V, etc. are functions of degrees $n - 2$, $n - 3$, etc., while M is a function of degree $n - 1$, it follows that $\dfrac{S(by + ax)}{M} = C,$ $\dfrac{T(by + ax)^2}{M} = D,$ $V\dfrac{(by + ax)^3}{M} = E,$ etc.

so that

$$(ay - bx)^2 + A(by + ax) + B + \frac{C}{by + ax}$$

$$+ \frac{D}{(by + ax)^2} + \frac{E}{(by + ax)^3} + \cdots = 0.$$

This equation expresses the nature of a curve whose parts reaching to infinity coincide with the curve expressed by $P + Q + R + S + \cdots = 0,$ when we let $by + ax$ go to infinity. When the curve goes to infinity $(ay - bx)^2$ can have either a finite or an infinite value, but if infinite it will be less than ∞^2, although $by + ax$ has infinite value.

183. Now we change the axis to the asymptote we found earlier and choose for our abscissa $t = \dfrac{ax + by}{(a^2 + b^2)^{\frac{1}{2}}}$ and for ordinate we let $u = \dfrac{ay - bx}{(a^2 + b^2)^{\frac{1}{2}}}.$ For the sake of brevity we let $g = \sqrt{a^2 + b^2}$, so that we have $u^2 + \dfrac{At}{g} + \dfrac{B}{g^2} + \dfrac{C}{g^3 t} + \dfrac{D}{g^4 t^2} + \dfrac{E}{g^5 t^3} + \cdots = 0.$ In the present case we have $A = 0$ and $B = 0$ so that $u^2 + \dfrac{C}{g^3 t} + \dfrac{D}{g^4 t^2} + \dfrac{E}{g^5 t^3} + \cdots = 0.$ As long as C is not equal to zero, when t goes to infinity the terms $\dfrac{D}{g^4 t^2} + \dfrac{E}{g^5 t^3} + \cdots$ will vanish when compared to $\dfrac{C}{g^3 t}.$ We are left with $u^2 + \dfrac{C}{g^3 t} = 0,$ which represents the curve which coincides with the original curve when t goes

to infinity. Since $u = \pm \sqrt{-C/g^3 t}$, the curve has two branches, both of which approach the same axis from either side.

184. If it should happen that $C = 0$, then we consider the equation $u^2 + \dfrac{D}{g^4 t^2} = 0$, and again we have three cases, depending on whether D is positive, negative, or zero. In the first case, due to the impossible equation, the curve has no branch going to infinity and is contained in some bounded region. In the second case, if $\dfrac{D}{g^4} = -f^2$, since $u^2 = \dfrac{f^2}{t^2}$, when we let t approach either $+\infty$ or $-\infty$, the ordinate u has two values, both negative and positive which vanish at infinity. Hence the curve has four branches two of which converge to the axis from either side at $+\infty$ and the other two from either side at $-\infty$. The third case is when $D = 0$, so that the equation takes the form $u^2 + \dfrac{E}{g^5 t^3} = 0$ but this is like the case considered in the previous section, so that we should continue our consideration of the equation $P + Q + R + S + \cdots = 0$ by examining further terms.

185. Suppose now that P, the highest member of the equation $P + Q + R + S + \cdots = 0$, has three real linear factors. It is clear that if these three factors are all unequal, then what was said about a single such factor remains true for each of the three factors. In this case the curve has six branches going to infinity, which converge to the three straight line asymptotes. If two of the factors are equal, then about the third factor which is unequal, the same holds as before. Concerning the two equal factors the treatment given above also holds. There remains only to consider the case when all three factors are equal. Let $P = (ay - bx)^3 M$. Since the equation $P + Q + R + S + \cdots = 0$ cannot hold at infinity unless $(ay - bx)^3$ has either a finite value or at

least a value less than ∞^3 so that the power of infinity to which the highest member goes is less than ∞^n, and so at infinity $\dfrac{y}{x} = \dfrac{b}{a}$.

186. In order to explain this case we should first consider the second member Q to see whether or not it has $ay - bx$ as a factor. First we suppose that Q is not divisible by $ay - bx$. Since Q is a function of degree $n - 1$ and N is a function of degree $n - 3$, it follows that $\dfrac{Q}{(ax + by)^2 M}$ is a function of degree zero, so that when we let $\dfrac{y}{x} = \dfrac{b}{a}$, the expression becomes a constant equal to A. Then $(ay - bx)^3 + A(ax + by)^2 = 0$ since the following members give terms which will vanish at infinity in comparison with $A(ax + by)^2$.

187. The curve which this equation expresses is such that when extended to infinity it will coincide with the curve expressed by the equation $P + Q + R + S + \cdots = 0$. For a closer examination of that curve we make the change of coordinates by choosing as abscissa $t = \dfrac{ax + by}{g}$ and as ordinate $u = \dfrac{ay - bx}{g}$, where $g = \sqrt{a^2 + b^2}$. Then $u^2 + \dfrac{At^2}{g} = 0$, and this equation, when $t = \infty$, give the part of the desired curve $P + Q + R + S + \cdots = 0$ extended to infinity. Hence if the shape of the curve $u^3 + \dfrac{At^2}{g} = 0$ is known, then we will also know the shape at infinity of $P + Q + R + S + \cdots = 0$. In the next chapter we will consider these asymptotic curves.

188. Now we suppose that the second member Q has $ay - bx$ as a factor. We consider two cases, either it is divisible also by $(ay - bx)^2$ or it is not. Suppose first that it is not divisible by $(ay - bx)^2$. We take the zero degree function $\dfrac{Q}{(ay - bx)(ax + by)M}$, and when we let $\dfrac{y}{x} = \dfrac{b}{a}$, we obtain the required constant A, so that

$(ay - bx)^3 + A(ay - bx)(ax + by) + \dfrac{R}{M} + \dfrac{S}{M} + \cdots = 0.$ When

we let $\dfrac{y}{x} = \dfrac{b}{a}$, $\dfrac{R}{M}$ is equal to either $B(ay - bx)$ or $B(ax + by)$

depending on whether R is divisible by $ay - bx$ or not. In any case $\dfrac{S}{M}$

is equal to a constant C. When we change the axis the equation in u

and t takes either the form $u^3 + \dfrac{Atu}{g} + \dfrac{Bu}{g^2} + \dfrac{C}{g^3} = 0$ or

$u^3 + \dfrac{Atu}{g} + \dfrac{Bt}{g^2} + \dfrac{C}{g^3} = 0.$ Since we are interested only in the case

when $t = \infty$, and in that case the last term vanishes. Then the first form

becomes $u^3 + \dfrac{Atu}{g} + \dfrac{Bu}{g^2} = 0$, which has two asymptotes: $u = 0$ and

$u + \dfrac{At}{g} = 0$ The first of these is a straight line, and the second is a

parabola. In the second form, when $t = \infty$, if u has a finite value, due

to the vanishing of the finite compared to the infinite, we have

$\dfrac{Atu}{g} + \dfrac{Bt}{g^2} = 0.$ This gives the straight line $u = \dfrac{-B}{Ag}$. If u takes an

infinite value, due to the vanishing of the third term, we obtain the para-

bola $u^2 + \dfrac{At}{g} = 0.$ Hence, in both cases we obtain two asymptotes, one

of which is a straight line and the other a parabola. For this reason it is

not necessary to distinguish these cases.

189. We now suppose that Q is divisible by $(ay - bx)^2$ and distin-

guish between the cases where R is divisible by $ay - bx$ or is not. Fol-

lowing the same procedures as before we obtain the following equations

in t and u: $u^3 + \dfrac{Au^2}{g} + \dfrac{Bu}{g^2} + \dfrac{C}{g^3} = 0$ and $u^3 + \dfrac{Au^2}{g} + \dfrac{Bt}{g^2} = 0.$

In the first form if all of the roots are real, then we have three parallel

straight lines. If two of the roots are complex, then there is one straight

line asymptote. There is some possible variety when we have the three lines since they could coincide to give two distinct parallel lines or even only one line. In the second form, $u^3 + \dfrac{Au^2}{g} + \dfrac{Bt}{g^2} + 0$, we cannot have $t = \infty$ unless at the same time u becomes infinite, so that the term $\dfrac{Au^2}{g}$ will vanish when compared to the first term u^3. The resulting equation is $u^3 + \dfrac{Bt}{g^2} = 0$, and this is the equation of a curve of the third order which is the asymptote.

190. Now we suppose that $A = 0$, $B = 0$, $C = 0$, so that the equation $\quad P + Q + {}+R + S + \cdots = 0 \quad$ takes the form $u^3 + \dfrac{D}{g^4 t} + \dfrac{E}{g^5 t^2} + \dfrac{F}{g^6 t^3} + \cdots = 0$. In this case, unless $D = 0$, the third and all following terms vanish, so that we have $u^3 + \dfrac{D}{g^4 t} = 0$. If $D = 0$, we have $u^3 + \dfrac{E}{g^5 t^2}$. If in addition $E = 0$, then we have $u^3 + \dfrac{F}{g^6 t^3} = 0$, and so forth. All of these equations denote curves which, when $t = \infty$, coincide with the curves represented by the equation $P + Q + R + S + \cdots = 0$. Since all of these equations have the term u^3, there are always real roots with branches going to infinity. Indeed, in all of these cases the straight line $u = 0$ is an asymptote since it is an asymptote for the curves $u^3 + \dfrac{D}{g^4 t} = 0$, $u^3 + \dfrac{E}{g^5 t^2} = 0$, etc.

191. Since the branches which converge to straight lines can differ from each other very much, we should ponder the differences very carefully. We will accomplish this by defining the simplest curves which converge to the same straight line which is an asymptote for the given curve. For example, although the equation $u^3 + \dfrac{Au^2}{g} + \dfrac{Bu}{g^2} + \dfrac{C}{g^3} = 0$ manifests three parallel straight lines, provided the roots are all real, we still

do not know whether the branches of the curve which go to infinity are hyperbolic, that is of the form $u = \dfrac{C}{t}$ or some other type, such as $u = \dfrac{C}{t^2}$ or $u = \dfrac{C}{t^3}$ etc. In order to obtain this information we take the next terms suggested by the equation, that is $\dfrac{D}{g^4 t}$, or if this has vanished, we take $\dfrac{E}{g^5 t^2}$ or if that term also vanishes, we take $\dfrac{F}{g^6 t^3}$. In order to treat the cases in general we will take the following term to be $\dfrac{K}{t^k}$. From the nature of the equation $P + Q + R + \cdots = 0$, which is of degree n, it is clear that k is no greater than $n - 3$. Let the factors of the expression $u^3 + \dfrac{Au^2}{g} + \dfrac{Bu}{g^2} + \dfrac{C}{g^3}$ be $(u - \alpha)(u - \beta)(u - \gamma)$, so that $(u - \alpha)(u - \beta)(u - \gamma) - \dfrac{K}{t^k} = 0$.

Let $u - \alpha = \dfrac{I}{t^\mu}$. This equation expresses the nature of one of the asymptotes, and we have $\dfrac{I}{t^\mu}(\alpha - \beta + \dfrac{I}{t^\mu})(\alpha - \gamma + \dfrac{I}{t^\mu}) = \dfrac{K}{t^k}$. When we let t go to infinity we have $\dfrac{(\alpha - \beta)(\alpha - \gamma)I}{t^\mu} = \dfrac{K}{t^k}$.

192. This equation holds if the root α is not equal to the other roots β and γ. In this case we have $I = \dfrac{K}{(\alpha - \beta)(\alpha - \gamma)}$ and $\mu = k$. It follows that the root $u = \alpha$ provides the curvilinear asymptote $u - \alpha = \dfrac{K}{(\alpha - \beta)(\alpha - \gamma)t^k}$. Now if all three roots are unequal, then each provides an asymptote of this kind. If two of the roots are equal, for instance $\beta = \alpha$, then the two asymptotes coalesce into one, with $\dfrac{I^2(\alpha - \gamma)}{t^{2\mu}} = \dfrac{K}{t^k}$, so that $I^2 = \dfrac{K}{\alpha - \gamma}$ and $2\mu = k$. Hence, the nature of this double asymptote is expressed by the equation

$(u - \alpha)^2 = \dfrac{K}{(\alpha - \gamma)t^k}$. If all three roots are equal, then all three

asymptotes come together into one, whose nature is expressed by the

equation $(u - \alpha)^3 = \dfrac{K}{t^k}$.

193.   If   the   highest   member P of   the   equation
$P + Q + R + S + \cdots = 0$ has four real linear factors and all four

are unequal, or two are equal, or even three are equal, we can gather

from what has gone before, what the branches going to infinity and their

asymptotes will be. The only case which needs to be explained is that

when all the roots are equal. Hence, we let $P = (ay - bx)^4 M$, where M

is a function with degree $n - 4$. As has been done before, we let

$\dfrac{y}{x} = \dfrac{b}{a}$ to obtain the constants and we change the axis so that

$t = \dfrac{ax + by}{g}$ and $u = \dfrac{ay - bx}{g}$, where $g = \sqrt{a^2 + b^2}$. We obtain

the following equations in t and u to express the asymptotes. The first

equation, obtained if Q is not divisible by $ay - bx$, is $u^4 + \dfrac{At^3}{g} = 0$.

194. If Q is divisible by $ay - bx$ but not by $(ay - bx)^2$, we

obtain $u^4 + \dfrac{At^2 u}{g} + \dfrac{Bt^2}{g^2} = 0$. When we let $t = \infty$, the ordinate u

can be either finite or infinite. We obtain two asymptotes: the straight

line $u + \dfrac{B}{gA}$, and the curve $u^3 + \dfrac{At^2}{g} = 0$. As for the straight line, in

order to have better information, we take the next term, which is $\dfrac{K}{t^k}$.

The resulting equation is $u + \dfrac{B}{gA} + \dfrac{gK}{At^{k+2}} = 0$, which is the equation

for a curve which coincides with the desired curve when the abscissa

$t = \infty$.

195. Now suppose that Q is divisible by $(ay - bx)^2$ but not by $(ay - bx)^3$. We have to see whether or not R is divisible by $ay - bx$. If R is divisible by $ay - bx$, we obtain $u^4 + \dfrac{Atu^2}{g} + \dfrac{Btu}{g^2} + \dfrac{Ct}{g^3} = 0$.

If it is not divisible, we have $u^4 + \dfrac{Atu^2}{g} + \dfrac{Bt^2}{g^2} + \dfrac{Ct}{g^3} = 0$. In the first case we have two equations, depending on whether u is finite or infinite, and these are $u^2 + \dfrac{Bu}{gA} + \dfrac{C}{g^2A} = 0$ and $u^2 + \dfrac{At}{g} = 0$. If in the first of these the two roots are unequal and real, we obtain two parallel straight lines. If the roots are complex, we have no branch going to infinity. The equation $u^2 + \dfrac{At}{g} = 0$, gives a parabolic asymptote. The other equation, $u^4 + \dfrac{Atu^2}{g} + \dfrac{Bt^2}{g^2} = 0$ (since $\dfrac{Ct}{g^3}$ vanishes in comparison to $\dfrac{Bt^2}{g^2}$ when $t = \infty$,) contains two equations of the form $u^2 + \alpha t = 0$. It follows that we have two parabolic asymptotes provided A^2 is greater than $4B$. The two become one if $A^2 = 4B$; if A^2 is less than $4B$, then the roots are complex and there is no branch going to infinity.

196. If Q is divisible by $(ay - bx)^3$, and depending on whether or not R and S are divisible by $ay - bx$, we obtain the following equations.

$$u^4 + \frac{Au^3}{g} + \frac{Bu^2}{g^2} + \frac{Cu}{g^3} + \frac{D}{g^4} = 0$$

$$u^4 + \frac{Au^3}{g} + \frac{Bu^2}{g^2} + \frac{Ct}{g^3} = 0$$

$$u^4 + \frac{Au^3}{g} + \frac{But}{g^2} + \frac{Ct}{g^3} = 0$$

$$u^4 + \frac{Au^3}{g} + \frac{Bt^2}{g^2} = 0$$

The first of these is for four parallel straight lines, provided all of the roots are real and distinct. If two or more roots are equal, then two or more lines coalesce into one. If the roots are complex, either all or only two, then all or two of the lines are removed. In the second equation, since $t = \infty$, the ordinate u must become infinite, so that the equation becomes $u^4 + \frac{Ct}{g^3} = 0$, which is an asymptotic curve of the fourth order. From the third equation a finite value is possible, so that we have $u + \frac{C}{gB} = 0$; we also have $u^3 + \frac{Bt}{g^2} = 0$, which is an asymptotic third order curve. Finally, the fourth equation becomes $u^4 + \frac{Bt^2}{g^2} = 0$ when $t = \infty$. If B is positive, the equation is impossible; if B is negative, the equation gives two parabolas with a common vertex but opening in opposite directions. When extended to infinity this coincides with the original curve.

197. From these discussions we see the way to progress further, if more real linear factors of the highest member P are equal. As for factors which are not equal, each of them can be considered separately, and from each of them we define a straight line asymptote. When two factors are equal, we can define some characteristics of the curve according to the treatment in sections 178 and following. Similarly, if there are three equal factors we consult sections 185 and following. The case when four factors are equal we have just discussed, and from these discussions we should be able to treat the cases when more than four factors are

equal. In general we have seen the great multiplicity and variety in curves, just in regard to branches which go to infinity. The variety which occurs in the bounded regions we still have to discuss.

CHAPTER VIII

Concerning Asymptotes.

198. In the preceding chapter we have seen several different kinds of asymptotes. Besides the straight line asymptotes, we have found asymptotic curves with the equation $U^\mu = Ct^\nu$. Even the straight line asymptotes have given rise to an abundance of curvilinear asymptotes to which the curves more closely approximate than to the straight lines. Whenever a straight line has been found to be an asymptote to some curve, we can always find another curve which has this straight line for an asymptote and is itself an asymptote for the originally given curve. Curvilinear asymptotes of this kind much more accurately express the nature of the curve for which it is an asymptote. It not only gives the number of branches which converge to the straight line, but also the pattern, in that it may lie above or below the straight line, or it may wind back and forth across the straight line as it approaches it.

199. This infinite variety of asymptotes can be most conveniently treated in some order, if we follow that source from which we have taken them. That is, some asymptotes come from the individual factors of the highest members which are not equal to any other factor, other asymptotes come from two equal factors, still others from three equal factors, others from four equal factors, and so forth. We suppose that the given

equation is of degree n in the coordinates x and y, and is expressed as $P + Q + R + S + \cdots = 0$, where P is the highest member containing all terms of degree n. Q is the second member containing all terms of degree $n - 1$, and similarly in regard to R, S, and so forth.

200. Now suppose that $ay - bx$ is a linear factor of P and that there is no other linear factor of P equal to it. We let $P = (ay - bx)M$, where M is a homogeneous function of degree $n - 1$ which is not divisible by $ay - bx$. Let AZ be the axis, as in *figure 35*, from which we choose the abscissa $AP = x$ and let the ordinate $PM = y$. In order that the factor $ay - bx$ may be more succinctly expressed, we take another straight line AX as an axis which intersects the first axis at the origin A and makes an angle XAZ whose tangent is equal to $\dfrac{b}{a}$. Then

$$\sin XAZ = \frac{b}{\left(a^2 + b^2\right)^{\frac{1}{2}}}$$

and

$$\cos XAZ = \frac{a}{\left(a^2 + b^2\right)^{\frac{1}{2}}}.$$

On the new axis we take the abscissa $AQ = t$ and ordinate $QM = u$. When we draw Pg and Pf parallel to the new coordinates u and t, we have

$$Pg = Qf = \frac{bx}{\left(a^2 + b^2\right)^{\frac{1}{2}}}, \ Ag = \frac{ax}{\left(a^2 + b^2\right)^{\frac{1}{2}}}, \ Mf = \frac{ay}{\left(a^2 + b^2\right)^{\frac{1}{2}}},$$

and $Pf = Qg = \dfrac{by}{\left(a^2 + b^2\right)^{\frac{1}{2}}}$. It follows that

$$t = Ag + Qg = \frac{ax + by}{\left(a^2 + b^2\right)^{\frac{1}{2}}}$$

and

$$u = Mf - Qf = \frac{ay - bx}{\left(a^2 + b^2\right)^{\frac{1}{2}}}.$$

We now have the ordinate u as a factor of the highest member P.

201. Conversely we have $y = \dfrac{au + bt}{\left(a^2 + b^2\right)^{\frac{1}{2}}}$ and $x = \dfrac{at - bu}{\left(a^2 + b^2\right)^{\frac{1}{2}}}$.

If we substitute these values into the equation $P + Q + R + \cdots = 0$, we obtain an equation in t and u for the same curve related to the AX axis. In order to avoid a multitude of coefficients we will keep α, β, γ, δ, etc. instead of all the coefficients. After the substitutions have been made, the individual members have the values:

$$M = \alpha t^{n-1} + \alpha t^{n-2} u + \alpha t^{n-3} u^2 + \cdots$$

$$Q = \beta t^{n-1} + \beta t^{n-2} u + \beta t^{n-3} u^2 + \cdots$$

$$R = \gamma t^{n-2} + \gamma t^{n-3} u + \gamma t^{n-4} u^2 + \cdots$$

$$S = \delta t^{n-3} + \delta t^{n-4} u + \delta t^{n-5} u^2 + \cdots$$

$$T = \epsilon t^{n-4} + \epsilon t^{n-5} u + \epsilon t^{n-6} u^2 + \cdots$$

etc. Since we are looking for asymptotes, we should let the abscissa go to infinity, so that in each member all the terms vanish when compared to the first. Hence, if the first term of any member is present, then the following terms can be neglected. However, if the first term is lacking, we take the second; if the first two are missing, then we begin with the third.

202. Since the function M is not divisible by u, its first term must be present. Hence we let $\alpha t^{n-1}u + \beta t^{n-1} = 0$ so that u has a finite value, which we let equal c. That is, the straight line parallel to and at a distance equal to c from the axis AX is an asymptote. Now in order to find a curvilinear asymptote more closely approximating the curve, we put everywhere, except in the first term, $u = c$. The equation we obtain is

$$\alpha t^{n-1}u + \beta t^{n-1} + t^{n-2}(\alpha c^2 + \beta c + \gamma)$$

$$+ t^{n-3}(\alpha c^3 + \beta c^2 + \gamma c + \delta) + \cdots = 0.$$

Since $\alpha u + \beta = u - c$ we have

$$(u - c)t^{n-1} + t^{n-2}(\alpha c^2 + \beta c + \gamma)$$

$$+ t^{n-3}(\alpha c^3 + \beta c^2 + \gamma c + \delta) + \cdots = 0.$$

Unless the second term is lacking, all the following terms can be neglected, so that we have $(u - c) + \dfrac{A}{t} = 0$. If the second term is lacking, the third is taken and we have $(u - c) + \dfrac{A}{t^2} = 0$. If the third term is also lacking, then $(u - c) + \dfrac{A}{t^3} = 0$, and so forth. If all terms except for the final constant term are lacking, then $(u - c) + \dfrac{A}{t^{n-1}} = 0$. If every term were missing, the whole equation would be divisible by $u - c$, then the straight line $u - c = 0$ would be a part of the curve.

203. If we let $u - c = z$, that is, if we choose the abscissas on this straight line asymptote, then all of the curvilinear asymptotes which this one factor of the highest member provides are covered by the general

equation $z = \dfrac{C}{t^k}$, where k is some integer less than the exponent n. Let us compare the curvilinear asymptotes as t becomes infinite. We let XY be the straight line asymptote, which we take as the axis, with the origin at the point A, as in *figure 36*. When the straight line CD is drawn we obtain the four quadrants which are designated P, Q, R, and S. First we let $z = \dfrac{C}{t}$. When we take t to be negative, z also is negative, so that the curve has two branches, EX and FY, in the two opposite quadrants P and S, converging to the straight line XY. The same thing happens if k is any odd integer. On the other hand, if $k = 2$, that is $z = \dfrac{C}{t^2}$, since for either positive or negative values of t, z is always positive. It follows that the curve has two branches EX and FY as in *figure 37,* in the quadrants P and Q which converge to the straight line XY. The same thing happens if k is any even number, but the convergence is quicker the larger k is.

204. Suppose that the principal member P has two equal factors $ay - bx$. When we change coordinates as before, we have the following:

$$P = \cdots + \alpha t^{n-2} u^2 + \alpha t^{n-3} u^3 + \cdots$$

$$Q = \beta t^{n-1} + \beta t^{n-2} u + \beta t^{n-3} u^2 + \beta t^{n-4} u^3 + \cdots$$

$$R = \gamma t^{n-2} + \gamma t^{n-3} u + \gamma t^{n-4} u^2 + \gamma t^{n-5} u^3 + \cdots$$

$$S = \delta t^{n-3} + \delta t^{n-4} u + \delta t^{n-5} u^2 + \delta t^{n-6} u^3 + \cdots$$

and so forth. Hence, depending on which of the first terms of Q are present or lacking, we have the two equations

I

$$\alpha t^{n-2} u^2 + \beta t^{n-1} = 0$$

that is

$$\alpha u^2 + \beta t = 0$$

II

$$\alpha t^{n-2} u^2 + \beta t^{n-2} u + \gamma t^{n-2} = 0$$

that is

$$\alpha u^2 + \beta u + \gamma = 0$$

If the first equation $\alpha u^2 + \beta t = 0$ holds, the asymptote is a parabola, whose two branches coincide with two branches of the curve at infinity. Hence the curve has branches in the two quadrants P and R, as in *figure 38,* which finally coincide with the branches of the parabola EAF.

205. If the second equation $\alpha u^2 + \beta u + \gamma = 0$ results, then we have to determine whether the equation has real roots or not. If the roots are not real, then this equation indicates no branches going to infinity. On the other hand, if both roots are real and unequal, one $u = c$ and the other $u = d$, then the curve has asymptotes which are two parallel straight lines. We will now investigate the peculiarities of each. For instance, since $\alpha u^2 + \beta u + \gamma = (u - c)(u - d)$, when we let $u = c$ everywhere except in the factor $u - c$, we obtain

$$(c - d) t^{n-2}(u - c) + t^{n-3}(\alpha c^3 + \beta c^2 + \gamma c + \delta)$$

$$+ t^{n-4}(\alpha c^4 + \beta c^3 + \gamma c^2 + \delta c + \epsilon) + \cdots = 0.$$

Unless the second term vanishes, all of the succeeding terms vanish when we let $t = \infty$. We then have the asymptote $(u - c) + \dfrac{A}{t} = 0$. If the second term vanishes, then we have $(u - c) + \dfrac{A}{t^2} = 0$, and so forth. If all the terms except for the final constant vanish, then we have $(u - c) + \dfrac{A}{t^{n-2}} = 0$. We have already described the configuration of all of these curves when $t = \infty$.

206. If both roots of the equation $\alpha u^2 + \beta u + \gamma = 0$ are equal, that is, $\alpha u^2 + \beta u + \gamma = (u - c)^2$, when we let $u = c$ in all the other terms, we obtain the equation

$$t^{n-2}(u - c)^2 + t^{n-3}(\alpha c^3 + \beta c^2 + \gamma c + \delta)$$

$$+ \ t^{n-4}(\alpha c^4 + \beta c^3 + \gamma c^2 + \delta c + \epsilon) + \ \cdots \ = 0.$$

If the second term does not vanish, or if the second term vanishes but the third does not, or if the second and third term vanish, but not the fourth, we obtain the following equations for the asymptotes.

$$(u - c)^2 + \frac{A}{t} = 0$$

$$(u - c)^2 + \frac{A}{t^2} = 0$$

$$(u - c)^2 + \frac{A}{t^3} = 0$$

all the way to

$$(u - c)^2 + \frac{A}{t^{n-2}} = 0,$$

if all of the terms except the constant last term vanish. If the constant term is also lacking, then $(u - c)^2 = 0$ and the straight line is a part of the curve, so that the curve is complex.

207. Although it would seem that we have considered all of the cases when there are two equal factors, still this last equation can take different forms, from which we find different asymptotes. If the coefficient of t^{n-3} is divisible by $u - c$, then there remains the factor $u - c$ as in the first term, and we must add the next following term. In this way we obtain the following equations.

$$(u - c)^2 + \frac{A(u - c)}{t} + \frac{B}{t^2} = 0$$

$$(u - c)^2 + \frac{A(u - c)}{t} + \frac{B}{t^3} = 0$$

all the way to

$$(u - c)^2 + \frac{A(u - c)}{t} + \frac{B}{t^{n-2}} = 0.$$

If the second term is lacking or is divisible by $(u - c)^2$, then we consider the third term, and if this is divisible by $u - c$, we keep it with the $u - c$ and we add the next following term. In this case we have the following equations.

$$(u - c)^2 + \frac{A(u - c)}{t^2} + \frac{B}{t^3} = 0$$

$$(u - c)^2 + \frac{A(u - c)}{t^2} + \frac{B}{t^4} = 0$$

all the way to

$$(u - c)^2 + \frac{A(u - c)}{t^2} + \frac{B}{t^{n-2}} = 0.$$

If the third term is lacking and the fourth is divisible by $u - c$, or if the fifth is lacking, and so forth, we can obtain an equation for an asymptotic curve

$$(u - c)^2 + \frac{A(u - c)}{t^p} + \frac{B}{t^q} = 0,$$

where the exponent p is always less than q and q is always less than $n - 1$.

208. When we let $u - c = z$, these equations all have the form $z^2 - \frac{Az}{t^p} + \frac{B}{t^q} = 0$. At this point we consider three cases: if q is greater than $2p$, if q is equal to $2p$, or if q is less than $2p$.

In the first case, when q is greater than $2p$, we have two equations, $z - \frac{A}{t^p} = 0$ and $Az - \frac{B}{t^{q-p}} = 0$. Both of these equations hold when $t = \infty$. Indeed, if we let $z = \frac{A}{t^p}$, the first equation becomes $\frac{A^2}{t^{2p}} - \frac{A^2}{t^{2p}} + \frac{B}{t^q}$, that is $A^2 - A^2 + \frac{B}{t^{q - 2p}} = 0$, which is satisfied since q is greater than $2p$ and p is less than $\frac{n - 2}{2}$. If $z = \frac{B}{At^{q-p}}$, then $\frac{B^2}{A^2 t^{2q - 2p}} - \frac{B}{t^q} + \frac{B}{t^q} = 0$, that is $\frac{B^2}{A^2 t^{q - 2p}} - B + B = 0$. This equation is satisfied since the first term vanishes when $t = \infty$. It follows that in this case the straight line asymptote is approximated by two different curvilinear asymptotes, so that there are four branches which go to infinity.

In the second case, when $q = 2p$, we have the equation $z^2 - \frac{Az}{t^p} + \frac{B}{t^{2p}} = 0$. The roots of this equation are complex if A^2 is less than $4B$ and then there is no asymptote. There are two similar asymptotes $z = \frac{C}{t^p}$ if A^2 is greater than $4B$.

In the third case, when q is less than $2p$, the middle term always vanishes when $t = \infty$. Then $z^2 + \frac{B}{t^q} = 0$ is the equation for one asymptote. We have already discussed the forms of the other asymptotes, so now we will consider asymptotes with the form $z^2 = \dfrac{C}{t^k}$.

209. If we take the axis to be the straight line asymptote $u = c$ and we take the ordinate $u - c = z$, all of these curvilinear asymptotes have an equation of the form $z^2 = \dfrac{C}{t^k}$, where k is an integer less than $n - 1$. The branches of these curves go to infinity, that is, when $t = \infty$, the following takes place. If $k = 1$, that is, $z^2 = \dfrac{C}{t}$, since t is never negative, the curve has two branches EX and FX which go to infinity in quadrants P and R, as in *figure 39*. The same thing happens if k is any odd number. If k is even, for example 2, that is $z^2 = \dfrac{C}{t^2}$, then we see whether C is negative or positive. If it is negative, then there are no real roots, so that the curve has no branch going to infinity. If C is positive, then the curve has four branches going to infinity and approaching the asymptote XY, namely EX, FX, GY, and HY in each of the quadrants P, Q, R, and S, as in *figure 40*.

210. We suppose now that the principal member P has three equal factors and that the equations have been expressed in the coordinates t and u, so that u is the triple factor of P. We have

$$P = \cdots + \alpha t^{n-3}u^3 + \alpha t^{n-4}u^4 + \cdots$$

$$Q = \beta t^{n-1} + \beta t^{n-2}u + \beta t^{n-3}u^2 + \beta t^{n-4}u^3 + \beta t^{n-5}u^4 + \cdots$$

$$R = \gamma t^{n-2} + \gamma t^{n-3}u + \gamma t^{n-4}u^2 + \gamma t^{n-5}u^3 + \gamma t^{n-6}u^4 + \cdots$$

$$S = \delta t^{n-3} + \delta t^{n-4}u + \delta t^{n-5}u^2 + \delta t^{n-6}u^2 + \delta t^{n-7}u^4 + \cdots$$

and so forth. Hence, depending on the constitution of the members Q and R we will have the following

$$\text{I.}$$

$$\alpha\, t^{n-3}u^3 + \beta\, t^{n-1} = 0$$

$$\text{II.}$$

$$\alpha\, t^{n-3}u^3 + \beta\, t^{n-2}u + \gamma t^{n-2} = 0$$

$$\text{III.}$$

$$\alpha\, t^{n-3}u^3 + \beta\, t^{n-3}u^2 + \gamma\, t^{n-2} = 0$$

$$\text{IV.}$$

$$\alpha\, t^{n-3}u^3 + \beta\, t^{n-3}u^2 + \gamma\, t^{n-3}u + \delta\, t^{n-3} = 0.$$

211. The first equation becomes $\alpha u^3 + \beta t^2 = 0$. This asymptote is a third order line, whose configuration is shown in *figure 41*. We choose the abscissa t on the axis XY with the point A as origin. It has two branches E and F which go to infinity in quadrants P and Q.

The second equation can be expressed as $\alpha u^3 + \beta tu + \gamma t = 0$. When we let $t = \infty$, u can take two values, either finite or infinite, so we resolve it into the two equations $\beta u + \gamma = 0$ and $\alpha u^2 + \beta t = 0$. The latter equation is of a parabola, as we have seen before, so that the curve has two branches which go to infinity while approaching the parabola. The former equation can be expressed as $u - c = 0$, which is a straight line asymptote whose properties can be seen if, except for $\beta u + \gamma = u - c$, we substitute c for u. We then have

$$t^{n-2}(u - c) + t^{n-3}(\alpha c^3 + \beta c^2 + \gamma c + \delta)$$

$$+ t^{n-4}(\alpha c^4 + \beta c^3 + \gamma c^2 + \delta c + \epsilon) + \cdots = 0.$$

It follows, as above, that we have either $(u - c) + \dfrac{A}{t} = 0$ or $(u - c) + \dfrac{A}{t^2} = 0$, etc. The last equation which can arise is

$(u - c) + \dfrac{A}{t^{n-2}} = 0$. In this case the curve has a double asymptote; one of which is a straight line with the property just described, and the other is a parabola.

212. The third equation can be written as $\alpha u^3 + \beta u^2 + \gamma t = 0$. When we let $t = \infty$, the equation cannot hold unless $u = \infty$. It follows that the term βu^2 vanishes in comparison to the term αu^3. We obtain as an asymptote the third degree equation $\alpha u^3 + \gamma t = 0$, whose configuration is shown in *figure 42*. There are two branches AE and AF which go to infinity in the opposite quadrants P and S.

The fourth equation takes the form $\alpha u^3 + \beta u^2 + \gamma u + \delta = 0$. This gives either one or three parallel straight line asymptotes, unless two or all three are equal to each other. In order to discover the characteristic of the lines we first let $u = c$, which we suppose is a root of the equation different from all other roots. Then

$$\alpha u^3 + \beta u^2 + \gamma u + \delta = (u - c)(fu^2 + gu + h).$$

We substitute c for u except in the factor $(u - c)$, to obtain the following equation

$$t^{n-3}(u - c) + At^{n-4} + Bt^{n-5} + Ct^{n-6} + \cdots = 0.$$

form $u - c = \dfrac{K}{t^k}$, where k is less than $n - 2$.

213. If two roots of the equation $\alpha u^3 + \beta u^2 + \gamma u + \delta = 0$ are equal, so that the expression becomes equal to $(u - c)^2(fu + g)$. Then when we substitute c for u, except in the factors $u - c$, we obtain $(u - c)^2 + \dfrac{A(u - c)}{t^p} + \dfrac{B}{t^q} = 0$, where q is less than $n - 2$ and p is less than q. We have already considered this case. There remains the case in which the equation $\alpha u^3 + \beta u^2 + \gamma u + \delta = 0$ has three equal

roots, namely $(u - c)^3$. From this we obtain the equation $(u - c)^3 t^{n-3} + Pt^{n-4} + Qt^{n-5} + Rt^{n-6} + St^{n-7} + \cdots = 0$. As long as P is not divisible by $u - c$, we substitute c for u to obtain $(u - c)^3 + \dfrac{a}{t} = 0$. If P has $u - c$ as a divisor, then we substitute c for u everywhere except in this factor, to obtain an equation of the form $(u - c)^3 + \dfrac{A(u - c)}{t} + \dfrac{B}{t^q} = 0$, where q is less than $n - 2$ and $\dfrac{B}{t^q}$ is the next term which does not vanish when $u = c$. If P is divisible by $(u - c)^2$, and Q has no factor $u - c$, then we obtain an equation of the form $(u - c)^3 + \dfrac{A(u - c)^2}{t} + \dfrac{B}{t^2} = 0$. If the second term is divisible by $(u - c)^3$, then we proceed in order to find the first term which is not divisible by $(u - c)^3$. If this is divisible by $u - c$, then we must proceed further until we arrive at a term not divisible by $u - c$. If that term is divisible by $(u - c)^2$, then we go further until we obtain a term not divisible by $u - c$, where we stop. In this way we can always obtain a equation in the general form $(u - c)^3 + \dfrac{A(u - c)^2}{t^p} + \dfrac{B(u - c)}{t^q} + \dfrac{C}{t^r}$, where r is less than $n - 2$, q is less than r and p is less than q.

214. In this equation there are three of the form $(u - c) = \dfrac{K}{t^k}$, or one of this form and one of the form $(u - c)^2 = \dfrac{K}{t^k}$, or even one of the form $(u - c)^3 = \dfrac{K}{t^k}$. This last case occurs if both $3p$ is greater than r and $3q$ is greater than $2r$. It can also happen that two of the equations are complex, and then no asymptote is indicated. We have already considered the forms of the asymptotes except the last equation, with the form $(u - c)^3 = \dfrac{K}{t^k}$. If k is odd, then the configuration is that

given in *figure 36,* with two branches *EX* and *FY* going to infinity in the opposite quadrants *P* and *S*. If *k* is even then we have the configuration given in *figure 37,* with two branches *EX* and *EY*, with the same straight line asymptote *XY*, going to infinity in the quadrants *P* and *Q*.

215. Since from what has already been done, it is easy to see what the form of the asymptotes will be when four or more real linear factors of the principal member of the equation are equal, we will go no further. Rather I will conclude this chapter with an example in which the given rules are applied.

EXAMPLE

Let the given curve have the equation

$$y^3 x^2 (y - x) - xy(y^2 + x^2) + 1 = 0,$$

whose principal member $y^3 x^2 (y - x)$ has a single factor $y - x$, two equal factors x^2, and three equal factors y^3.

First we consider the single linear factor $y - x$. When we substitute x for y we obtain $y - x - \dfrac{2}{x} = 0$. When $x = \infty$ we have $y - x = 0$, which is the equation of the straight line asymptote *BAC* in *figure 43.* This makes half of the right angle *BAY* with the axis *XY* at the origin. In order to change coordinates so as to have the asymptote for an axis we let $y = \dfrac{u + t}{2^{\frac{1}{2}}}$ and $x = \dfrac{t - u}{2^{\frac{1}{2}}}$. With these substitutions we obtain the equation

$$\frac{(u + t)(t^2 - u^2)u}{4} + \frac{(t^2 - u^2)(t^2 + u^2)}{2} + 1 = 0.$$

When we multiply the equation by 4, we obtain

$$0 = t^5 u + t^4 u^2 - 2t^3 u^3 - 2t^2 u^4 + tu^5 + u^6 - 2t^4 + 2u^4 + 4.$$

From this equation, when $t = \infty$, we find $u = 0$, so that except for the two terms $t^5 u - 2t^4$ the other terms vanish. It follows that the curvilinear asymptote is $u = \dfrac{2}{t}$. Because of this factor the curve has two branches bB and cC which go to infinity.

216. Now we consider the two equal factors x^2. Since $x^2 = \dfrac{xy(y^2 + x^2) - 1}{y^3(y - x)}$, we choose as the new axis the straight line AD perpendicular to the old axis XY; we let $y = t$ and $x = u$ so that the resulting equation is

$$0 = t^4 u^2 - t^3 u^3 - t^3 u - t u^3 + 1.$$

When we let $t = \infty$ the equation becomes $t^4 u^2 - t^3 u + 1 = 0$. From this we obtain the two equation $u = \dfrac{1}{t}$ and $u = \dfrac{1}{t^3}$. From this factor we obtain four branches going to infinity. The first two dD and eE from the equation $u = \dfrac{1}{t}$, and the other two δD and ϵE from the equation $u = \dfrac{1}{t^3}$.

217. The three equal factors y^3 are referred to the axis XY, where we let $t = x$ and $y = u$. Then we obtain the equation

$$0 = -t^3 u^3 + t^2 u^4 - t^3 u - t u^3 + 1.$$

When we let $t = \infty$, what remains is $t^3 u^3 + t^3 u = 0$, that is, $u(u^2 + 1) = 0$. Since $u^2 + 1$ has no real roots, the only asymptote is the straight line $u = 0$, which coincides with the axis XY. The special nature of the asymptote is expressed by the equation $t^3 u = 1$, that is, $u = \dfrac{1}{t^3}$. We conclude that this triple factor provides only two branches yY and xX which go to infinity. Altogether the curve has eight branches going to infinity. This is not the place to explain how these branches all come together in the bounded region.

218. From this and the preceding chapter we see clearly the variety of branches which go to infinity. First there are branches of curves which converge to a straight line as an asymptote, an example of which is the hyperbola. Then there are branches which have no straight line asymptote, for instance, parabolas. Branches of curves of the first kind are called *hyperbolic*, the others are called *parabolic*. There are innumerable species in each of the two classes. The species of hyperbolic branches are expressed by the following equations in t and u, in which we let t go to infinity.

$$u = \frac{A}{t}, \; u = \frac{A}{t^2}, \; u = \frac{A}{t^3}, \; u = \frac{A}{t^4}, \cdots$$

$$u^2 = \frac{A}{t}, \; u^2 = \frac{A}{t^2}, \; u^2 = \frac{A}{t^3}, \; u^2 = \frac{A}{t^4}, \cdots$$

$$u^3 = \frac{A}{t}, \; u^3 = \frac{A}{t^2}, \; u^3 = \frac{A}{t^3}, \; u^3 = \frac{A}{t^4}, \cdots,$$

etc. The species of parabolic branches are given by the following equations.

$$u^2 = At, \quad u^3 = At, \quad u^4 = At, \quad u^5 = At, \cdots$$

$$u^3 = At^2, \quad u^4 = At^2, \quad u^5 = At^2, \quad u^6 = At^2, \cdots$$

$$u^4 = At^3, \quad u^5 = At^3, \quad u^6 = At^3, \quad u^7 = At^3, \cdots,$$

etc. Each of these equations provides at least two branches which go to infinity, provided the exponents of t and u are not both even. However, if both exponents are even, then there may be no branches going to infinity, or there may be four such branches, depending on whether the roots of the equation are all complex or there are real roots.

CHAPTER IX

On the Species of Third Order Lines.

219. The nature and the number of branches going to infinity are quite properly considered to be the essential criteria for assigning a curve to its species. On this basis we can most conveniently give the subdivision of lines of any order. Indeed the species of second order lines have been divided according to these criteria already.

Let the general equation of a second order line be

$$\alpha y^2 + \beta yx + \gamma x^2 + \delta y + \epsilon x + \zeta = 0,$$

whose principal member $\alpha y^2 + \beta xy + \gamma x^2$ we consider especially as to whether it has real linear factors or not. If it has no such factors, then we have the first species, called an *ellipse*. If the factors are real, then we see whether they are unequal or equal. If they are unequal, then we have a *hyperbola;* if they are equal, then the curve is a *parabola*.

220. In case the factors of the principal member are real and unequal, the curve has two straight line asymptotes. In order to investigate their nature we let

$$\alpha y^2 + \beta yx + \gamma x^2 = (ay - bx)(cy - dx),$$

so that

$$(ay - bx)(cy - dx) + \delta y + \epsilon x + \zeta = 0$$

We consider the first factor $ay - bx$, which at infinity gives $\dfrac{y}{x} = \dfrac{b}{a}$, so that

$$ay - bx + \frac{\delta b + \epsilon a}{bc - ad} + \frac{\zeta}{cy - dx} = 0.$$

It follows that the equation $ay - bx + \dfrac{\delta b + \epsilon a}{bc - ad} = 0$ defines the position of one of the straight line asymptotes. In a similar way the equation

$$cy - dx + \frac{\delta d + \epsilon c}{ad - bc} = 0 \text{ gives the other asymptote.}$$

221. In order to scrutinize the nature of each asymptote, we change coordinates by letting $y = \dfrac{au + bt}{(a^2 + b^2)^{\frac{1}{2}}}$ and $x = \dfrac{at - bu}{(a^2 + b^2)^{\frac{1}{2}}}$.

We let $\sqrt{a^2 + b^2} = g$, so that

$$u((ac + bd)u + (bc - ad)t) + \frac{(\delta a - \epsilon b)u + (\delta b + \epsilon a)t}{g} + \zeta = 0.$$

From this we have

$$g(bc - ad)tu + g(ac + bd)u^2 + (\delta b + \epsilon a)t + (\delta a - \epsilon b)u + \zeta g = 0.$$

If now we substitute $-\dfrac{\delta b - \epsilon a}{g(bc - ad)}$ for u in the other members, we obtain

$$\left(g(bc - ad)u + \delta b + \epsilon a\right)t + \frac{(ac + bd)(\delta b + \epsilon a)^2}{g(bc - ad)^2}$$

$$+ \frac{(\delta a - \epsilon b)(\delta b - \epsilon a)}{g(bc - ad)} + \zeta g = 0.$$

That is

$$g(bc - ad)u + \delta b + \epsilon a + \frac{g(\delta b + \epsilon c)(\delta b + \epsilon a)}{(bc - ad)^2 t} + \frac{\zeta g}{t} = 0.$$

We conclude that the asymptote is hyperbolic with the genus $u = \frac{A}{t}$. In a similar way the other asymptote arising from the factor $cy - dx$ is defined, so that the curve has two pair of branches going to infinity, each of which is expressed by an equation of the form $u = \frac{A}{t}$.

222.   Suppose that both factors are equal, that is, $\alpha y^2 + \beta xy + \gamma x^2 = (ay - bx)^2$. When we make a change of coordinates given by $y = \frac{au + bt}{g}$ and $x = \frac{at - bu}{g}$, we have

$$g^2 u^2 + \frac{(\delta a - \epsilon b)u}{g} + \frac{(\delta b + \epsilon a)t}{g} + \zeta = 0.$$

When we let t go to infinity, the remaining equation is $u^2 + \frac{(\delta b + \epsilon a)t}{g^3} = 0$. This equation shows two parabolic branches of the species $u^2 = At$. This means that the curve itself is a parabola as well as its asymptote. However, if $\delta b + \epsilon a = 0$, then the equation becomes $g^2 u^2 + \frac{\delta g u}{a} + \zeta = 0$. This is the equation for two parallel straight lines, and this is the case when the whole second degree equation can be expressed as two linear factors.

Thus we would have found all of the species of second order lines, even if we had not already discovered them.

223.   In the same way now we proceed to the third order lines, which have the general equation

$$\alpha y^3 + \beta y^2 x + \gamma yx^2 + \delta x^3 + \epsilon y^2$$

$$+ \zeta yx + \eta x^2 + \theta y + \iota x + \kappa = 0.$$

The principal member $\alpha y^3 + \beta y^2 x + \gamma y x^2 + \delta x^3$ has an odd degree, so it has either one real linear factor or three such factors. The following are the cases which we will treat.

I

If there is one real linear factor.

II

If all three are real and unequal.

III

If two factors are equal.

IV

If all three factors are equal.

Since in each case it is sufficient to confine our calculations to one factor, we let that factor be $ay - bx$, whether it is the only factor, or there are others either equal to it or not. As we have done in the past we adapt the position of the axis to obtain another equation, which we use, since it is just as general as the original:

$$\alpha t^2 u + \beta t u^2 + \gamma u^3 + \delta t^2 + \epsilon t u + \zeta u^2 + \eta t + \theta u + \iota = 0,$$

where the principal member is $\alpha t^2 u + \beta t u^2 + \gamma u^3$, and this has at least one factor u.

CASE 1.

224. The principal member has a unique real factor u, and this will occur when β^2 is less than $4\alpha\gamma$. When we let $t = \infty$, we have $\alpha u + \delta = 0$, and this is the equation for a straight line asymptote. We write this equation as $u = c$, then we have

$$\alpha t^2 (u - c) + (\beta c^2 + \epsilon c + \eta) + \gamma c^3 + \zeta c^2 + \theta c + \iota = 0.$$

This is the equation for the nature of the asymptotes. Depending on whether $\beta c^2 + \epsilon c + \eta$ is equal to zero or not we have two kinds of asymptotes, namely $u - c = \dfrac{A}{t^2}$ or $u - c = \dfrac{A}{t}$. It follows that there

are two species of third order lines under the first case.

The FIRST species has a single asymptote of the species $u = \dfrac{A}{t}$.

The SECOND species has a single asymptote of the species

$u = \dfrac{A}{t^2}$.

CASE 2.

225. Suppose all three of the real linear factors are distinct. This occurs in the equation

$$\alpha t^2 u + \beta t u^2 + \gamma u^3 + \delta t^2 + \epsilon t u + \zeta u^2 + \eta t + \theta u + \iota = 0$$

when β^2 is greater than $4\alpha\gamma$. In this case whatever is true of one factor is also true of the other two. That is, each one provides two hyperbolic

branches which are either of the species $u = \dfrac{A}{t}$ or of the species

$u = \dfrac{A}{t^2}$. Hence in this case there are four different species of third order

lines, each of which has three different straight line asymptotes, with various angles between the straight lines. These are the species.

The THIRD species has three asymptotes of the species $u = \dfrac{A}{t}$.

The FOURTH species has two asymptotes of the species $u = \dfrac{A}{t}$

and one of the species $u = \dfrac{A}{t^2}$.

The fifth species has one asymptote of the species $u = \dfrac{A}{t}$ and two

of the species $u = \dfrac{A}{t^2}$. See section 227.

The sixth species has three asymptotes of the species $u = \dfrac{A}{t^2}$.

226. We have to consider whether all of these species are actually possible. For this purpose we take this most general equation

$$y(\alpha y - \beta x)(\gamma y - \delta x) + \epsilon xy + \zeta y^2 + \eta x + \theta y + \iota = 0$$

whose principal member has three real factors. Although the term x^2 has been eliminated, still the equation is no less general. From what has gone before we know that the factor y gives an asymptote of the form $u = \dfrac{A}{t}$ as long as η is not equal to zero. Now let us see what kind of asymptote is given by the factor $\alpha y - \beta x$. To this end we let $y = \alpha u + \beta t$ and $x = \alpha t - \beta u$. For the sake of brevity we suppose that $\alpha^2 + \beta^2 = 1$, which we can always assume. The equation takes the form

$$\beta(\beta\gamma - \alpha\delta)t^2 u + (2\alpha\beta\gamma - (\alpha^2 - \beta^2)\delta)tu^2 + \alpha(\alpha\gamma + \beta\delta)u^3$$

$$+ \beta(\alpha\epsilon + \beta\zeta)t^2 + (2\alpha\beta\zeta + (\alpha^2 - \beta^2)\epsilon)tu$$

$$+ \alpha(\alpha\zeta - \beta\epsilon)u^2 + (\alpha\eta + \beta\theta)t + (\alpha\theta - \beta\eta)u + \iota = 0.$$

The factor $\alpha y - \beta x$ becomes u. Then when we let $t = \infty$, we have $u = \dfrac{\alpha\epsilon + \beta\zeta}{\alpha\delta - \beta\gamma} = c$. We substitute this value for u in the second member containing t, and this shows that from the factor $u = \alpha y - \beta x$ we have an asymptote of the form $u = \dfrac{A}{t}$ unless

$$\frac{\alpha\eta + \beta\theta}{\beta} + \frac{(\alpha\epsilon + \beta\zeta)(\gamma\epsilon + \delta\zeta)}{(\alpha\delta - \beta\gamma)^2} = 0.$$

In a similar way the factor $\gamma y - \delta x$ gives an asymptote of the form $u = \dfrac{A}{t}$ unless

$$\frac{\gamma\eta + \delta\theta}{\delta} + \frac{(\alpha\epsilon + \beta\zeta)(\gamma\epsilon + \delta\zeta)}{(\alpha\delta - \beta\gamma)^2} = 0.$$

227. From this it is clear that it can happen that neither η nor the most recent expression vanishes, so that the third species can indeed exist. As for the fourth species, first let $\eta = 0$, so that we have one asymptote of the form $u = \dfrac{A}{t^2}$. Then the other expressions are the same, so that the two remaining asymptotes have the form $u = \dfrac{A}{t}$, unless $\theta + \dfrac{(\alpha\epsilon + \beta\zeta)(\gamma\epsilon + \delta\zeta)}{(\alpha\delta - \beta\gamma)^2} = 0$. We conclude that the fourth species is possible. However, if $\eta = 0$, and also one of the other expressions vanishes, then the other expression also vanishes. For this reason, in order that two asymptotes have the form $u = \dfrac{A}{t^2}$, it is necessary that the third also have the same form. It follows that the fifth species is impossible. The sixth species however, is possible, since if $\eta = 0$ and $\theta = \dfrac{-(\alpha\epsilon + \beta\zeta)(\gamma\epsilon + \delta\zeta)}{(\alpha\delta - \beta\gamma)^2}$, then we have the sixth species. From these two cases we actually have only five species of third order lines. What was called the fifth species above must be eliminated. We say that the FIFTH species has three asymptotes of the species $u = \dfrac{A}{t^2}$.

CASE 3.

228. We suppose that the principal member has two equal factors u. This occurs when the first term $\alpha t^2 u$ in the previous case vanishes. Then we have the general equation

$$\alpha t u^2 - \beta u^3 + \gamma t^2 + \delta t u + \epsilon u^2 + \zeta t + \eta u + \theta = 0,$$

so that the principal member has two equal factors u and the third factor $\alpha t - \beta u$ is unequal to the others. This third factor produces an asymptote with the form either $u = \dfrac{A}{t}$ or $u = \dfrac{A}{t^2}$ depending on whether the expression

$$(\alpha\delta + 2\beta\gamma)(\alpha^2\epsilon + \alpha\beta\delta + \beta^2\gamma) - \alpha^3(\alpha\eta + \beta\zeta)$$

does not vanish or does vanish.

229. As for the two equal factors, the first case occurs when γ is not equal to zero. In this case when $t = \infty$ we have $\alpha u^2 + \gamma t = 0$, and this is the equation for a parabolic asymptote of the species $u^2 = At$. From this discussion we see two new species of third order lines.

The SIXTH species has one asymptote of the species $u = \dfrac{A}{t}$ and one asymptote of the species $u^2 = At$.

The SEVENTH species has one asymptote of the species $u = \dfrac{A}{t^2}$ and one parabolic asymptote of the species $u^2 = At$.

230. We now suppose that $\gamma = 0$, and the third factor $\alpha t - \beta u$ gives an asymptote of the form $u = \dfrac{A}{t^2}$ provided that

$$\delta(\alpha\epsilon + \beta\delta) = \alpha(\alpha\eta + \beta\zeta).$$

In this equation does not hold, then the asymptote has the form $u = \dfrac{A}{t}$. Hence we have the equation

$$\alpha t u^2 - \beta u^3 + \delta t u + \epsilon u^2 + \zeta t + \eta u + \theta = 0.$$

When we let $t = \infty$ we have $\alpha u^2 + \delta u + \zeta = 0$. If $\delta^2 < 4\alpha\zeta$, then there is no asymptote, and we have from this case the following two species.

The EIGHTH species has a single asymptote of the species $u = \dfrac{A}{t}$.

The NINTH species has a single asymptote of the species $u = \dfrac{A}{t^2}$.

231. If both of the roots of the equation $\alpha u^2 + \delta u + \zeta = 0$ are real and distinct, that is $\delta^2 > 4\alpha\zeta$, then we have two parallel straight line asymptotes, each of the form $u = \dfrac{A}{t}$. It follows that this case again gives two species.

The TENTH species has one asymptote of the species $u = \dfrac{A}{t}$ and two parallel asymptotes of the species $u = \dfrac{A}{t}$.

The ELEVENTH species has one asymptote of the species $u = \dfrac{A}{t^2}$ and two parallel asymptotes of the species $u = \dfrac{A}{t}$.

232. Suppose the equation $\alpha u^2 + \delta u + \zeta = 0$ has two equal roots, that is $\delta^2 = 4\alpha\zeta$. We then have $\alpha u^2 + \delta u + \zeta = \alpha(u - c)^2$, so that $\alpha t(u - c)^2 = \beta c^3 - \epsilon c^2 - \eta c - \theta$. From this we obtain an asymptote of the species $u^2 = \dfrac{A}{t}$. From this case we obtain two new species.

The TWELFTH species has one asymptote of the species $u = \dfrac{A}{t}$ and one of the species $u^2 = \dfrac{A}{t}$.

The THIRTEENTH species has one asymptote of the species $u = \dfrac{A}{t^2}$ and one of the species $u^2 = \dfrac{A}{t}$.

CASE IV.

233. If all three factors of the principal member are equal, then the equation has the following form.

$$\alpha u^3 + \beta t^2 + \gamma tu + \delta u^2 + \epsilon t + \zeta u + \eta = 0.$$

First we consider the term βt^2. If this term does not vanish, then the curve has a parabolic asymptote of the species $u^3 = At^2$, and we have another species

The FOURTEENTH species has a single parabolic asymptote of the form $u^3 = At^2$.

234. Suppose now that the term $\beta t^2 = 0$, so that

$$\alpha u^3 + \gamma tu + \delta u^2 + \epsilon t + \zeta u + \eta = 0.$$

When we let $t = \infty$, we have $\alpha u^3 + \gamma tu + \epsilon t = 0$ unless both $\gamma = 0$ and $\epsilon = 0$. Hence we suppose that $\gamma \neq 0$ and we have the two equations $\alpha u^2 + \gamma t = 0$ and $\gamma u + \epsilon = 0$. The first of these is for a parabolic asymptote of the species $u^2 = At$, and the second, if we let $\dfrac{-\epsilon}{\gamma} = c$, gives this equation

$$\gamma t(u - c) + \alpha c^3 + \delta c^2 + \zeta c + \eta = 0,$$

which is for a hyperbolic asymptote of the species $u = \dfrac{A}{t}$. It follows that we have another species.

The FIFTEENTH species has one parabolic asymptote of the species $u^2 = At$, and one straight line asymptote of the species $u = \dfrac{A}{t}$, where the axis of the parabola is parallel to the other straight line asymptote.

235. Now we suppose that $\gamma = 0$, so that the equation has the form

$$\alpha u^3 + \delta u^2 + \epsilon t + \zeta u + \eta = 0,$$

where ϵ cannot vanish if the equation is to represent a curve. When we let $t = \infty$, u must necessarily also become infinite so that we have

$\alpha u^3 + \epsilon t = 0$, and this is our final species.

The SIXTEENTH species has one parabolic asymptote of the species $u^3 = At$.

236. We conclude that all third order lines can be classified as one of these *sixteen species,* so that all of the *seventy-two* species in NEWTON'S classification can be reduced to these. It is not to be wondered at that there is such a difference between the Newtonian classification and ours, since we used only the criteria of branches going to infinity and their species, while Newton considered also that part of the curve which is in a bounded region. This is the source of the difference in the species. Although this criterion for classification may seem arbitrary, still Newton was able to find many more species by following his criteria, while using my method I am able to classify no more nor less curves than he.

237. In order that the nature and complexity of each species may be better understood, I will exhibit a general equation for each species, and that in its simplest form. For each species I will give the Newtonian species which pertain to it.

FIRST SPECIES

$$y(x^2 - 2mxy + n^2y^2) + ay^2 + bx + cy + d = 0,$$

where $m^2 > n^2$ and $b \neq 0$. The Newtonian species 33, 34, 35, 36, 37, and 38 belong to the first species.

SECOND SPECIES.

$$y(x^2 - 2mxy + n^2y^2) + ay^2 + cy + d = 0,$$

where $m^2 < n^2$. The Newtonian species 39, 40, 41, 42, 43, 44, and 45 belong to the second species.

THIRD SPECIES.

$$y(x - my)(x - ny) + ay^2 + bx + cy + d = 0,$$

where

$$b \neq 0,\ mb + c + \frac{a^2}{(m - n)^2} \neq 0,\ nb + c + \frac{a^2}{(m - n)^2} \neq 0,\ m \neq n.$$

The Newtonial species 1, 2, 3, 4, 5, 6, 7, 8, and 9 belong to the third species, as do also Newtonian species 24, 25, 26, 27, if $a = 0$.

FOURTH SPECIES.

$$y(x - my)(x - ny) + ay^2 + cy + d = 0,$$

where $c + \dfrac{a^2}{(m - n)^2} \neq 0,\ m \neq n$. Newtonian species 10, 11, 12, 13, 14, 15, 16, 17, 18, 19, 20, 21 belong to the fourth species, as well as 28, 29, 30, 31 if $a = 0$.

FIFTH SPECIES.

$$y(x - my)(x - ny) + ay^2 - \frac{a^2 y}{(m - n)^2} + d = 0,$$

where $m \neq n$. Newtonian species 22, 23, 32 belong to this species.

SIXTH SPECIES.

$$y^2(x - my) + ax^2 + bx + cy + d = 0,$$

where $a \neq 0,\ 2m^3 a^2 - mb - c \neq 0$. Newtonian species 46, 47, 48, 49, 50, 51, 52 belong to this species.

SEVENTH SPECIES.

$$y^2(x - my) + ax^2 + bx + m(2m^2 a^2 - b)y + d = 0,$$

where $a \neq 0$. Newtonian species 53, 54, 55, 56 belong to this species.

EIGHTH SPECIES.

$$y^2(x - my) + b^2x + cy + d = 0,$$

where $c \neq - mb^2$, $b \neq 0$. Newtonian species 61, 62 belong to this species.

NINTH SPECIES.

$$y^2(x - my) + b^2x - mb^2y + d = 0,$$

where $b \neq 0$. Newtonian species 63 belongs to this species.

TENTH SPECIES.

$$y^2(x - my) - b^2y + cy + d = 0,$$

where $c \neq mb^2$, $b \neq 0$. Newtonian species 57, 58, 59 belong to this species.

ELEVENTH SPECIES.

$$y^2(x - my) - b^2x + mb^2y + d = 0,$$

where $b \neq 0$. Newtonian species 60 belongs to this species.

TWELFTH SPECIES

$$y^2(x - my) + cy + d = 0,$$

where $c \neq 0$. Newtonian species 64 belongs to this species.

THIRTEENTH SPECIES

$$y^2(x - my) + d = 0.$$

Newtonian species 65 belongs to this species.

FOURTEENTH SPECIES

$$y^3 + ax^2 + bxy + cy + d = 0,$$

where $a \neq 0$. Newtonian species 67, 68, 69, 70, 71 belong to this species.

FIFTEENTH SPECIES

$$y^3 + bxy + cx + d = 0,$$

where $b \neq 0$. Newtonian species 66 belongs to this species.

SIXTEENTH SPECIES

$$y^3 + ay + bx = 0,$$

where $b \neq 0$. Newtonian species 72 belongs to this species.

238. If we consider the form which curves have in a bounded region, then for the most part each species has notable varieties. It is for this reason that Newton has such a large number of species, since he distinguished the curves which are notably different in a bounded region. It would have been advantageous to have called *genera* what we have called *species,* and the varieties within each could have been referred to as species. This is especially applicable if one wishes to classify the lines of the fourth or higher order. In that case there is so much more variety in each species thus found.

CHAPTER X

On the Principal Properties of Third Order Lines.

239. Just as we have already deduced the principal properties of second order lines from the general equation, so we can also know the principal properties of third order lines from a general equation. In a similar way we could conclude to properties of fourth and higher order lines from an equation. For this reason we will consider the most general equation for a third order line, which is

$$\alpha y^3 + \beta y^2 x + \gamma y x^2 + \delta x^3 + \epsilon y^2 + \zeta y x + \eta x^2 + \theta y + \iota x + \kappa = 0.$$

This expresses the nature of any third order line in the coordinates x and y with any straight line as axis and any angle of inclination for the ordinates.

240. Unless $\alpha = 0$, to each abscissa x there correspond either one or three real values for the ordinate. We suppose that there are three real values for the ordinate; it is clear that their interrelationship can be defined from the equation. We let $\alpha = 1$, so that the equation becomes

$$y^3 + (\beta x + \epsilon)y^2 + (\gamma x^2 + \zeta x + \theta)y + \delta x^3 + \eta x^2 + \iota x + \kappa = 0$$

The sum of the three values of the ordinate is equal to $-\beta x - \epsilon$, also the sum of the three products of the ordinates taken two at a time is equal to $\gamma x^2 + \zeta x + \theta$ and finally the product of all these values of the ordinate is equal to $-\delta x^3 - \eta x^2 - \iota x - \kappa$. If two of the ordinates are complex, then these facts are still true but cannot give information about the configuration of the line, since we have no useful information from the sum of the complex ordinates or from the sum of their products.

241. Let any third order line be related to the axis AZ, with ordinates drawn with a given angle, where LMN and lmn are secants which cut the curve in three points, as in *figure 44*. We let the abscissa $AP = x$ and the ordinate y has the three values PL, PM, and $-PN$. It follows that $PL + PM - PN = -\beta x - \epsilon$. If we take

$$PO = z = \frac{PL + PM - PN}{3}$$

then the point O will be in the middle, in the sense that $LO = MO + NO$. Since $z = -\dfrac{\beta x - \epsilon}{3}$, this point O lies on the straight line OZ, and this straight line has the property that it intersects each chord lmn parallel to LMN in a point o such that $lo + mo = no$. This property is analogous to the property of a diameter which each second order line possesses. If two parallel chords, each of which intersects the curve in three points, and are themselves cut by points O and o so that the sum of two ordinates on one side are equal to the third on the other side, then the straight line drawn through O and o will cut any other such parallel chord in the same way. Hence this line is a kind of diameter for the third order line.

242. Since all diameters of a second order line intersect in the same point, we must see how different diameters of third order lines are related to each other. We think of ordinates with a different angle to the same axis AP. We let the abscissa be t and the ordinate be equal to u. Then we have $y = nu$ and $x = t - mu$. We substitute these values in the general equation

$$y^3 + \beta y^2 x + \gamma yx^2 + \delta x^3 + \epsilon y^2 + \zeta yx + \eta x^2 + \theta y + \iota x + \kappa = 0$$

to obtain the equation

$$n^3 u^3 + \beta n^2 u^2 t + \gamma nut^2 + \delta t^3 + \epsilon n^2 u^2$$

$$+ \zeta nut + \eta t^2 + \theta nu + \iota t + \kappa - \beta mn^2 u^3 - 2\gamma mnu^2 t$$

$$- 3\delta mut^2 - \zeta mnu^2 - 2\eta mut - \iota mu + \gamma m^2 nu^3 + 3\delta m^2 u^2 t$$

$$+ \eta \ m^2 u^2 - \delta m^3 u^3 = 0.$$

From this equation we see that if we call the ordinate of the point on the new diameter v when the abscissa is t, then we have

$$3v = \frac{-\beta n^2 t + 2\gamma mnt - 3\delta m^2 t - \epsilon n^2 + \zeta mn - \eta m^2}{n^3 - \beta \ mn^2 + \gamma \ m^2 n - \delta m^3}.$$

243. Let O be the intersection of these two diameters, and from this point to the axis AZ we draw the line OP parallel to the first set of ordinates and the line OQ parallel to the second set of ordinates, as in *figure 45*. Then we have $AP = x$, $PO = z$, $AQ = t$, and $OQ = v$. Then $z = nv$ and $x = t - mv$ so that $v = \dfrac{z}{n}$ and $t = x + \dfrac{m}{n}z$. First we have $3z = -\beta x - \epsilon$, furthermore $3v = \dfrac{-\beta x}{n} - \dfrac{\epsilon}{n}$ and $t = x - \dfrac{\beta mx}{3n} - \dfrac{\epsilon m}{3n}$. When we substitute these values into the equation found above, we have

$$- \beta n^2 x + \beta^2 mnx - \beta \gamma m^2 x + \frac{\beta \delta m^3 x}{n}$$

$$- \epsilon n^2 + \beta \epsilon mn - \gamma \epsilon m^2 + \frac{\delta \epsilon m^3}{n}$$

$$+ \beta n^2 x - \frac{\beta^2 mnx}{3} - \frac{\beta \epsilon mn}{3} + \epsilon n^2$$

$$- 2\gamma mnx + \frac{2\beta \gamma m^2 x}{3} + \frac{2\gamma \epsilon m^2}{3}$$

$$- \zeta m^2 + 3\delta m^2 x - \frac{\beta \delta \ m^3 x}{n} - \frac{\delta \epsilon \ m^3}{n} + \eta m^2 = 0,$$

that is

$$\frac{2}{3}\beta^2 mnx - \frac{1}{3}\beta \gamma \ m^2 x - 2\gamma mnx + 3\delta m^2 x$$

$$+ \frac{2}{3}\beta \epsilon \ mn - \frac{1}{3}\gamma \epsilon \ m^2 - \zeta mn + \eta m^2 = 0.$$

244. Hence the intersection of the two diameters in the point O depends on the inclination of the two ordinates, since this information is contained in the quantities m and n. For this reason we cannot say that every third order line has a center (if we want to call the intersection of the diameters a center). However, we can find cases where all of the diameters intersect in the same fixed point. To accomplish this we take separately the terms of the equation which contain mn and mm respectively and set each equal to zero. Then the two values of x which we obtain are set equal to each other, to obtain

$$x = \frac{3\zeta - 2\beta \epsilon}{2\beta^2 - 6\gamma} = \frac{3\eta - \gamma \epsilon}{\beta \gamma - 9\delta}.$$

In order for these two values to be equal, it is necessary that

$$6\beta^2 \eta - 2\beta^2 \gamma \epsilon - 18\gamma \eta + 6\gamma \epsilon = 3\beta \gamma \zeta - 2\beta^2 \gamma \epsilon - 27\delta \zeta + 18\beta \delta \epsilon,$$

that is

$$\beta \gamma \zeta - 2\beta^2 \eta - 9\delta \zeta + 6\gamma \eta + 6\beta \delta \epsilon - 2\gamma^2 \epsilon = 0.$$

It follows that $\eta = \dfrac{\beta \gamma \zeta - 9\delta \zeta + 6\beta \delta \epsilon - 2\gamma^2 \epsilon}{2\beta^2 - 6\gamma}$. Whenever η has this value, it will be true that all of the diameters intersect in the same point. Third order lines of this type have a center which can be found from the coordinates

$$AP = \frac{3\zeta - 2\beta\epsilon}{2\beta^2 - 6\gamma}$$

$$PO = \frac{-3\beta\zeta + 6\gamma\epsilon}{2\beta^2 - 6\gamma}$$

245. The determination of the center takes the following form if, instead of unity, we have α for the coefficient of the first term. Suppose that the most general equation of the third order line is

$$\alpha y^3 + \beta y^2 x + \gamma y x^2 + \delta x^3 + \epsilon y^2 + \zeta xy + \eta x^2 + \theta y + \iota x + \kappa = 0,$$

and this curve has a center if

$$\eta = \frac{\beta\gamma\zeta - \alpha\delta\zeta + 6\beta\delta\epsilon - \gamma^2\epsilon}{2\beta^2 - 6\alpha\gamma}.$$

The center will be at O with coordinates $AP = \dfrac{3\alpha\zeta - 2\beta\epsilon}{2\beta^2 - 6\alpha\gamma}$ and $PO = \dfrac{6\gamma\epsilon - 3\beta\zeta}{2\beta^2 - 6\alpha\gamma}$. Wherefore, if one chord which intersects the curve in three points is divided in such a way that two of the ordinates on one side are equal to the third on the other side, then the straight line through this division point and the center will so divide all other such chords which are parallel to the first.

246. If we apply these results to the equations of the species enumerated above, it becomes clear that the first five species have centers provided $\alpha = 0$. In this case we take the center to be the origin. The sixth and seventh species have no center since the coefficient α cannot vanish. The eight, ninth, tenth, eleventh, twelfth, and thirteenth species have a center which is always put at the origin. The fourteenth, fifteenth

and sixteenth species have a center at infinity, so that all the triameters of the curve are parallel.

247. Now that we have these results from a consideration of the sum of the values of the three coordinates, we next consider the product of these three values, since there is nothing of interest to be learned from the sum of products two at a time. From the general equation in section 239 we see that $-PM \cdot PL \cdot PN = -\delta x^3 - \eta x^2 - \iota x - \kappa$. In order to obtain some insight, we note that when we let $y = 0$, we have $\delta x^3 + \eta x^2 + \iota x + \kappa = 0$. The roots of this equation give the intersections of the axis AZ with the curve. If we call these points B, C, and D, then

$$\delta x^2 + \eta x^2 + \iota x + \kappa = \delta(x - AB)(x - AC)(x - AD).$$

It follows that

$$PL \cdot PM \cdot PN = \delta PB \cdot PC \cdot PD.$$

Hence, if we take any other chord lmn parallel to the previous one, we will have $\dfrac{PL \cdot PM \cdot PN}{PB \cdot PC \cdot PD} = \dfrac{pl \cdot pm \cdot pn}{pB \cdot pC \cdot pD}$. This property is completely analogous to the property we found above for second order lines from the product of their ordinates. There are similar properties for fourth, fifth, and higher order lines.

248. Suppose now that a third order line has three straight line asymptotes, FBf, GDg, and HCh as in *figure 46*. Since that third order line would become these three asymptotes if the equation for the curve could be expressed as three linear factors of the form $py + qx + r$. The special equation for the asymptotes, as a complex line, could be expressed and its principal member would be the same as the principal member of the curve. Then since the position of the asymptotes is determined by the second member, the equation for the asymptotes and the equation for

the curve would have the second member also in common. Hence, if the curve is related to the axis AP, then the equation of the curve in the coordinates with the abscissa $AP = x$ and ordinate $PM = y$ is

$$y^3 + (\beta x + \epsilon)y^2 + (\gamma x^2 + \zeta x + \theta)y + \delta x^3 + \eta x^2 + \iota x + \kappa = 0.$$

For the asymptotes the equation related to the same axis AP in the coordinates with abscissa $AP = x$ and ordinate $PG = z$ is

$$z^3 + (\beta x + \epsilon)z^2 + (\gamma x^2 + \zeta x + B)z + \delta x^3 + \eta x^2 + Cx + D = 0,$$

in which the coefficients ζ, B, C, and D are defined in such a way that the equation can be expressed as the product of the linear factors.

249. If we draw any ordinate PN which intersects the curve in the three points L, M, and N, then it also intersects the asymptotes in three points F, G, and H. From the equation for the curve we have $PL + PM + PN = -\beta x - \epsilon$. From the equation for the asymptotes we also have $PF + PG + PH = -\beta x - \epsilon$. For this reason we have $PL + PM + PN = PF + PG + PH$, that is $FL - GM + HN = 0$. Furthermore, if some other ordinate pf is drawn, in the same way we have $fn - gm - hl = 0$. Hence, if any straight line intersects both the asymptotes and the curve in three points, then the sum of two of the intervals between the curve and the asymptotes it approaches on one side is equal to the third such interval in which the curve is approaching from the other side.

250. If follows that for any third order line with three straight line asymptotes, the three branches which approach the asymptotes cannot all approach from the same side, but if two converge from one side, then the third necessarily converges from the other side. For this reason the third order line represented in *figure 47* is impossible, since the straight line cutting the asymptotes in the points f, g, and h and the curve in

the points l, m and n gives the intervals fn, gm, and hl, all approaching from the same side, so that the sum cannot be equal to zero. The intervals approaching from the same side have the same sign, for example $+$, while the one from the other side has the sign $-$. It follows that the sum of the three intervals cannot vanish unless there are different signs.

251. From this it should be clear why in a third order line there cannot be two straight line asymptotes with the species $u = \dfrac{A}{t^2}$ while the third would be with the species $u = \dfrac{A}{t}$, since the two hyperbolic branches would approach their asymptotes infinitely more rapidly, then the other hyperbolic branch with the species $u = \dfrac{A}{t}$. Now consider the straight line fl, as in *figure 46,* when it is moved out to infinity so that the intervals fn, gm, and hl have become infinitely small. If two of the branches, nx and my were of the species $u = \dfrac{A}{t^2}$, and the third branch lz of the species $u = \dfrac{A}{t}$, then the intervals fn and gm would be infinitely smaller than the interval hl, so that $gm \neq fn + hl$.

252. It follows that in lines of higher order which have the same number of asymptotes as their order, it is impossible for there to be a single asymptote of the species $u = \dfrac{A}{t}$, while the others are of higher species, such as $u = \dfrac{A}{t^2}$, $u = \dfrac{A}{t^3}$, etc. If there is one asymptote of the species $u = \dfrac{A}{t}$, then there are necessarily others of the same species. For the same reason, if there is no asymptote of the species $u = \dfrac{A}{t}$, then there cannot be a single asymptote of the species $u = \dfrac{A}{t^2}$, but there

must be at least two of such asymptotes. The reason is that branches of the hyperbolic species $u = \dfrac{A}{t^3}$ and $u = \dfrac{A}{t^4}$, etc. converge to their asymptotes infinitely more rapidly than those of the species $u = \dfrac{A}{t^2}$. Therefore in the classification of the species of any higher order, the impossible cases can easily be excluded and in this way some significantly difficult calculations can be avoided.

253. We suppose now that the third order line is intersected by some straight line in only two points, and that any other straight line parallel to the first intersects the curve in either two or no points. Now if we choose any axis and take ordinates parallel to the original straight line, then the equation has the form

$$y^2 + \frac{(\gamma x^2 + \zeta x + \theta)y}{\beta x + \epsilon} + \frac{\delta x^3 + \eta x^2 + \iota x + \kappa}{\beta x + \epsilon} = 0.$$

For example, if the abscissa $AP = x$ there are two ordinates y, namely PM and $-PN$. From the nature of the equation we have $PM - PN = \dfrac{-\gamma x^2 - \zeta x - \theta}{\beta x + \epsilon}$. Let the chord MN be bisected in the point O, as in *figure 48*. Then $PO = \dfrac{\gamma x^2 + \zeta x + \theta}{\beta x + \epsilon}$. If we let $PO = z$, then $z(\beta x + \epsilon) = \gamma x^2 + \zeta x + \theta$. From this it follows that every O which bisects a chord parallel to MN lies on a hyperbola, unless $\gamma x^2 + \zeta x + \theta$ is divisible by $\beta x + \epsilon$, in which case the point O lies on a straight line.

254. If $\gamma x^2 + \zeta x + \theta$ is divisible by $\beta x + \epsilon$, then the curve has a diameter, that is, all chords parallel to MN are bisected by a straight line. But this is a property of all second order lines. Indeed, if $\gamma x^2 + \zeta x + \theta$ is divisible by $\beta x + \epsilon$, then it should vanish when we let $x = \dfrac{-\epsilon}{\beta}$. It follows that if $\gamma \epsilon^2 - \beta \epsilon \zeta + \beta^2 \theta = 0$, then the third

order line has a diameter.

255. Hence we can determine in the most general way all of the cases in which a third order line has a diameter. Let the general equation be

$$\alpha y^3 + \beta y^2 x + \gamma yx^2 + \delta x^3 + \epsilon y^2 + \zeta yx + \eta x^2 + \theta y + \iota x + \kappa = 0,$$

whose ordinate y, since it has three values or only one value, cannot have the property of a diameter. For this reason we draw some ordinate u which makes a different angle with the same axis, so that $y = nu$ and $x = t - mu$. When these values are substituted we obtain

$$\alpha n^3 u^3 + \beta n^2 u^2 t + \gamma nut^2 + \epsilon n^2 u^2 + \zeta nut + \eta t^2 + \theta nu$$

$$+ \iota t + \kappa - \beta\, mn^2 u^3 - 2\gamma mnu^2 t - 3\delta mut^2 - \zeta nmu^2 - 2\eta mut$$

$$- \iota mu + \gamma m^2 nu^3 + \delta m^2 u^2 t + \eta m^2 u^2 - \delta\, m^3 u^3 = 0.$$

In the first place, in order that the new ordinate be rendered apt to become a diameter, it is necessary that no higher degree than the second be present. It follows that

$$\alpha n^3 - \beta mn^2 + \gamma m^2 n - \delta m^3 = 0.$$

256. Besides this, it is necessary that the quantity which multiplies u, namely

$$(\gamma n - 3\delta m)t^2 + (\zeta n - 2\eta m)t + \theta n - \iota m,$$

should be divisible by that which multiplies u^2, which is

$$(\beta n^2 - 2\gamma mn + 3\delta m^2)t + \epsilon n^2 - \zeta mn + \eta m^2.$$

That means that the previous expression should vanish when we substitute

$$t = \frac{-\epsilon n^2 + \zeta mn - \eta m^2}{\beta n^2 - 2\gamma mn + 3\delta m^2}.$$

From this we obtain

$$\iota = \frac{\theta n}{m} \frac{(\zeta n - 2\eta m)(\epsilon n^2 - \zeta mn + \eta m^2)}{(\beta n^2 - 2\gamma mn + 3\delta m^2)m} +$$

$$\frac{(\gamma n - 3\delta m)(\epsilon n^2 - \zeta mn + \eta m^2)^2}{(\beta n^2 - 2\gamma mn + 3\delta m^2)^2 m}.$$

257. If we apply this criterion to the species we have enumerated we see that in the first species there is no place for a diameter. In the second species the chords which are parallel to the axis from which the abscissas x are taken will be bisected by a diameter. The third species admits no diameter. The fourth species always has one diameter which bisects chords which are parallel to one of the asymptotes. The fifth species has three diameters which bisect chords which are parallel to each of the asymptotes. The sixth species can have no diameter. The seventh species always has one diameter for the chords parallel to the asymptote arising from the factor $x - my$. The eighth has one diameter for chords parallel to the axis. The ninth species has two diameters; one for chords parallel to the axis, and the other for chords parallel to the other asymptote. The tenth is like the eighth, and the eleventh is like the ninth. The twelfth is like the eighth and the thirteenth is like the ninth. The fourteenth has one diameter for chords parallel to the axis. The fifteenth and sixteenth species have no chords which intersect the curve in two points, so they have no diameters. These properties of diameters were well known to Newton, and for that reason it has been a pleasure to commemorate his work in this place.

258. Although the equations which we have given above for each of species of third order lines were given in rectangular coordinates x and

y, the nature of the species is not changed if the coordinates are oblique. Whenever an equation in rectangular coordinates gives branches going to infinity, the same number of such branches remain if oblique coordinates are used. Nor is the nature of the branch going to infinity changed when coordinates are changed. A branch which is hyperbolic remains hyperbolic, and those which are parabolic remain parabolic. Nor does the species of either a hyperbolic or a parabolic branch change. It follows that any curve given by an equation for the first species remains in the first species whether rectangular coordinates or oblique coordinates are used. The same is true for all of the species of third order lines.

259. If we allow arbitrary oblique coordinates, the equations which we gave above for the various species remain valid when we substitute νu for y and $t - \mu u$ for x, where $\mu^2 + \nu^2 = 1$. When we take the angle of obliquity to be arbitrarily chosen, the equations given above can be simplified. What follows is a list of the simplest form of the equations in the oblique coordinates t and u.

FIRST SPECIES

$$u(t^2 + n^2 u^2) + au^2 + bt + cu + d = 0,$$

where $n \neq 0$ and $b \neq 0$.

SECOND SPECIES

$$(t^2 + n^2 u^2) + au^2 + cu + d = 0,$$

where $n \neq 0$.

THIRD SPECIES

$$u(t^2 + n^2 u^2) + au^2 + bt + cu + d = 0,$$

where $n \neq 0$, $b \neq 0$, and $\pm nb + c + \dfrac{a^2}{4n^2} \neq 0$.

FOURTH SPECIES

$$u(t^2 - n^2 u^2) + au^2 + cu + d = 0,$$

where $n \neq 0$ and $c + \dfrac{a^2}{4n^2} \neq 0$.

FIFTH SPECIES

$$u(t^2 - n^2 u^2) + au^2 - \dfrac{a^2 u}{4n^2} + d = 0,$$

where $n \neq 0$.

SIXTH SPECIES

$$tu^2 + at^2 + bt + cu + d = 0,$$

where $a \neq 0$, and $c \neq 0$.

SEVENTH SPECIES

$$tu^2 + at^2 + bt + d = 0,$$

where $a \neq 0$.

EIGHTH SPECIES

$$tu^2 + b^2 t + cu + d = 0,$$

where $b \neq 0$, and $c \neq 0$.

NINTH SPECIES

$$tu^2 + b^2 t + d = 0,$$

where $b \neq 0$.

TENTH SPECIES

$$tu^2 - b^2 t + cu + d = 0,$$

where $b \neq 0$, and $c \neq 0$.

ELEVENTH SPECIES

$$tu^2 - b^2t + d = 0,$$

where $b \neq 0$.

TWELFTH SPECIES

$$tu^2 + cu + d = 0,$$

where $c \neq 0$.

THIRTEENTH SPECIES

$$tu^2 + d = 0.$$

FOURTEENTH SPECIES

$$u^3 + at^2 + cu + d = 0.$$

FIFTEENTH SPECIES

$$u^3 + atu + bt + d = 0,$$

where $a \neq 0$.

SIXTEENTH SPECIES

$$u^3 + at = 0.$$

CHAPTER XI

On Fourth Order Lines.

260. The general equation for fourth order lines is

$$\alpha y^4 + \beta y^3 x + \gamma y^2 x^2 + \delta y x^3 + \epsilon x^4 + \zeta y^3 + \eta y^2 x + \theta y x^2 + \iota x^3 +$$

$$\kappa y^2 + \lambda y x + \mu x^2 + \nu y + \xi x + o = 0.$$

By changing the inclination of the coordinates and by changing the axis and the origin, the equation can be simplified in many ways for different cases. In order that we might follow the method given above for finding the various species, or rather the *genera* of the lines contained in this order, we consider the principal member, which leads to the following different cases.

I. If all of the linear factors are complex.

II. If only two of the linear factors are real and they are distinct.

III. If only two factors are real, and they are equal.

IV. If all four factors are real and distinct.

V. If two of the factors are equal, and the other two are distinct.

VI. If besides two equal factors, there are two others equal to each other.

VII. If three linear factors are equal to each other.

VIII. If all four factors are equal.

CASE I

261. If all of the factors of the principal member are complex, then there will be no branches going to infinity. Since we are using the criterion of diversity of branches for assigning the genus, this case provides one genus. Hence we have:

GENUS I.

There are no branches going to infinity and the simplest equation by which it may be expressed is

$$(y^2 + m^2x^2)(y^2 - 2pxy + q^2x^2) + ay^2x + byx^2 +$$

$$cy^2 + dyx + ex^2 + fy + gx + h = 0,$$

where $p^2 < q^2$. Since it is necessary that the terms y^4 and x^4 be present in the principal number, for a given quantity the coordinates x and y can be increased or decreased so that the terms y^3 and x^3 can be absent from the second member.

CASE II

If only two factors of the principal member are real and unequal, by choosing oblique coordinates and a change of axis we can make one of the factors be y and the other be x, so that the equation is

$$yx(y^2 - 2myx + n^2x^2) + ay^2x + byx^2 + cy^2 + dyx +$$

$$ex^2 + fy + gx + h = 0,$$

where $m^2 < n^2$. Since y^3x and yx^3 are necessarily present, the terms y^3 and x^3 from the second member can be omitted. The curve has two straight line asymptotes, one of which has the equation $y = 0$, and the other has the equation $x = 0$. The species of the first is given by the equation $n^2yx^3 + ex^2 + gx + h = 0$. That of the second is given by $xy^3 + cy^3 + fy + h = 0$. Hence we have

GENUS II.

With two straight line asymptotes, each with the property $u = \dfrac{A}{t}$, where $c \neq 0$ and $e \neq 0$.

GENUS III.

This has two straight line asymptotes, one with the species $u = \dfrac{A}{t}$ and the other the species $u = \dfrac{A}{t^2}$, expressed by the equation

$$yx(y^2 - 2myx + n^2x^2) + ay^2x + byx^2 + cy^2 + dyx + fy + gx + h = 0,$$

where $e \neq 0$ and $g \neq 0$.

GENUS IV.

This has two straight line asymptotes, one with the species $u = \dfrac{A}{t}$ and the other with $u = \dfrac{A}{t^3}$, and contained in the equation

$$yx(y^2 - 2myx + n^2x^2) + ay^2x + byx^2 + cy^2 + dyx + fy + h = 0,$$

where $c \neq 0$.

GENUS V.

This has two straight line asymptotes, both with species $u = \dfrac{A}{t^2}$, and is contained in the equation

$$yx(y^2 - 2myx + n^2x^2) + ay^2x + byx^2 + dyx + fy + gx + h = 0,$$

where $f \neq 0$ and $g \neq 0$.

GENUS VI.

This has two straight line asymptotes, one with species $u = \dfrac{A}{t^2}$ and the other with species $u = \dfrac{A}{t^3}$, and is contained in the equation

$$yx(y^2 - 2myx + n^2x^2) + ay^2x + byx^2 + dyx + fy + h = 0,$$

where $f \neq 0$.

GENUS VII.

This has two straight line asymptotes, both with species $u = \dfrac{A}{t^3}$ and is contained in the equation

$$yx(y^2 - 2myx + n^2x^2) + ay^2x + byx^2 + dyx + h = 0,$$

where $n^2 > m^2$ in all of the above.

CASE III

263. Let both of the factors of the principal member which are real be equal, so that the equation becomes

$$y^2(y^2 - 2myx + n^2x^2) + ayx^2 + bx^3 + cy^2 + dyx + ex^2 +$$

$$fy + gx + h = 0,$$

where again $n^2 > m^2$. Unless $b = 0$, the equation gives

GENUS VIII,

which has one parabolic asymptote with the species $u^2 = At$.

If $b = 0$, when we let $x = \infty$, we obtain $y^2 + \dfrac{ay}{n^2} + \dfrac{g}{n^2x} + \dfrac{b}{n^2x^2} = 0$. Hence, if $a^2 < 4n^2e$, then we have

GENUS IX,

which has no branch going to infinity.

If $b = 0$, $a^2 > 4n^2e$ and $g \neq 0$, we have

GENUS X,

which has two parallel asymptotes with the species $u = \dfrac{A}{t}$.

If $b = 0$, $g = 0$, and $a^2 > 4n^2e$, we have

GENUS XI,

which has parallel asymptotes with the species $u = \dfrac{A}{t^2}$.

If $b = 0$, $a^2 = 4n^2e$, and $g \neq 0$, then we have

GENUS XII,

which has one hyperbolic asymptote with species $u^2 = \dfrac{A}{t}$.

If $b = 0$, $g = 0$, $a^2 = 4n^2e$, and $h < 0$, then we have

GENUS XIII,

which has one hyperbolic asymptote with the species $u^2 = \dfrac{A}{t^2}$.

If $b = 0$, $g = 0$, $a^2 = 4n^2e$, and $h > 0$, then we have

GENUS XIV,

which has no branches going to infinity.

CASE IV.

264. Let all four linear factors of the principal member be real and distinct, so that the equation has the form

$$yx(y - mx)(y - nx) + ay^2x + byx^2 + cy^2 + dyx + ex^2 +$$

$$fy + gx + h = 0.$$

The curve has four straight line asymptotes with the species $u = \dfrac{A}{t}$, $u = \dfrac{A}{t^2}$, or $u = \dfrac{A}{t^3}$. Following the rule discussed in section 251, we have the following genera.

GENUS XV

has four hyperbolic asymptotes, all with the species $u = \dfrac{A}{t}$.

GENUS XVI

has four hyperbolic asymptotes, three with species $u = \dfrac{A}{t}$ and one with species $u = \dfrac{A}{t^2}$.

GENUS XVII

has four hyperbolic asymptotes, three with species $u = \dfrac{A}{t}$ and one with

species $u = \dfrac{A}{t^3}$.

GENUS XVIII

has four hyperbolic asymptotes, two with species $u = \dfrac{A}{t}$ and two with

species $u = \dfrac{A}{t^2}$.

GENUS XIX

has four hyperbolic asymptotes, two with species $u = \dfrac{A}{t}$, one with

species $u = \dfrac{A}{t^2}$, and one with species $u = \dfrac{A}{t^3}$.

GENUS XX

has four hyperbolic asymptotes, two with species $u = \dfrac{A}{t}$, and two with

species $u = \dfrac{A}{t^3}$.

GENUS XXI

has four hyperbolic asymptotes, all of the species $u = \dfrac{A}{t^2}$.

GENUS XXII

has four hyperbolic asymptotes, three with species $u = \dfrac{A}{t^2}$, and one with

species $u = \dfrac{A}{t^3}$.

GENUS XXIII

has four hyperbolic asymptotes, two with species $u = \dfrac{A}{t^2}$ and two with

species $u = \dfrac{A}{t^3}$.

GENUS XXIV

has four hyperbolic asymptotes, all with species $u = \dfrac{A}{t^3}$.

CASE V

265. Let two factors of the principal member be equal and the other two be distinct, so that the equation is of the form

$$y^2 x(y + nx) + ayx^2 + bx^3 + cy^2 + dyx + ex^2 + fy + gx + h = 0.$$

In the first place, due to the equality of the factors, all of the genera which occurred in *case III,* and each of these with the varieties suggested by the unequal factors, that is, all of the genera contained in the second case. All together we have six times seven, that is forty-two genera from this case. However, two of these genera are impossible, namely those with two parallel asymptotes with species $u = \dfrac{A}{t^2}$ while the third is one with species $u = \dfrac{A}{t}$, and the fourth either with species $u = \dfrac{A}{t^2}$, or species $u = \dfrac{A}{t^3}$. Hence this case gives forty genera, which with the previous genera gives a total of *sixty-four.* It would take too long to describe all of these here. Since there is not time to treat in detail all of these genera, we cannot say for sure that all of them are real. Whoever wishes, can take it upon himself to perform this task, and if necessary, emend this classification.

CASE VI

266. This is the case where two pair of equal factors are present, so that the equation is

$$y^2 x^2 + ay^3 + bx^3 + cy^2 + dyx + ex^2 + fy + gx + h = 0.$$

Each pair of equal factors considered by themselves give rise to seven genera, so that taken together we have forty-nine genera. However, since

b cannot be both positive and negative at the same time, two of the genera are impossible, so that from this case we obtain forty-seven genera. This number also is too large for us to discuss the individual genera. Hence we have up to this point found *one hundred eleven* genera.

CASE VII

267. If three of the factors are equal the equation is as follows.

$$y^3 x + ayx^2 + bx^3 + cy^2 + dyx + ex^2 + fy + gx + h = 0.$$

This factor x gives an asymptote with the species $u = \dfrac{A}{t}$ as long as $c \neq 0$. If $c = 0$, and $f \neq 0$, then the asymptote has the species $u = \dfrac{A}{t^2}$. If $c = 0$ and $f = 0$, then the asymptote has the species $u = \dfrac{A}{t^3}$. Then the factor y^3, unless $b = 0$, gives a parabolic asymptote with the species $u^3 = At^2$. If $b = 0$, then when we let $x = \infty$, we have

$$y^3 + ayx + dy + ex + g + \frac{cy^2 + fy + h}{x} = 0.$$

Here, if $e \neq 0$ we have $y^3 + ayx + ex = 0$, so that if $a \neq 0$ the equation becomes $y^2 + ax = 0$ and $ay + e = 0$. Hence we have besides the parabolic asymptote with species $u^2 = At$, also the hyperbolic asymptote with the equation

$$(ay + e)x - \frac{e^3}{a^3} - \frac{de}{a} - g + \frac{ce^2 - afe + a^2 h}{a^2 x}.$$

If $e^3 + a^2 de + a^3 g \neq 0$, this asymptote has the species $u = \dfrac{A}{t}$. If this expression is equal to zero, then the species is $u = \dfrac{A}{t^2}$. If $a = 0$ and $e \neq 0$, then $y^3 + ex = 0$ which gives a parabolic asymptote with species $u^3 = At$. If $e = 0$, and $a = 0$, then we have $y^3 + dy + g = 0$, which gives either a single asymptote with species $u = \dfrac{A}{t}$ or three of the same species, or one with species $u = \dfrac{A}{t}$ and one with species $u^2 = \dfrac{A}{t}$, or one with species $u^3 = \dfrac{A}{t}$. All together we have eight different varieties which occur. When we multiply these by the three from the x factor, we have twenty-four genera. Hence all of the cases treated so far give *one hundred thirty five* genera.

CASE VIII

268. If all of the factors are equal, then we have the equation

$$y^4 + ay^2x + byx^2 + kx^3 + cy^2 + dyx + ex^2 + fy + gx + h = 0.$$

If $k \neq 0$. then we have

GENUS CXXXVI.

This has only one parabolic asymptote with species $u^4 = At^3$. If $k = 0$. and $b \neq 0$, then we have $y^4 + byx^2 + ex^2 = 0$. Hence $y^3 + bx^2 = 0$ and $by + e = 0$. Hence for the straight line asymptote $by + e = 0$, we have

$$(by + e)x^2 + \frac{e^4}{b^4} + \frac{ae^2x}{b^2} + \frac{ce^2}{b^2} - \frac{dex}{b} - \frac{ef}{b} + gx + h = 0.$$

Hence, unless $ae^2 - bde + b^2g = 0$, the asymptote is of the species $u = \dfrac{A}{t}$. If the expression is equal to zero, then the asymptote has the species $u = \dfrac{A}{t^2}$. Thus we have

GENUS CXXXVII.

This has one parabolic asymptote with the species $u^3 = At^2$ and one hyperbolic asymptote with the species $u = \dfrac{A}{t}$. We likewise have

GENUS CXXXVIII.

This has one parabolic asymptote with the species $u^3 = At^2$ and one hyperbolic asymptote with the species $u = \dfrac{A}{t^2}$.

269. Suppose now that $k = 0$ and $b = 0$, so that

$$y^4 + ay^2x + cy^2 + dyx + ex^2 + fy + gx + h = 0.$$

If $e \neq 0$, then $y^4 + ay^2x + ex^2 = 0$. This equation is impossible if $a^2 < 4e$. If $a^2 > 4e$, then there are two parabolic asymptotes related to the same axis with the species $u^2 = At$. If $a^2 = 4e$, these two parabolas become one, so that we have GENERA CXXXIX, CXL, and CXLI.

If $e = 0$, then the equation becomes

$$y^4 + ay^2x + cy^2 + dyx + fy + gx + h = 0.$$

If $a \neq 0$, we have

$$y^4 + ay^2x + cy^2 + dyx + gx = 0,$$

so that both $y^2 + ax = 0$, and $y =$ constant, so that $ay^2 + dy + g = 0$ and we conclude that y has either two different values, or they are equal, or there are no real values. In the first case the curve has, besides one parabolic asymptote, two parallel asymptotes with species $u = \dfrac{A}{t}$. In the second case, one with the species $u^2 = \dfrac{A}{t}$. In the third case there is no other asymptote. Hence, again we have the three GENERA CXLII, CXLIII, and CXLIV.

270. We now suppose that $a = 0$. so that the equation becomes

$$y^4 + cy^2 + dyx + fy + gx + h = 0.$$

If $d \neq 0$, the curve has a parabolic asymptote with the species $u^3 = At$, and one straight line equation $dy + g = 0$, with the species $u = \dfrac{A}{t}$. Finally, if $d = 0$, the curve has one parabolic asymptote with the species $u^4 = At$. We see that the total of all genera of fourth order lines is *one hundred forty-six*. Each of these genera contain many species which may differ from each other notably.

271. From this investigation it should be clear what a large number of genera there must be in the fifth and higher order lines, when we recall the number of genera in the third order. Indeed, one could not undertake to answer this question, unless a whole volume were to be dedicated to this one task. As to the primary properties of lines of the fourth and higher orders, these can be derived from the general equation in a way similar to that which we used for third order lines. We will not delay any longer with this question.

CHAPTER XII

On the Investigation of the Configuration of Curves.

272. The topics we have been discussing in the previous chapters have been for the sake of understanding those parts of the curve which go to infinity. It is quite often very difficult to know from the equation the configuration of the curve in a bounded region. In order to find all of the ordinate values for any finite abscissa we would have to solve the equation and distinguish real from complex roots. This task, when the equation is of a higher degree, is beyond the powers of the known principles of analysis. If some particular value is given to the abscissa, then the ordinate takes the role of the unknown. Hence the solution of the equation depends on the degree of the ordinate in the resulting equation. This task can be greatly simplified by reducing the equation to a simpler form, both by the choice of the most convenient axis and the most apt inclination of the coordinates. Since either coordinate can be chosen for the abscissa, the labor can be greatly reduced by a choice which leaves the ordinate with the lowest degree in the equation.

273. Hence, if we want to investigate the configuration of the third order line which belongs to the first species, we will take the simplest equation, which is given in section 258. Between the coordinates t and u, we choose u for the abscissa and t for the ordinate since the equation

is second degree in t. Hence we have the equation in the form

$$y^2 = \frac{2by + ax^2 + cx + d - n^2x^3}{x},$$

which can be solved to give

$$y = \frac{b \pm \sqrt{b^2 + dx + cx^2 + ax^3 - n^2x^4}}{x},$$

where $b \neq 0$ and $n \neq 0$.

274. There are two values of the ordinate which correspond to the values of x which make the function $b^2 + dx + cx^2 + ax^3 - n^2x^4$ take a positive value. There is a single value of the ordinate, that is, both values equal each other, and this corresponds to values of x which make the function vanish. There is no ordinate corresponding to the values of the abscissa which make the function take negative values. The values of this function, if they are positive, cannot become negative unless the function first vanishes, where the values of y become equal. The cases which are most important to consider are those where the function $b^2 + dx + cx^2 + ax^3 - n^2x^4$ vanishes. It is certain that this will happen at least twice. There are certain limits for x beyond which whether in the negative or in the positive direction, the value of this function becomes negative. Hence the whole curve lies between this determined interval of the abscissas, since beyond this interval the ordinates become imaginary.

275. We suppose first that the expression $b^2 + dx + cx^2 + ax^3 - n^2x^4$ has only two real factors, that is, it vanishes for only two values of x. If this happens at the points P and S, then corresponding to these points the ordinate has only a single value. Throughout the interval PS the ordinate will have two real values, and outside this interval all of the ordinates will be complex. Hence all of the

curve lies between the ordinates Kk and Nn in *figure 49*. The ordinate at the origin A is an asymptote for the curve, but the curve also intersects this ordinate in a point. If we let $x = 0$, then

$$\sqrt{b^2 + dx + cx^2 + ax^3 - n^2x^4} = b + \frac{dx}{2b},$$

so that $y = \dfrac{b \pm \left(b + dx/2b\right)}{x}$. That is, either $y = \infty$ or $y = \dfrac{-d}{2b}$. Hence the curve has the form represented in *figure 50*.

276.   Now   we   suppose   that   the   expression $b^2 + dx + cx^2 + ax^3 - n^2x^4$ has four real linear factors which are distinct. Hence the expression vanishes at the four points $P,Q,R,$ and S, so that the ordinate has a single value at these points. Since the ordinates corresponding to values of x between X and P in *figure 51* will be complex, then for the interval PQ they will be real. For the interval QR the ordinates will be complex, and then real again for the interval RS. Beyond S towards Y they become imaginary again. Hence the curve consists of two parts which are separated; one lies between the straight lines Kk and Ll, while the other lies between the straight lines Mm and Nn. Since the ordinates for the origin are real, the origin A must lie on the interval PQ or RS. In this case the curve has the configuration shown in *figure 51,* called a CONJUGATE OVAL, that is, it consists of an oval, called a conjugate oval, separated from the rest of the curve which is related to the asymptote DE.

277. If two of the roots are equal, either P and Q or Q and R or R and S coincide. If the first happens, since A lies between P and Q, both roots must be x; since $b \neq 0$, this cannot happen. If R and S coincide, then the conjugate oval becomes infinitely small and becomes the CONJUGATE POINT. If Q and R coincide, the oval in joined to the rest of the curve to produce a curve with a NODE, as in *figure 52*. If

three of the roots are equal, that is the points Q, R, and S coincide, then
the node evanesces into the CUSP, as represented in *figure 53*. Hence
there are five different varieties which find a place in the first species.
Newton considered each of these to be a different species.

278. In a similar way Newton finds the subdivisions of the other
species, since all of the equations are so constituted that one or the other
of the coordinates has a degree no more than two. When one of the
coordinates has only one degree, the shape of the curve is quite easily
recognized. In this case the equation has the form $y = P$, where P is
some non-irrational function of the abscissa x. For each value of x there
is a corresponding single value of the ordinate, so that the curve follows
the axis to infinity in both directions. If the function P is a rational
function, then it can happen that the ordinate in one or several places
becomes infinite, so that the curve has an asymptote when the denominator of the function P vanishes.

279. If we let $y = \dfrac{P}{Q}$, then the ordinates become infinite at all of
the roots of $Q = 0$. That is, any root of this equation, for example
$x = f$, indicates that if the abscissa $x = f$, then the ordinate y becomes
infinite, since $Q = 0$. Furthermore, it is clear that if the ordinates y are
positive for $x > f$ and that for $x < f$ the ordinates are negative, then
the ordinate is an asymptote of the species $u = \dfrac{A}{t}$. This is true for all
factors which are distinct. If the denominator Q has two equal factors,
for instance $(x - f)^2$, then the ordinate will be positive whether $x < f$
or $x > f$. When $x = f$, the asymptote is of the species $u^2 = \dfrac{A}{t}$. If the
denominator Q has three equal factors, namely $(x - f)^3$, then the ordinate changes signs as the abscissa passes from less than f to greater than
f, as in the first case.

280. Next, the equations with the form $y^2 = \dfrac{2Py - R}{Q}$, where P, Q, and R are any polynomial functions in the abscissa x, are quite easily treated. For any value of the abscissa x there corresponds either two or no values for the ordinate; there are two values when $P^2 > QR$, and there are none when $P^2 < QR$. At the point where the two values become imaginary, we have $P^2 = QR$, so that $y = \dfrac{P}{Q}$ and the ordinate draws the curve together or is tangent in the one point. In order to understand the form of the curve, we should consider the equation $P^2 - QR = 0$, whose real roots will give the places where the ordinate constricts the curve to a single point. We note these points on the axis, and if all of the roots are distinct, then the intervals on the axis between these points alternate in having two real and two complex values for the ordinate. Thus the curve consists of as many separated parts as there are alterations of this kind. This is the source of the conjugate ovals.

281. If the equation $P^2 - QR = 0$ has two equal roots, then two of the points noted on the axis coincide, so that an interval having complex ordinates, or one having real ordinates will disappear. In the former case, we obtain a node, as in *figure 52;* in the latter case a conjugate oval will vanish to a point. If that equation has three equal roots, the node shrinks to a point and becomes a cusp, as in *figure 53.* If four roots of the equation become equal, then either two separated ovals come together in a point, or a node is added to the cusp, or two cusps are joined together on opposite sides. If five equal roots are present, we obtain almost nothing new. There is a cusp with not one, but two ovals joined to it. When there are even more equal roots, there are no really new configurations.

282. A node, or the intersection of two branches of a curve, is also

called a DOUBLE POINT, since a straight line intersecting the curve in that point must be considered to cut the curve in two points. Furthermore, if another branch of the curve passes through the node, then we obtain a *triple* point if two doube points coinside. Hence we understand the nature and source of any *multiple* points. That is, we may have a vanishing oval, or a conjugate point; there can be a double point; or there can be a cusp, which arises when a conjugate point is adjoined to the rest of the curve.

283. If the equation by which the ordinate y is expressed in terms of the abscissa x is a cubic or of a higher degree, so that y is equal to a multiple-valued function of x, then for each value of the abscissa, there will be the same number of values of y, as there are degrees of y in the equation, or that number diminished by 2, 4, 6, etc. Hence two values of the ordinate always become complex simultaneously, and at the point where this happens, the two values become equal. Hence in the transition from complex to real, there are many forms, but they are all either of the form already discussed, or they are a combination of these. Hence if we know all of the values of the ordinate at many values of the abscissa, both negative and positive, then we can easily draw the curve through each of these points, and the configuration of the curve is known.

284. We will illustrate these remarks with an example in which the degree is higher, but the values of y can still be expressed by quadratic roots. Let

$$2y = \pm \sqrt{6x - x^2} \pm \sqrt{6x + x^2} \pm \sqrt{36 - x^2}.$$

From this equation, for any abscissa, we obtain eight values for the ordinate. It is clear that if $x < 0$, then the ordinate is complex, while the same thing occurs when $6 < x$. Hence the whole curve lies between

$x = 0$ and $x = 6$. We let x have the successive values 0, 1, 2, 3, 4, 5, and 6, to obtain the following

if	$x=0$	$x=1$	$x=2$	$x=3$	$x=4$	$x=5$	$x=6$
$\sqrt{6x-x^2}$	0.000	2.235	2.828	3.000	2.828	2.235	0.000
$\sqrt{6x+x^2}$	0.000	2.645	4.000	5.196	6.324	7.416	8.484
$\sqrt{36-x^2}$	6.000	5.916	5.656	5.196	4.470	3.316	0.000
sum	6.000	10.796	12.484	13.392	13.622	12.967	8.484
$y =$ if							
$+ + +$	3.000	5.398	6.242	6.696	6.811	6.483	4.242
$- + +$	3.000	3.163	3.414	3.696	3.983	4.248	4.242
$+ - +$	3.000	2.753	2.242	1.500	0.487	0.933	-4.242
$+ + -$	-3.000	-0.518	0.586	1.500	2.341	3.167	4.242

The remaining four permutations of the signs differ from these only by a change of all signs, so that to each value of the abscissa we have corresponding eight values of the ordinate, which are shown in *figure 54*. The curve consists of two parts: $AFBEcagbcDA$ and $afbECAGBCDa$. It has two cusps, at A and a, it has two double points or self-intersections at four points, D, E, C and c.

CHAPTER XIII

On the Dispositions of Curves.

285. Just as we have previously described the species of branches going to infinity by assigning straight lines or simpler curves which coincide with the given curve at infinity, so in this chapter we will examine portions of the curve contained in a bounded region. We will investigate straight lines or simpler curves which coincide with the given curve for at least a small bit. In the first place it is clear that every straight line which is tangent to the curve, at least at the point of tangency coincides with the curve, that is, it has at least two points in common with the curve. There are also other curves that we can find which coincide with a portion of the given curve more accurately, and, as it were, kiss the curve. When we know these, the condition of the curve in any location and its dispositions are seen most clearly.

286. Suppose we have an equation in x and y for some curve. Let the abscissa x be given some value $AP = p$, and we look for the corresponding values of the ordinate y. If there are several of these, we arbitrarily choose one of these $PM = q$, so that M is a point on the curve, that is, a point through which the curve passes. Then if in the equation in x and y we substitute p for x and u for y, then all the terms of the equation cancel so that nothing remains. In order to

discover the nature of that portion of the curve which passes through M, we draw the straight line through M parallel to the axis AP, which now is taken as the axis. We call the new abscissa $Mq = t$, and the corresponding ordinate $qm = u$, as in *figure 55*. Since the point m is also on the curve, if the straight line mq is extended to the original axis at p, so that, when $Ap = p + t$ is substituted for x, and $pm = q + u$ is substituted for y, the equation still should be satisfied.

287. When this substitution is made in the equation in x and y, all of the terms which contain neither t nor u should cancel each other and the terms which contain the new coordinates t and u are the only remaining terms. Hence we have an equation of the form

$$0 = At + Bu + Ct^2 + Dtu + Eu^2 + Ft^3 + Gt^2u + Htu^2 + \cdots,$$

where A, B, C, D, etc. are constants composed of the constants of the original equation, with p and q, which we now take to be constants. Hence this new equation expresses the nature of the curve, but with respect to the axis Mq with the origin at M.

288. In the first place, it is obvious that if we let $Mq = t = 0$, then $qm = u = 0$, since then the point m coincides with M. Then, since we want to investigate only a small part of the curve near M, we accomplish this by taking very small values of t. In this case the value of $qm = u$ will also be quite small. What we want is the nature of the arc Mm as it almost disappears. If we take values for t and u as small as possible, the terms t^2, tu, and u^2 are even smaller, as are also t^3, t^2u, tu^2, u^3, etc. much smaller, and so forth. For this reason, we can omit all except the terms of the form $At + Bu$, which is an equation of the straight line $M\mu$ passing through the point M, and as m approaches M the straight line coincides with the curve.

289. Hence this straight line $M\mu$ is the tangent to the curve at the point M. It follows that at any point on the curve M we can draw the tangent μMT. That is, from the equation $At + Bu = 0$, we have $\dfrac{u}{t} = \dfrac{-A}{B} = \dfrac{q\mu}{Mq}$, so that $\dfrac{q\mu}{Mq} = \dfrac{MP}{PT} = \dfrac{-A}{B}$. Hence, since $PM = q$, we have $PT = \dfrac{-Bq}{A}$. This portion of the axis PT is called the SUBTANGENT. From this we deduce the following

RULES

for finding subtangents. In the equation for the curve, after finding that the ordinate $y = q$ corresponds to the abscissa $x = p$, we substitute $x = p + t$ and $y = q + u$. After the substitution, we keep only the terms in which t and u appear with the exponent 1, and we neglect all other terms, to obtain the equation $At + Bu = 0$. From this we know A and B, and so the subtangent $PT = \dfrac{-Bq}{A}$.

EXAMPLE I

Let the given curve be the parabola with the equation $y^2 = 2ax$, *where AP is the principal axis and A is the vertex.*

We let $AP = p$, and $PM = q$, so that $q^2 = 2ap$, that is, $q = \sqrt{2ap}$. Now we let $x = p + t$, and $y = q + u$, so that $q^2 + 2qu + u^2 = 2ap + 2at$. By the rule, we keep only the terms $2qu = 2at$. This gives $at - qu = 0$, and $\dfrac{u}{t} = \dfrac{a}{q} = \dfrac{-A}{B}$. Then the subtangent $PT = \dfrac{q^2}{a} = 2p$, since $q^2 = 2ap$. Hence the subtangent PT is twice the abscissa AP.

EXAMPLE II.

Let the curve be the ellipse with center at A and with the equation $y^2 = \dfrac{b^2}{a^2}(a^2 - x^2)$. *That is* $a^2y^2 + b^2x^2 = a^2b^2$.

We take $AP = p$ and let $PM = q$, so that $a^2q^2 + b^2p^2 = a^2b^2$. Then we let $x = p + t$ and $y = q + u$. Since we retain only the terms in which t and u appear with the exponent 1 and omit the others, the resulting equation is $2a^2qu + 2b^2pt = 0$. Hence,

$\dfrac{u}{t} = \dfrac{-b^2p}{a^2q} = \dfrac{-A}{B}$. Then the subtangent

$$PT = \frac{-B}{A}p = \frac{-a^2q^2}{b^2p} = \frac{-a^2 + p^2}{p}.$$

Since this expression is negative it indicates that the point T falls on the opposite side. In addition, this expression is a perfect confirmation of the determination of the tangent for an ellipse, which we gave above.

EXAMPLE III

Let the given curve be a third order line of the seventh species with equation $y^2x = ax^2 + bx + c$.

When we take $AP = p$ and let $PM = q$, then $pq^2 = ap^2 + bp + c$. Now let $x = p + t$ and $y = q + u$, so that

$$(p + t)(q^2 + 2qu + u^2) = a(p^2 + 2pt + t^2) + b(p + t) + c.$$

When we omit all of the superfluous terms we have $2pqu + q^2t = 2apt + bt$. Hence

$$\frac{u}{t} = \frac{2ap + b - q^2}{2pq} = \frac{-A}{B}.$$

Then the subtangent

$$PT = \frac{-B}{A}q = \frac{2pq^2}{2ap + b - q^2} = \frac{2ap^2 + 2bp + 2c}{2ap + b - q^2}$$

$$= \frac{2ap^3 + 2bp^2 + 2cp}{ap^2 - c}.$$

That is $PT = \dfrac{2p^2q^2}{ap^2 - c}$.

290. When we know the tangent to the curve in this way, we simultaneously know the direction which the curve follows at the point M. Indeed, a most apt way of considering a curve is that of a path which a point follows with a continuous change of direction. Hence the point M on the curve Mm moves in the direction of the tangent $M\mu$. If the direction did not change, the straight line $M\mu$ would describe the curve, while the direction must change if the track is to curve. Hence, in order to know the path of a curve we should know the position of the tangent at each point, and this is easily done by the method just given. Nor should there be any difficulty, provided the equation for the curve is non-irrational and contains no fractions. We can always express our equation in that form. If an equation happens to be irrational or involved in fractions and cannot be put into the non-irrational form with no fractions, then the same method, with certain modifications, can be used, where the modifications give us *differential calculus*. For this reason, the method for finding tangents when the equation for the curve is irrational, or not free of fractions, we reserve for differential calculus.

291. Hence we know the inclination of the tangent $M\mu$ to the axis AP or to its parallel Mq. Since $\dfrac{q\mu}{Mq} = \dfrac{-A}{B}$, if the coordinates are orthogonal, so that the angle $Mq\mu$ is a right angle, then the tangent of the angle $qM\mu$ is equal to $\dfrac{-A}{B}$; if the coordinates are oblique, then from the given angle $Mq\mu$ and the ratio of the sides Mq and $q\mu$, by trigonometry we can find the angle $qM\mu$. It is clear that if in the resulting equation $At + Bu = 0$ we have $A = 0$, then the angle $qM\mu$ vanishes, so that the tangent $M\mu$ is parallel to the axis AP. On the other hand, if $B = 0$, then the tangent $M\mu$ is parallel to the ordinate PM, that is, the ordinate PM is tangent to the curve at the point M.

292. After having found the tangent MT, if we draw the line MN normal to the tangent at the point of tangency, this line will also be normal to the curve. Hence this line is also easy to find in any case. It is most conveniently expressed when the coordinates are rectangular. In that case the triangles $Mq\mu$ and MPN are similar, so that $\dfrac{Mq}{q\mu} = \dfrac{MP}{PN}$, that is $\dfrac{-B}{A} = \dfrac{q}{PN}$, so that $PN = \dfrac{-Aq}{B}$. That part of the axis PN, between the ordinate and the normal MN, is usually called the SUB-NORMAL. This subnormal, when the coordinates are rectangular, is easily found from the subtangent PT. Since $\dfrac{PT}{PM} = \dfrac{PM}{PN}$, we have $PN = \dfrac{PM^2}{PT}$. Furthermore if the angle APM is a right angle, then the tangent $MT = \sqrt{PT^2 + PM^2}$ and the normal $MN = \sqrt{PM^2 + PN^2}$, that is, since $\dfrac{PT}{TM} = \dfrac{PM}{MN}$, we have

$$MN = \frac{PM \cdot TM}{PT} = \frac{PM}{PT}\sqrt{PT^2 + PM^2}.$$

293. We have seen that if in the equation $At + Bu = 0$, either $A = 0$ or $B = 0$, then the tangent is parallel to either the axis or the ordinates. There remains the consideration of the case when both A and B simultaneously vanish. If this occurs, then we consider the following terms in the equation found in section 286, that is, those with the second degree in t and u, which can no longer be neglected since $At + Bu$ vanishes. For this reason we consider the equation $Ct^2 + Dtu + Eu^2 = 0$, where we neglect the following terms, since, when t and u become infinitely small, the higher degree terms vanish when compared to the second degree terms. From this equation, as is true in general, it is clear that when $t = 0$, then also $u = 0$, so that M is a point on the curve, which fits the hypothesis.

294. Since the equation $Ct^2 + Dtu + Eu^2 = 0$ indicates that a point M of the curve is nearby, it is clear that if $D^2 < 4CE$, then the equation is complex unless $t = 0$ and $u = 0$. In this case the point M actually belongs to the curve, but it is isolated from the rest of the curve. It is a conjugate oval which has shrunk to a point, as in the case we noted in the previous chapter. Hence there is no place for the concept of a tangent in this case, since a tangent is a straight line which has two adjacent points in common with a curve, while in this case there is but a single point. In this way we know a conjugate point, if the curve has one, and distinguish it from the other points of the curve.

295. If $D^2 > 4CE$, then the equation $Ct^2 + Dtu + Eu^2 = 0$ can be resolved into two equations of the form $\alpha t + \beta u = 0$, each of which gives information about the nature of the curve. Since each gives the position of a tangent, that is the direction of the curve, at the point M, it follows that two branches of the curve cross each other at the point M, which makes M a double point. For instance, if we take $Mq = t$, in *figure 56*, then $q\mu$ and $q\nu$ are both values of u, given by that equation, and the straight lines $M\mu$ and $M\nu$ are both tangents to the curve at point M. Hence there is the intersection of two branches of the curve at the point M, one goes in the direction of $M\mu$, and the other in the direction of $M\nu$. Since a conjugate point should be considered to be a double point, the equation $Ct^2 + Dtu + Eu^2 = 0$ always indicates a double point just as the equation $At + Bt = 0$ indicates a simple point whenever it holds.

296. If $D^2 = 4CE$, then both of the tangents, $M\mu$ and $M\nu$, coincide, so that the angle $\mu M\nu$ vanishes. From this we know that the two branches not only meet, but they also have the same direction, and so they are tangent to each other. In this case the point M is still a double point, since the straight line through this point in the direction of the

curve is considered to cut the curve in two points. Hence whenever the equation we obtained in section 286 has both of the first coefficients, A and B equal to zero, we conclude that the curve has a double point, of which there are three species: either an oval shrinks to a point, that is, a conjugate point, or two branches of the curve intersects in a node, or two branches of the curve are mutually tangent, and these three species of double points are determined by the equation $Ct^2 + Dtu + Eu^2 = 0$.

297. If in addition to the coefficients A and B also the three C, D, and E all vanish, then the next sum of terms in which t and u have the third degree, $Ft^3 + Gt^2u + Htu^2 + Iu^3 = 0$. If this equation has only one real linear factor, then this indicates the tangent or direction of the curve passing through the point M, while the two complex factors point to a vanishing oval. If all three factors are real, then we know that three branches of the curve intersect in the point M or are mutually tangent there, depending on whether the roots are unequal or equal. Whichever of these occurs, the curve has a triple point at M and a straight line through M in the direction of the curve should be considered as having three points in common with the curve.

298. If besides all the preceding coefficients the four F, G, H, and I also vanish, then in order to understand the nature of the point M, it is necessary to consider the following terms in which t and u have the fourth degree. In this case the point is declared to be a quadruple point. Here either two conjugate oval coalesce, which happens when the fourth degree equation has only complex roots. Or M is a point of intersection or mutual tangency of two branches with a conjugate point, which occurs if two roots are real and two are complex. Finally, M can be the intersection of four branches of the curve if all roots of the equation are real. The intersection or tangency of two, three, or all four branches depends on whether two, three, or all four of the roots are equal. In a similar

way if all of the coefficients of terms less that the fifth degree vanish, then we proceed to the terms of the fifth degree or higher degree.

299. After having thought these matters through, it is easy to find a general equation for all curves which not only pass through M, but have M as a simple, or double, or triple, or whatever desired point. If we let $AP = p$, and $PM = q$, and also let P, Q, R, S, etc. be any functions of the coordinates x and y, it is clear that the equation $P(x - p) + Q(y - q) = 0$ expresses a curve through the point M, since if we let $x = AP = p$, then $y = PM = q$, provided neither P is divisible by $y - q$, nor Q is divisible by $x - p$; that is, provided the factors $x - p$ and $y - q$, upon which the passing of the curve through M depends, cannot be eliminated from the equation by division. It is clear that all curves which pass through M are contained in this equation $P(x - p) + Q(y - q) = 0$. The point M will be simple if this equation is not of the form which we are about to show for multiple points.

300. If M is to be a double point, the equation for the curve is contained in the general form

$$P(x - p)^2 + Q(x - p)(y - q) + R(y - q)^2 = 0,$$

provided the form is not lost through division. From this point of view, it becomes clear that a second order line cannot have a double point, since this can be the equation of a second order line only if P, Q, and R are constants. In that case we would have an equation, not of a curve, but of two straight lines. If P, Q, and R are first degree functions, such as $\alpha x + \beta y + \gamma$, then the line is of the third order and has a double point at M. A third order line, unless it is the equation for three straight lines, can have no more than one double point. In order to clarify this we suppose that there are two double points. Through these points we draw a straight line, which then intersects the curve in four points. But this contradicts the nature of a third order line. Likewise a fourth order line has no more than two double points. A fifth order line can have no more than three double points, and so forth.

301. Let M be a triple point of a curve, and let the nature of the curve be expressed by the equation

$$P(x - p)^3 + Q(x - p)^2(y - q)$$

$$+ R(x - p)(y - q)^2 + S(y - q)^3 = 0.$$

If this equation is to define a curve, the order must be greater than three, since if P, Q, R, and S were constants, as required by the nature of a third order line, then the equation would have three factors of the form $\alpha(x - p) + \beta(y - q)$ and would be an equation for three straight lines. Hence there are no triple points on curves simpler than the fourth order, nor can a fifth order equation have more than one triple point, since otherwise there would be a straight line which intersected a fifth order line in six points. There is nothing to prevent a sixth order line from having two triple points.

302. If an equation is of the form

$$P(x - p)^4 + Q(x - p)^3(y - q) + R(x - p)^2(y - q)^2$$

$$+ S(x - p)(y - q)^3 + T(y - q)^4 = 0,$$

then the curve has a quadruple point at M. Hence the simplest curve which has a quadruple point must be of the fifth order. Two quadruple points cannot occur on a curve unless it has order at least eight. In a similar way general equations can be found for lines with a quintuple point at M, or with any arbitrary multiple point.

If M is a double point or a triple point, or of any multiplicity, then either that number of the branches of the whole curve intersect or are tangent at M, or if the number of branches meeting there is less, then there are one or more conjugate points at M. This information can be found by methods already discussed. That is, in the functions P, Q, R, S, etc. we substitute p for x and instead of $x - p$ and $y - q$ we write t and u. Then from these equations we can find the state of the curve at M and the tangents to the branches there.

CHAPTER XIV

On the Curvature of a Curve.

304. Just as we have previously investigated straight lines which indicate at a point the direction of a curve, so now we investigate simpler curves which at some place coincide with a given curve so closely, that, at least for a very small distance, the two are the same. In this way, from a knowledge of the nature of the simpler curve, we gather at the same time something of the nature of the proposed curve. We will use a method which is similar to that which we used when investigating the nature of branches which go to infinity. First we consider the tangent line, then we consider a simpler curve which is much closer to the proposed curve, in that it is not only tangent, but rather kisses. Curves of this kind with such a tight fit are usually called OSCULATING.

305. Let any equation in the coordinates x and y be given. We want to investigate the nature of a very small portion of the curve Mm near the point M. When we find the abscissa $AP = p$ and the ordinate $PM = q$, then we choose a new axis MR, as in *figure 55*. We choose a very small abscissa $Mq = t$ with ordinate $qm = u$. Then $x = p + t$ and $y = q + u$. When we substitute these values into the equation we obtain

$$0 = At + Bu + Ct^2 + Dtu + Eu^2 + Ft^3 + Gt^2u + \cdots,$$

which expresses the nature of the curve with respect to the axis MR. Since we have chosen the new coordinate t and u very small, the following terms will be minuscule compared to the first two, so that they may be neglected without introducing an error.

306. Unless both coefficients A and B vanish, we eliminate all the following terms to obtain the equation $0 = At + Bu$. This gives the straight line $M\mu$ which is tangent to the curve at the point M, and has the same direction at that point as the curve. Hence, we have $\dfrac{Mq}{q\mu} = \dfrac{B}{-A}$, so that from a knowledge of the quantities A and B, the position of the tangent $M\mu$ is known. Since the tangent touches the curve only at the point M, we will see how much the curve Mm deviates from the straight line $M\mu$, at least for a short distance. For this purpose, we choose the normal MN as the axis, to which the perpendicular ordinate mr is dropped from m. We let $Mr = r$ and $rm = s$, so that

$$t = \frac{-Ar + Bs}{(A^2 + B^2)^{\frac{1}{2}}}, \quad u = \frac{-As - Br}{(A^2 + B^2)^{\frac{1}{2}}}, \quad r = \frac{-At - Bu}{(A^2 + B^2)^{\frac{1}{2}}},$$

and

$$s = \frac{Bt - Au}{(A^2 + B^2)^{\frac{1}{2}}}.$$

Then since

$$-At - Bu = Ct^2 + Dtu + Eu^2 + Ft^3 + Gt^2u + \cdots,$$

the quantity r will be infinitely smaller than t and u, with the result that also r is infinitely smaller than s. This is so since s is expressed in terms of t and u, while r is expressed by squares or higher powers of t and u.

307. We will know the nature of the curve much more accurately if we bring into our computations the terms $Ct^2 + Dtu + Eu^2$, and neglect only the terms which follow these. Hence we have the equation in t and u, $-At - Bu = Ct^2 + Dtu + Eu^2$. When we substitute for t and u the expressions found above, we have

$$r\sqrt{A^2 + B^2} = \frac{(A^2C + ABD + B^2E)r^2}{A^2 + B^2}$$

$$+ \frac{(A^2D - B^2D - 2ABC + 2ABE)rs}{A^2 + B^2}$$

$$+ \frac{(A^2E - ABD + B^2C)s^2}{A^2 + B^2}.$$

Since r is infinitely smaller than s, the terms r^2 and rs will vanish in relation to s^2. Hence $s^2 = \dfrac{(A^2 + B^2)r\sqrt{A^2 + B^2}}{A^2E - ABD + B^2C}$. This is the equation which expresses the nature of the osculating curve for the given curve at the point M.

308. Hence the very short arc of the curve Mm coincides with the vertex of the parabola constructed on the axis MN with latus rectum, or a parameter, equal to $\dfrac{(A^2 + B^2)\sqrt{A^2 + B^2}}{A^2E - ABD + B^2C}$. We say that the curvature of the given curve at the point M is equal to the curvature of this parabola at its vertex. Since a circle is the the curve whose curvature is known most clearly; its curvature is the same at every point, and is inversely proportional to the radius, it is most convenient to define the curvature of a curve by means of a circle with the same curvature. This circle is usually called the *osculating circle*. For this reason we should define a circle whose curvature is the same as the curvature of the proposed parabola at its vertex. Then we can reasonably substitute that circle for the osculating parabola.

309. In order to accomplish this, let us suppose that the curvature of a circle is unknown, and express it in the way we have just suggested, through the curvature of the parabola. Then conversely, we can substitute the osculating circle for the osculating parabola. Hence let the proposed curve Mm be a circle with radius equal to a. The nature of this curve is given by the equation $y^2 = 2ax - x^2$. We let $AP = p$, and $PM = q$, so that $q^2 = 2ap - p^2$. Next we let $x = p + t$ and $y = q + u$, to obtain the equation

$$q^2 + 2qu + u^2 = 2ap + 2at - p^2 - 2pt - t^2.$$

Since $q^2 = 2ap - p^2$, the equation simplifies to

$$0 = 2at - 2pt - 2qu - t^2 - u^2.$$

When we compare this with the form given above we find $A = 2a - 2p$, $B = -2q$, $C = -1$, $D = 0$, and $E = -1$. Then

$$A^2 + B^2 = 4(a^2 - 2ap + p^2 + q^2) = 4a^2,$$

$$(A^2 + B^2)\sqrt{A^2 + B^2} = 8a^3,$$

and

$$A^2E - ABD + B^2C = -A^2 - B^2 = -4a^2.$$

It follows that the circle with radius a has at any point the osculating parabola $s^2 = 2ar$. Conversely, if a curve has an osculating parabola with the equation $s^2 = br$, then the same curve has an osculating circle with radius $\tfrac{1}{2}b$.

310. Since we have already found that the curve Mm has an osculating parabola with equation $s^2 = \dfrac{(A^2 + B^2)\sqrt{A^2 + B^2}}{A^2E - ABD + B^2C}r$, it is clear that the same curve has curvature at M which is the same as the curvature of the circle with radius equal to $\dfrac{(A^2 + B^2)\sqrt{A^2 + B^2}}{2(A^2E - ABD + B^2C)}$. This

expression gives the radius of the osculating circle. This radius is called the *osculating radius* or the *radius of curvedness* or even *radius of curvature*. Hence from the equation in t and u which we derived from the given equation in x and y, we can immediately define the radius of the osculating circle at the point M, that is the radius of the osculating circle of the curve at M. In the equation in t and u we eliminate the terms of degree higher than two, and then from the equation which has the form

$$0 = At + Bu + Ct^2 + Dtu + Eu^2,$$

we find the osculating radius equal to $\dfrac{(A^2 + B^2)\sqrt{A^2 + B^2}}{2(A^2 E - ABD + B^2 C)}$.

311. Since the sign of the radical $\sqrt{A^2 + B^2}$ involves an ambiguity, it is not clear whether this expression is positive or negative, that is, whether the curve is convex or concave with respect to the point N. In order to remove this doubt, we should inquire whether the point m on the curve lies on the same side of the tangent $M\mu$ as the axis AN or on the opposite side. In the former case the curve is concave with respect to N and the center of the osculating circle lies on the straight line MN, in the latter case, the center lies on the extension of the straight line NM beyond M. All doubt will be banished if we inquire whether qm is shorter than $q\mu$ or longer. In the former case the curve is concave with respect to N, in the latter case it is convex.

312. Since $q\mu = \dfrac{-At}{B}$, and $qm = u$, we have to see whether $\dfrac{-At}{B}$ is greater or less than u. Since $m\mu$ is a miniscule interval, we let $m\mu = w$ so that $u = \dfrac{-At}{B} - w$. When we make the substitution, we have

$$0 = -Bw + Ct^2 - \frac{ADt^2}{B} - Dtw + \frac{A^2Et^2}{B^2} + \frac{2AEtw}{B} + Ew^2.$$

Since w is so small with respect to t, the terms tw and w^2 vanish.

Hence we have $w = \dfrac{(B^2C - ABD + A^2E)t^2}{B^3}$. Now if w is positive,

which happens if $\dfrac{B^2C - ABD + A^2E}{B^3}$, or $\dfrac{A^2E - ABD + B^2C}{B}$ is

positive, then the curve is concave with respect to N. If

$\dfrac{A^2E - ABD + B^2C}{B}$ is negative, then the curve is convex with respect

to the point N.

313. In order that this may become more clear, let us consider separately some different cases which may occur. Suppose first that $B = 0$, so that the ordinate becomes the tangent to the curve Mm, as in *figure 57*. Then the osculating radius is equal to $\dfrac{A}{2E}$. Whether the curve is concave with respect to R, as in the figure, or convex can be discovered from the equation $0 = At + Ct^2 + Dtu + Eu^2$. Since $Mq = t$ and $qm = u$, and t is infinitely smaller than u, the terms t^2 and tu will vanish with respect to u^2, so that $At + Eu^2 = 0$. From this equation we know whether the coefficients A and E have opposite signs, that is, whether $\dfrac{E}{A}$ is negative, then the curve will be concave with respect to R. If the coefficients A and E have the same signs, so that $\dfrac{E}{A}$ is positive, then the curve is situated on the other side of the tangent line. This is so since the abscissa Mq should be negative so that the ordinate qm may be real.

314. Now let the tangent $M\mu$ be inclined with respect to the axis AP that is, with respect to the line parallel to the axis, as in *figure 55*, so that the angle $RM\mu$ is acute, and the normal MN intersects the axis

at the point N beyond P. In this case the abscissa t corresponds to a positive ordinate u. Hence the coefficients A and B have opposite signs, and the fraction $\frac{A}{B}$ is negative. In this case, as we saw before, the curve is concave with respect to N if $\frac{A^2E - ABD + B^2C}{B}$ is positive. That is, since $\frac{B}{A}$ is negative, if $\frac{A^2E - ABD + B^2C}{A}$ is negative. If on the other hand, $\frac{A^2E - ABD + B^2C}{B}$, is negative or $\frac{A^2E - ABD + B^2C}{A}$ is positive then the curve is convex with respect to N. In either case the osculating radius is equal to $\frac{(A^2 + B^2)\sqrt{A^2 + B^2}}{2(A^2E - ABD + B^2C)}$.

315. Now let $A = 0$ then the straight line MR is both parallel to the axis and tangent to the curve, as in *figure 58*. Furthermore, u is infinitely smaller than t, so that $0 = Bu + Ct^2$. Hence, if B and C have the same sign, that is, if $BC > 0$, then u will have a negative value, so that the curve is concave with respect to the point P, which coincides with N, as is shown by the rule given above, since $A = 0$. The osculating radius is equal to $\frac{B}{2C}$. The same rule given above is still valid if the tangent MT intersects the axis beyond P, as in *figure 59*. In that case the curve will be concave or convex with respect to N depending on whether the expression $\frac{A^2E - ABD + B^2C}{B}$ is positive or negative. The osculating radius is still equal to $\frac{(A^2 + B^2)\sqrt{A^2 + B^2}}{2(A^2E - ABD + B^2C)}$.

316. Let the given curve be an ellipse, or at least one of its arcs DMC, as in *figure 60*. Let the center be at A, the transverse semiaxis $AD = a$, and the conjugate semiaxis $AC = b$. We take the abscissa x on the axis AD from the center A. We have the equation for an ellipse $a^2y^2 + b^2x^2 = a^2b^2$. Take any abscissa $AP = p$, with the

corresponding ordinate $PM = q$, so that $a^2q^2 + b^2p^2 = a^2b^2$. We let $x = p + t$, and $y = q + u$, so that

$$a^2q^2 + 2a^2qu + a^2u^2 + b^2p^2 + 2b^2pt + b^2t^2 = a^2b^2.$$

That is,

$$2b^2pt + 2a^2qu + b^2t^2 + a^2u^2 = 0.$$

In the first place, because of the coefficients of t and u, the normal MN intersects the axis before P. We have $\dfrac{PM}{PN} = \dfrac{B}{A} = \dfrac{a^2q}{b^2p}$, and $PN = \dfrac{b^2p}{a^2}$, since $A = 2b^2p$ and $B = a^2q$. Furthermore, since $C = b^2$, $D = 0$, and $E = a^2$, we have

$$\frac{A^2E - ADB + B^2C}{B} = \frac{4a^2b^2(a^2q^2 + b^2p^2)}{2a^2q} = \frac{4a^4b^4}{2a^2q}.$$

This positive quantity indicates that the curve is concave with respect to N.

317. In order to find the osculating radius, we note that $A^2 + B^2 = 4(a^4q^2 + b^4p^2)$, and that $A^2E - ABD + B^2C = 4a^4b^4$. Hence the osculating radius is equal to $\dfrac{(a^4q^2 + b^4p^2)^{\frac{3}{2}}}{a^4b^4}$. Since $MN = \sqrt{q^2 + (b^4p^2/a^4)}$, it follows that $\sqrt{a^4q^2 + b^4p^4} = a^2 \cdot MN$, so that the osculating radius is equal to $\dfrac{a^2 \cdot MN^3}{b^4}$. If we extend the normal MN and drop the perpendicular AO from the center A, then, since $AN = p - \dfrac{b^2p}{a^2}$, and the triangles MNP and ANO are similar, we have

$$NO = \frac{a^2b^2p^2 - b^4p^2}{a^4 \cdot MN} \quad \text{and}$$

$$MO = NO + MN = \frac{a^2q^2 + b^2p^2}{a^2 \cdot MN} = \frac{b^2}{MN}.$$

Hence $MN = \dfrac{b^2}{MO}$ and the osculating radius is equal to $\dfrac{a^2b^2}{MO^3}$. This expression is accommodated to both of the axes, AD and AC.

318. If we know the osculating radius at every point of a curve, the nature of the curve is seen quite clearly. Indeed, if the curve is divided into miniscule arcs, each of these particles of the curve can be given as an arc of a circle whose radius is the osculating radius at that place. In this way we have a much more accurate description of the curve; if we know very many points on the curve, and at each point we find the tangent and normal with the osculating radius, then these small portions of the curve between the points can be drawn as parts of a circle. In this way the more points we have, the more accurate the curve will be.

319. Since a very small portion of the curve through M in *figure 55* coincides with an arc of the osculating circle, not only the arc Mm, but also the preceding arc Mn has the same curvature as the circle. Since the very small portion of the arc Mm is expressed by the equation $s^2 = \alpha r$ in the coordinates $Mr = r$ and $rm = s$, each of the very small abscissas $Mr = r$ has two corresponding ordinates, one positive and one negative. Hence the curve goes both in the direction of m and of n. It follows that whenever the osculating radius, which is equal to $\tfrac{1}{2}\alpha$, is finite, there the curvature is uniform for at least a very small space. It follows that in this case there will be no sudden formation of a cusp with the curve reflecting back, nor a sudden change of curvature with one portion Mn convex with respect to N and the other portion Mm concave with respect to N. Points where this kind of change of curvature occurs are called points of inflection or of contrary bending. Hence when the

osculating radius is finite, there will be neither a cusp nor an inflection point.

320. Since from the equation in t and u

$$0 = At + Bu + Ct^2 + Dtu + Eu^2$$

$$+ Ft^3 + Gt^2u + Htu^2 + \cdots$$

we find the osculating radius equal to $\dfrac{(A^2 + B^2)\sqrt{A^2 + B^2}}{2(A^2E - ABD + B^2C)}$, it is clear that if $A^2E - ABD + B^2C = 0$, then the osculating radius becomes infinite and the osculating circle becomes a straight line. Whenever this happens, the curvature of the curve is destroyed and two parts of the curve may not agree. In order to study this phenomenon more carefully we substitute $t = \dfrac{-Ar + Bs}{(A^2 + B^2)^{\frac{1}{2}}}$ and $u = \dfrac{-As - Br}{(A^2 + B^2)^{\frac{1}{2}}}$ also into the terms $Ft^3 + Gt^2u + Htu^2 + Iu^3$. Since all the other terms which contain r will vanish when compared to $r\sqrt{A^2 + B^2}$, we eliminate these terms and keep the equation

$$r\sqrt{A^2 + B^2} = \alpha s^2 + \beta s^3 + \gamma s^4 + \delta s^5 + \cdots.$$

321. From this equation we note immediately, as above, that the osculating radius is equal to $\dfrac{\sqrt{A^2 + B^2}}{2\alpha}$, unless $\alpha = 0$. In this case the radius is infinite, and we must take the following term, βs^3, for a more accurate knowledge of the nature of the curve. Hence we have $r\sqrt{A^2 + B^2} = \beta s^3$, where the other terms γs^4, δs^5, etc. vanish unless $\beta = 0$. In this case the osculating curve at M has the equation $r\sqrt{A^2 + B^2} = \beta s^3$, from which the configuration of the curve at the point M can be known. Since a negative value for the abscissa r corresponds to a negative value of the ordinate s, the curve has a

serpentine configuration $mM\mu$ as in *figure 61*. Hence M is a point of inflection.

322. If besides $\alpha = 0$ we also have $\beta = 0$, then the nature of the curve at M is expressed by the equation $r\sqrt{A^2 + B^2} = \gamma s^4$. Since for each value of the abscissa r there are two values of s, one positive and one negative, while the abscissa r cannot take both values, both parts of the curve Mm and $M\mu$ lie on the same side of the tangent, as in *figure 62*. If α, β, and γ all vanish, then the nature of the curve at M is expressed by the equation $r\sqrt{A^2 + B^2} = \delta s^5$. Then the curve has a point of inflection, as in *figure 61*. If we also have $\delta = 0$, so that $r\sqrt{A^2 + B^2} = \epsilon s^6$, then the curve has no point of inflection, as in *figure 62*. In general, if the exponent on s is odd, then the curve has a point of inflection at M, while if the exponent is even, then the curve has no point of inflection at M, as in *figure 62*.

323. So far we have discussed the phenomena at a simple point M, that is, when in the equation

$$0 = At + Bu + Ct^2 + Dtu + Eu^2 + Ft^3 \cdots,$$

the two coefficients A and B do not simultaneously vanish. However, if both $A = 0$ and $B = 0$, then the curve has two or more branches intersecting at the point M, as in *figure 56*. As before we will investigate the curvature and species of each branch individually. Suppose that $mt + nu = 0$ is the equation of the tangent to one of the branches. We look for the equation of this branch in the coordinates r and s, where r is taken on the normal MN, so that r is infinitely small when compared to s. Hence we let $t = \dfrac{-mr + ns}{(m^2 + n^2)^{\frac{1}{2}}}$ and $u = \dfrac{-ms - nr}{(m^2 + n^2)^{\frac{1}{2}}}$. When these substitutions are made and the terms which are infinitely small are eliminated, we obtain for a double point M the equation

$$rs = \alpha s^3 + \beta s^4 + \gamma s^5 + \delta s^6 + \cdots.$$

If M is a triple point, then the equation is

$$rs^2 = \alpha s^4 + \beta s^5 + \gamma s^6 + \cdots,$$

and so forth. All of these equations reduce to the following form:

$$r = \alpha s^2 + \beta s^3 + \gamma s^4 + \delta s^5 + \cdots.$$

324. From this equation we understand that the osculating radius at M is equal to $\dfrac{1}{2\alpha}$, and that if $\alpha = 0$, then the radius is infinite. In this case the nature of the curve is expressed by the curve $r = \beta s^3$, or $r = \gamma s^4$ or $r = \delta s^5$, etc. From these equations we can decide as before, whether or not there is a point of inflection at M. There is a point of inflection if the exponent of s is odd, and no point of inflection if the exponent is even. In this way we judge individually each branch passing through the point M provided each such branch has a different tangent.

325. If two or more branches have the same tangent at the point M, then both A and B vanish at M and the first member $Ct^2 + Dtu + Eu^2$ has two equal factors. Suppose that $Ct^2 + Dtu + Eu^2 = (mt + nu)^2$, and that we have changed to the coordinates $Mr = r$ and $rm = s$, as in *figure 55*. We accomplish this by letting $t = \dfrac{-mr + ns}{(m^2 + n^2)^{\frac{1}{2}}}$ and $u = \dfrac{-ms - nr}{(m^2 + n^2)^{\frac{1}{2}}}$. In this way we obtain the equation

$$r^2 = \alpha rs^2 + \beta s^3 + \gamma rs^3 + \delta s^4 + \epsilon rs^4 + \zeta s^5 + \cdots.$$

The terms in which r has two or more degrees will vanish when compared to the first r^2.

326. First we consider the term βs^3. If this term is present, then all other terms vanish, since r is infinitely smaller than s. Hence, unless $\beta = 0$, the nature of the curve at M is expressed by the equation $r^2 = \beta s^3$. Since $r = s\sqrt{\beta s} = s^2\sqrt{\beta/s}$, we know the osculating radius at M is equal to $\frac{1}{2}\sqrt{s/\beta}$. Since s vanishes at M, the osculating radius also becomes equal to zero. Hence the curvature at M becomes infinitely large, that is the portion of the curve at M becomes an arc of a circle which is infinitely small. Since the ordinate s has the same sign whether the abscissa r has positive or negative values, it is clear that the curve has a cusp at M, as in *figure 63,* and that the two branches Mm and $M\mu$ are separated by their mutual tangent Mt at M and they preserve their convexity.

327. Suppose that $\beta = 0$, and that the term δs^4 is present. Then $\gamma r s^3$ vanishes, and the nature of the curve near M is expressed by the equation $r^2 = \alpha r s^2 + \delta s^4$. If $\alpha^2 < -4\delta$, then M is a conjugate point due to the complex factors. If $\alpha^2 > -4\delta$, then we obtain the two equations $r = f s^2$ and $r = g s^2$. Hence at M two curves are mutually tangent, one of which has osculating radius equal to $\frac{1}{2f}$ and the other equal to $\frac{1}{2g}$. If both of these are concave in the same direction the configuration will be that of two internally tangent circular arcs, as in *figure 64.* If they are concave in different directions, then the configuration is that of two externally tangent circular arcs, as in *figure 65.*

328. If δ also vanishes, then the equation can either be resolved into two equations or it cannot be so resolved. In the former case, there are two branches mutually tangent at M, the nature of each branch is expressed by an equation like $r = \alpha s^m$. There are as many different

configurations as there are combinations of such pairs of branches, each of which has M as a simple point. All branches which are contained in equations of the form $r = \alpha s^m$ are called *branches of the first order*. The latter case, in which the equation cannot be resolved into two equations, has a curve whose nature is expressed by either $r^2 = \alpha s^5$, or $r^2 = \alpha s^7$, or $r^2 = \alpha s^9$, etc. Branches of this type are like $r^2 = \alpha s^3$, which we discussed above, are called *branches of the second order*, since they take the place of two branches of the first order which are mutually tangent at M. All of these second order branches have a cusp, as in *figure 63*, which was given by the equation $r^2 = \alpha s^3$. There is this difference, however, in that the osculating radius at M for the equation $r^2 = \alpha s^3$ is infinitely small, while in all other equations there is an infinitely large radius. Since from the equation $r^2 = \alpha s^5$, we have $r = s^2 \sqrt{\alpha s}$, so that the osculating radius at M is equal to $\dfrac{1}{2(\alpha s)^{\frac{1}{2}}}$, which is infinite since $s = 0$.

329. If three tangents of branches passing thorough M coincide, then there are either three branches of the first order which are mutually tangent at M, or there is one branch of the second order and one of the first order, or there is a *branch of the third order*. Branches of the third order are expressed by equations with the form $r^3 = \alpha s^4$, $r^3 = \alpha s^5$, $r^3 = \alpha s^7$, $r = \alpha s^8$, etc. That is, the general form is $r^3 = \alpha s^n$, where n is an integer larger than 3 and not divisible by 3. There are some differences in the configurations of these branches. If n is odd, then M is an inflection point, while it is not an inflection point , as in *figure 62*, if n is even. In addition, the osculating radius is infinitely small if $n < 6$ and it is infinitely large if $n > 6$.

330. In a similar way, if four tangents of branches passing through

M coincide, then there are either four branches of the first order, or two of the first order and one of the second, or two branches of the second order, or one of the first order and one of the third order which are mutually tangent at M, or finally, there may be one *branch of the fourth order* at M. Branches of the fourth order have their nature contained in the general equation $r^4 = \alpha s^n$, where n is an odd integer greater than 4. All of these equations have a cusp, just as those of the second order, as in *figure 63*. At M the osculating radius is infinitely small if $n < 8$, while it is infinitely large if $n > 8$.

331. In the same way *branches of the fifth* or higher order occur. As to their configuration, branches of the fifth, seventh, ninth, and all odd orders are similar to branches of the first order, with either a point of inflection, or no point of inflection. Branches of the sixth, eighth, and all even orders are like branches of the second and fourth orders in that they all have a cusp at M, as in *figure 63*. With regard to the osculating radius, since the nature of these arcs is expressed by the equation $r^m = \alpha s^n$, where $n > m$ and it is clear that if $n < 2m$, then the osculating radius is infinitely small, while if $n > 2m$, then the radius is infinitely large.

332. The phenomena which can occur are reduced to three types. The first is a curve with *continuous curvature* with no point of inflection, nor cusp. This happens when the osculating radius is everywhere finite. There are, however, cases when the radius becomes infinitely small or infinitely large and the curvature is not perturbed. This will occur if the nature of the curve is expressed by an equation $\alpha r^m = s^n$, where m is an odd number, n is even, and $n > m$. The second phenomenon is a *point of inflection* which can occur only when the osculating radius is either infinitely large or infinitely small. Its nature is expressed by an equation with the form $\alpha r^m = s^n$, where both m and n are odd and

$n > m$. The radius of curvature is infinitely large if $n > 2m$, or infinitely small if $n < 2m$. The third phenomenon is a *reflex point* or *cusp*, where two branches are convex with respect to each other seem to terminate in a point where they come together mutually tangent. Such a point is indicated by an equation $\alpha r^m = s^n$, where m is even and n is odd. At a cusp the radius of curvature is either infinitely small or infinitely large.

333. Since in these three genera all the varieties of phenomena are exhausted, we should understand that there will not occur in a branch of a continuous curve such a change of direction as given in *figure 66,* where the angle at C is a finite angle ACB. Since in any cusp both branches preserved convexity, there can be no cusp as given in *figure 67,* where the branches AC and BC have a common tangent at C but one is concave and the other convex. If this seems to occur, then we know that the curve is not complete. If a curve is given completely in accord with an equation, then there may be a configuration like that given in *figure 64.* There are methods of describing curves according to which cusps ACB arise, and these have been called *cusps of the second species* by L'HOPITAL. We should note that mechanical descriptions do not always give the whole curve which is contained in an equation, but frequently gives only a part. This remark should dispel any dispute over nothing but a difference in notation.

In spite of these arguments which would seem to discredit the existence of a cusp of the second species, there are innumerable algebraic curves endowed with such cusps. Among these is one which is even a line of the fourth order, contained in the equation

$$y^4 - 2y^2x - 4yx^2 - x^3 = 0,$$

which results from the formula $y = \sqrt{x} \pm x^{\frac{3}{4}}$. Although the first term is $\sqrt{x}$, the sign is not ambiguous; it is necessarily positive. If a negative

sign is assigned to it, then the second term, $x^{\frac{3}{4}} = (x\sqrt{x})^{\frac{1}{2}}$ becomes

complex. From this example, the extent to which the above examples are

affected, we should examine very carefully.

334. If two branches have a common tangent at M, so that there

are four arcs emanating from M, namely, Mm, $M\mu$, Mn, and Mv, as in

figure 64, expressed by different equations, then there is no doubt as to

which arc is a continuation of another. One is a continuation of another

if they are both covered by the same equation. Hence, the arc Mm is a

continuation of the arc Mn and $M\mu$ is a continuation of the arch Mv.

However, if both branches are expressed by the same equation, then the

previous argument no longer holds, so the arc Mm can be considered a

continuation of the arc vM and the arc nM. Since the arcs Mn and Mv

can both be considered continuations of the arc Mm, either can be con-

sidered the continuation of the other. Hence, the arcs mM and $M\mu$ can

be thought of as a continuous curve, just as any other pair. In this case

we then have at M two cusps of the second species, namely, $mM\mu$ and

nMv.

335. This phenomenon can occur not only when two branches

without a point of inflection have a common tangent at M and are

expressed by the same equation, but the same idea of continuity applies

no matter what the genus might be of the two branches which are mutu-

ally tangent at M, provided that they are both expressed by the same

equation. This occurs whenever the equation in r and s reduces to the

form $\alpha^2 r^{2m} - 2\alpha\beta r^m s^n + \beta^2 s^{2n} = 0$. In that case both branches are

expressed by the same equation $\alpha r^m = \beta s^n$. Now we have four arcs

emanating from the point M, any two of which may be taken as a con-

tinuous curve. From this there arise innumerable cusps of the second species. This idea of continuity would seem to bear the blame for some of the descriptions and mechanical constructions which frequently produce cusps of the second species. Nevertheless this cannot occur unless the description does not contain the whole curve expressed by the equation, but only one or a few of the branches.

CHAPTER XV

Concerning Curves with One or Several Diameters.

336. We have already seen that all second order lines have at least one orthogonal diameter, which cuts the whole curve into two similar and equal parts. A parabola has exactly one such diameter, and hence it consists of two equal and similar parts. An ellipse and a hyperbola have two such diameters which are normal to each other at the center. Hence these pairs divide the curve into four similar and equal arcs or branches. A circle is divided into two equal and similar parts by any line drawn through the center. Hence it has innumerable equal parts, namely all arcs which are subtended by equal chords. These equal chords are also similar to each other.

337. Now we turn our attention to this similarity of two or more parts of the same curve. We will recall the general equations of those curves which have such multiple parts which are similar to each other. First we consider an equation in the rectangular coordinates x and y, with the whole space divided into four quadrants, Q, R, S, and T, by the two straight line AB and EF which intersects at C, as in *figure 68*. We take both x and y to be positive for parts of the curve in quadrant Q. The abscissa x is positive, while the ordinate y is negative for the curve in quadrant R. The abscissa x becomes negative and the ordinate y positive for the portion of the curve in the quadrant S. Finally the

portion in quadrant T has both coordinates x and y negative.

338. The parts of the curves in quadrants Q and R are equal and similar if the equation is such that it is not changed when $-y$ is substituted for y. Since all even powers of y have this property, it is clear that if the equation for the curve contains no odd powers of y, then the parts of the curve which lie in the quadrants Q and R will be equal and similar. Hence the straight line AB, from which the abscissas $CP = x$ are taken, will be a diameter for the curve. All curves of this kind, provided they are algebraic, will be contained in general equations

$$0 = \alpha + \beta x + \gamma x^2 + \delta y^2 + \epsilon x^3 + \zeta x y^2 + \eta x^4 + \theta x^2 y^2 + \iota y^4 + \cdots.$$

Such an expression could be described as a non-irrational function in x and y^2. If Z is any non-irrational function in x and y^2, then the equation $Z = 0$ expresses a curve which the straight line AB divides into two similar and equal parts. Note that the portions in quadrants S and T will also be equal and similar.

339. Portions of the curve in quadrants Q and S are equal and similar if the equation is such that when $-x$ is substituted for x, the equation is not changed. Hence if Z is any non-irrational function in x^2 and y, then the equation $Z = 0$ expresses a curve which is bisected by the straight line EF into two similar and equal parts. An equation of this kind has the form

$$0 = \alpha + \beta y + \gamma x^2 + \delta y^2 + \zeta y^3 + \eta x^4 + \theta x^2 y^2 + \iota y^4 + \cdots.$$

The portions in quadrants S and Q of the curve expressed by this equation are similar and equal, as are also the portions in quadrants T and R.

340. The portions of a curve in the opposite quadrants Q and T will be similar and equal if the equation in the coordinates x and y is

such that, when both $-x$ is substituted for x and $-y$ for y, the equation is not changed. Let $Z = 0$ be an equation for such a curve. In the first place, it is clear that if Z is a function of x and y which has only even powers in each variable, or more generally, if it is the sum of homogeneous functions of even degrees, then the equation $Z = 0$ will have the desired property. On the other hand, if Z is the sum of homogeneous functions of odd degree, then when both x and y are taken to be negative, then Z becomes $-Z$. Hence, if $Z = 0$, then also $-Z = 0$. Hence we have two different general equations which have equal and similar parts in the opposite quadrants Q and T as well as R and S. That is, either

$$0 = \alpha + \beta x^2 + \gamma xy + \delta y^2 + \epsilon x^4 + \zeta x^3 y$$

$$+ \eta x^2 y^2 + \theta xy^3 + \iota y^4 + \kappa x^6 + \cdots$$

or

$$0 = \alpha x + \beta y + \gamma x^3 + \delta x^2 y + \epsilon xy^2$$

$$+ \zeta y^3 + \eta x^5 + \theta x^4 y + \iota x^3 y^2 + \cdots.$$

341. Hence the curves which have two similar and equal parts are of two sorts. Either the two parts are situated on both sides of a straight line in such a way that all the orthogonal chords are bisected by that straight line. In this case the straight line is an *orthogonal diameter* of the curve. This kind was treated in sections 337 and 338. The other sort have their equal and similar parts in opposite quadrants Q and R or T and S, so that every straight line drawn through the point C divides the curve into two parts which are alternately equal. Curves of this kind are treated in the preceding section. These equal parts are described differently. The first sort are said to be *diametrically equal,* while the

second are said to be *alternately equal*. In the second kind there is a
point C through which each straight line, extended in both directions to
the curve, is bisected at C. It is convenient to call this point the *center*,
so that curves with parts which are alternately equal are said to have a
center. On the other hand, curves with two diametrically equal parts are
said to have a diameter.

342. Since the equation $Z = 0$ provides a curve whose diameter is
the straight line AB when the ordinate y appears only with even
exponents in the function Z. The same equation $Z = 0$ provides a curve
whose diameter is the straight line EF when the abscissa x appears only
with even exponents. It follows that if Z is a function in x and y such
that all exponents on both x and y are even, then both AB and EF are
orthogonal diameters. Therefore all four quadrants Q, R, S and T have
equal and similar parts of the curve. Curves of this type have the general equation

$$0 = \alpha + \beta x^2 + \gamma y^2 + \delta x^4 + \epsilon x^2 y^2 + \zeta y^4 + \eta x^6 + \theta x^4 y^2 + \cdots .$$

343. Curves contained in this equation have two orthogonal diameters AB and EF which are mutually perpendicular and intersect at C.
These curves all belong to either the lines of second order, or fourth
order, or sixth order, etc. Hence no curve with two mutually perpendicular diameters can be a line of an odd order. Then, since this equation is
contained in the previous equation given in section 339, these curves also
have a center C, so that every straight line through this point, extended
in both directions to the curve, is bisected by that point. Such curves
with two diameters are given by the equation $Z = 0$ if Z is any non-
irrational function in x^2 and y^2.

344. Since we were brought to the topic by our consideration of
curves with two diameters, let us now inquire into equations for curves

which have several diameters. In the first place, it is easily shown that if some curve has only two diameters, then they must be mutually perpendicular. It follows that every curve with only two diameters is contained in the equation we have already considered. Suppose that some curve has two diameters AB and EF which intersect at C, but are not perpendicular, as in *figure 69*. Since EC is a diameter, the curve is similar on both sides. Because on one side the curve has a diameter AC, it must also have a diameter GC on the other side also, with the angles $GCE = ACE$. In a similar way, since GC is a diameter, there must be a straight line IC with $GCI = GCE$ which is also a diameter with the same properties as EC. Furthermore, there is a straight line LC with the angles $ICL = ICG$. We proceed it in this way, finding new diameters, until we return to the first one AC. This will occur if the angle ACE has a rational ratio to a right angle.

345. Unless the angle ACE has a rational ratio to a right angle, the number of diameters would be infinite, and in this case the curve would be a circle. Conversely, in a circle every line drawn through the center is an orthogonal diameter. In this discussion we restrict the name diameter to orthogonal diameters, since it is only these which divide a curve into equal and similar parts. From what we have said it can be understood that no algebraic curve can have two parallel diameters, since by the same kind of argument, it would have an infinite number of parallel diameters all separated by the same distance. In this case a straight line could intersect the curve in an infinite number of points, but this is impossible for an algebraic curve.

346. If some curve has several diameters, they all intersect in a common point C and they are separated by equal angles. These diameters are of two kinds, which alternate in succession. Diameter CG will have the same properties as diameter CA. Furthermore the equation for

the curve when diameter CG is taken for the axis will be the same as the equation for the curve when diameter CA is taken for the axis. The alternate diameters, CA, CG, CL, etc. have this same relationship to the curve. In like manner the diameters CE, CI, etc. have the same relationship to the curve. For this reason, if the number of diameters is finite, then the angle ACG will be some aliquot part of four right angles, that is, the angle ACE will be some aliquot part of 180 degrees or the arc of the semicircle, which we call π.

347. If the angle ACE is equal to 90 degrees or $\frac{1}{2}\pi$, we have the case already treated above, in which the curve has two diameters perpendicular to each other, as in *figure 70*. We will investigate these curves again, but with a method, different from the previous one, which is adapted to the investigation of curves with several diameters. Suppose the curve has two diameters AB and EF. We take any point M on the curve and draw the straight line CM from the center C, and we let $CM = z$, and the angle $ACM = s$. We seek an equation in z and s. First we note that since the straight line AC is a diameter, z should be a function of s, which remains the same when $-s$ is substituted for s. When we take the angle ACm as the negative of the angle $ACM = s$, the straight line Cm should be equal to SCM. But $\cos s$ is a function of s which remains the same when $-s$ is substituted for s. For this reason the requirement will be met if z is some non-irrational function of $\cos s$.

348. We let $CP = x$ and the ordinate $PM = y$, then $z = \sqrt{x^2 + y^2}$ and $\cos s = \dfrac{x}{z}$. Let $Z = 0$ be the equation for the curve, whose diameter is the straight line CA. Then Z should be a non-irrational function of z and $\dfrac{x}{z}$, that is of z and x, that is, due to the

non-irrationality, of $x^2 + y^2$ and x. If Z is a function of $x^2 + y^2$ and x, then it will also be a function of y^2 and x. Let $x^2 + y^2 = u$. Since Z is function of x and u, we let $u = t + x^2$, so that $t = y^2$. Then Z is a function of t and x, that is, of y^2 and x. Whenever Z is a non-irrational function of y^2 and x, then the straight line CA is a diameter of the curve. But this is the same property of a curve with one diameter which we found before.

349. The curve we are considering is supposed to have two diameters, AB and EF. Hence, CB will be a diameter with the same properties as CA. If the straight line $CM = z$ is referred to the diameter CB, since the angle $BCM = \pi - s$, z must be the same kind of function of s, which will remain the same when $\pi - s$ is substituted for s. Such a function is $\sin s$, since $\sin s = \sin(\pi - s)$, but the preceding condition is not satisfied by this function. We have to find an expression which remains the same for s, $-s$, and $\pi - s$. Such an expression is $\cos 2s$, since $\cos 2s = \cos(-2s) = \cos 2(\pi - s)$. Hence the equation $Z = 0$ is for a curve with two diameters AB and EF, if Z is a non-irrational function in z and $\cos 2s$. Recall that $\cos 2s = \dfrac{x^2 - y^2}{z^2}$. It follows that Z must be a function of $x^2 + y^2$ and $x^2 - y^2$, that is of x^2 and y^2, as we found before.

350. We now proceed to investigate curves with three diameters, AB, EF, and GH. These diameters meet at the same point C with angles $ACE = ECG = GCB = \dfrac{1}{3}\pi = 60°$, as in *figure 71*. The alternate diameters CA, CG, and CF have the same properties. Hence, if we let $CM = z$ and the angle $ACM = s$, since $GCM = \dfrac{2}{3}\pi - s$, the equation of the curve, $Z = 0$ must be such that Z is a non-irrational function of z and some quantity w which depends on s in such a way that it

remains the same when either $-s$ or $\dfrac{2}{3}\pi - s$ is substituted for s. We

let $w = \cos 3s$, since $\cos 3s = \cos(-3s) = \cos(2\pi - 3s)$. When we

let the coordinates $CP = x$ and $PM = y$, we have

$\cos 3s = \dfrac{x^3 - 3xy^2}{z^3}$, so that Z must be a non- irrational function of

$x^2 + y^2$ and $x^3 - 3xy^2$.

351. If we let $x^2 + y^2 = t$ and $x^3 - 3xy^2 = u$, then the general

equation for a curve with three diameters is

$$0 = \alpha + \beta t + \gamma u + \delta t^2 + \epsilon t u + \zeta u^2 + \eta t^3 + \ \cdots$$

which gives the equation in x and y,

$$0 = \alpha + \beta(x^2 + y^2) + \gamma x(x^2 - 3y^2) + \delta(x^2 + y^2)^2 + \ \cdots.$$

Since $0 = \alpha + \beta x^2 + \beta y^2$ is the equation for a circle, which has

infinitely many diameters, the circle also satisfies the equation for a curve

with three diameters. The simplest curve having only three diameters is

the third order line expressed by the equation

$x^3 - 3xy^2 = ax^2 + ay^2 + b^3$. This curve has three asymptotes which

bound an equilateral triangle whose center is C. Each of the asymptotes

are of the species $u = \dfrac{A}{t^2}$, so that this curve belongs to the fifth species

in the classification we gave above.

352. If the curve has four diameters AB, EF, GH, and IK, all of

which intersect in the point C with angles which are half of right angles

equal to $\dfrac{1}{4}\pi$, then the diameters CA, CG, CB, and CH have the same

properties, as in *figure 72*. Hence, when we let $CM = z$ and the angle

$ACM = s$, we have to find a function of s which does not change when

either $-s$ or $\dfrac{2}{4}\pi - s$ is substituted for s. Such a function is $\cos 4s$.

Hence, if Z is a function of z and $\cos 4s$, or what comes to the same thing as $x^2 + y^2$ and $x^4 - 6x^2y^2 + y^4$, then equation $Z = 0$ gives a curve with four diameters. It follows that Z is a function of t and u, where $t = x^2 + y^2$ and $u = x^4 - 6x^2y^2 + y^4$. If we let $v = t^2 - u$, then Z is a function of t and v, that is of $x^2 + y^2$ and x^2y^2. We could also define Z to be a function of $x^2 + y^2$ and $x^4 + y^4$.

353. In order that the curve $Z = 0$ may express a curve with five diameters, Z must be a function of z and $\cos 5s$. When we take rectangular coordinates x and y, since $\cos 5s = \dfrac{x^5 - 10x^3y^2 + 5xy^4}{z^5}$, we see that Z must be a non-irrational function of $x^2 + y^2$ and $x^5 - 10x^3y^2 + 5xy^4$. The simplest curve, besides the circle, which has five diameters is the fifth order line expressed by the equation $x^5 - 10x^3y^2 + 5xy^4 = a(x^2 + y^2)^2 + b(x^2 + y^2) + c$. This curve, because of the real factors of the principal member, has five asymptotes which bound a regular pentagon whose center is C.

354. From what we have already discussed, it is clear that in general the curve expressed by an equation $Z = 0$. will have n diameters, each successive pair of which meet at an angle equal to $\dfrac{\pi}{n}$, if Z is a function of z and $\cos ns$, or in rectangular coordinates, Z is a non-irrational function in $x^2 + y^2$ and

$$ x^n - \frac{n(n-1)}{1\cdot 2} x^{n-2}y^2 + \frac{n(n-1)(n-2)(n-3)}{1\cdot 2\cdot 3\cdot 4} x^{n-4}y^4 - \cdots . $$

That is, the equation for the curve with n diameters will have the form

$$ 0 = \alpha + \beta t + \gamma u + \delta t^2 + \epsilon tu + \zeta u^2 + \eta t^3 + \theta t^2 u + \cdots , $$

where $t = x^2 + y^2$ and

$$u = x^n - \frac{n(n-1)}{1 \cdot 2} x^{n-2} y^2$$

$$+ \frac{n(n-2)(n-2)(n-3)}{1 \cdot 2 \cdot 3 \cdot 4} x^{n-4} y^4 - \cdots .$$

It follows that we can always find a curve with any desired number of diameters which meet in a single point C with equal angles. Conversely, all algebraic curves with a given number of diameters are obtained in this way.

355. Curves of this kind with several diameters have more than two parts which are similar and equal. A curve with two diameters, as in *figure 70,* has four similar and equal parts, AE, BE, AF, and BF. A curve with three diameters, as in *figure 71,* has six similar and equal parts, AE, GE, GB, FB, FH, and AH. A curve with four diameters, as in *figure 72,* has eight similar and equal parts, AE, AK, GE, GI, BI, BF, HF, and HK. In a similar way, the number of equal parts will always be at least twice the number of diameters. Insofar as we have already seen curves with two similar parts, but without any diameter, so too, there are curves with more parts which are similar and equal, but are still without a diameter.

356. Let us begin with two equal parts in opposite regions, such as AME and BKF in *figure 73.* If the curve is to have only two equal parts, then they must of necessity be opposite to each other. This becomes clearer when we consider more than two equal parts. As before, we let $CM = z$ and the angle $ACM = s$. It is clear that the angles s and $\pi + s$ should correspond to the same values for z. If we take the angle $ACM = \pi + s$, then $z = CK$, so that we should have $CK = CM$. We look for an expression which has the same value for the angles s and $\pi + s$. Such an expression is $\tan s$, since

$\tan s = \tan(\pi + s)$. the equation $Z = 0$ is for such a curve as we are seeking if Z is a function of z and $\tan s$, that is, a function of $x^2 + y^2$ and $\dfrac{x}{y}$. We let $\dfrac{x}{y} = t$, so that $x^2 + y^2 = y^2(1 + t^2)$. Hence Z must be a function of t and $y^2(1 + t^2)$, that is, of t and y^2. We thus obtain the same equations as we found before.

357. In order to avoid the fractions which the tangents involve, we can bring about the same result with sines and cosines. Since $\sin 2s = \sin 2(\pi + s)$ and $\cos 2s = \cos 2(\pi + s)$, we have what we want if Z is any non-irrational function of the three variables z, $\sin 2s$, and $\cos 2s$, that is of $x^2 + y^2$, $2xy$, and $x^2 - y^2$. We must note that if either of the expressions $\sin 2s$, $\cos 2s$ is omitted, then the curve also has a diameter. It follows that Z must be a non-irrational function of x^2, y^2, and xy, and the equation will be

$$0 = \alpha + \beta\, x^2 + \gamma xy + \delta y^2 + \epsilon x^4 + \zeta x^3 y$$

$$+ \eta x^2 y^2 + \theta x y^3 + \iota y^4 + \cdots .$$

If the terms in which there is no x vanish, the whole equation can be divided by x to obtain

$$0 = \beta x + \gamma y + \epsilon x^3 + \zeta x^2 y + \eta x y^2 + \theta y^3 + \kappa x^5 + \cdots ,$$

which are both equations we found above.

358. Now we look for a curve which has only three similar and equal parts, AM, BN, and DL, as in *figure 74*. We draw three straight lines CM, CN, and CL with equal angles. Henceforth angles of this kind will always be mutually equal. When we let the angle $ACM = s$ and the straight line $CM = z$, the straight line z is defined by s in such a way that the three angles s, $\dfrac{2}{3}\pi + s$, and $\dfrac{4}{3}\pi + s$ correspond to equal

values of z, since the angles $MCN = NCL = \dfrac{2}{3}\pi$. The expressions $\sin 3s$ and $\cos 3s$ both have the same values for these three angles. It follows that if Z is a non- irrational function of the three quantities $x^2 + y^2$, $3x^2y - y^3$, and $x^3 - 3xy^2$, then the equation $Z = 0$ gives all of the desired curves. The general equation for curves of this kind is

$$0 = \alpha + \beta(x^2 + y^2) + \gamma(3x^2y - y^3) + \delta(x^2 - 2xy^2) + \epsilon(x^2 + y^2)^2$$

$$+ \zeta(x^2 + y^2)(3x^2y - y^3) + \eta(x^2 + y^2)(x^3 - 3xy^2) + \cdots.$$

Third order lines with this property are contained in the equation

$$0 = \alpha + \beta x^2 + \beta y^2 + \delta x^3 + 3\gamma x^2 y - 3\delta xy^2 - \gamma y^3.$$

359. If the curve is to have four equal parts AM, EN, BK, and FL, then the four straight lines, for instance CM, CN, CK, and CL from the point C with equal angles, must be equal, as in *figure 73*. We let the angle $ACM = s$, and the straight line $CM = z$, then, since the angles $MCN = NCK = KCL = \frac{1}{2}\pi = 90°$, the straight line z must be expressed by the angle s in such a way that the angles s, $\frac{1}{2}\pi + s$, $\pi + s$, and $\dfrac{3}{2}\pi + s$ all give the same value. The expressions $\sin 4s$ and $\cos 4s$ have the desired property. Hence the equation $Z = 0$ gives a curve with four equal parts if Z is any non-irrational function of the three quantities $x^2 + y^2$, $4x^3y - 4xy^3$, and $x^4 - 6x^2y^2 + y^4$. Hence the general equation for a curve of this kind is

$$0 = \alpha + \beta x^2 + \beta y^2 + \gamma x^4 + \delta x^3 y + \epsilon x^2 y^2 - \delta xy^3 + \gamma y^4 + \cdots.$$

360. Similarly, if we want a curve with no diameters, but with five equal and similar parts, in the equation $Z = 0$, Z must be a non-irrational function of the three quantities $x^2 + y^2$, $5x^4y - 10x^2y^3 + y^5$, and $x^4 - 10x^3y^2 + 5xy^4$. In general, if the number of equal parts is n,

then Z must be a non-irrational function of $x^2 + y^2$,

$$nx^{n-1} - \frac{n(n-1)(n-2)}{1\cdot2\cdot3}x^{n-3}y^3$$

$$+ \frac{n(n-1)(n-2)(n-3)(n-4)}{1\cdot2\cdot3\cdot4\cdot5}x^{n-5}y^5 - \cdots,$$

and

$$x^n - \frac{n(n-1)}{1\cdot2}x^{n-2}y^2$$

$$+ \frac{n(n-1)(n-2)(n-3)}{1\cdot2\cdot3\cdot4}x^{n-4}y^4 - \cdots.$$

If either of these last two expressions are missing, then the curve has n diameters.

361. In the two classifications of curves which have several equal parts, namely, those with and those without diameters, we have absolutely all algebraic curves which have two or more similar and equal parts. In order to prove this, we suppose that some continuous curve has two similar and equal parts OAa and OBb, as in *figure 75*. Let AB join the two points, and construct an equilateral triangle ACB with AB as base and the angel at C equal to the angle at O. Now, since the angle OAC and the angle OBC are equal, so also the parts of the curve, CAa and CBb are similar and equal. From the law of continuity, if we take the angles BCD, DCE, etc. all equal to the angle ACB, and we let $CD = CE = CA = CB$, then the curve must also have besides these straight line parts, the parts Dd,Ee, etc. which are similar and equal to the parts Aa and Bb. Unless the ratio of the angle ACB to 360 degrees is irrational, the number of equal parts is finite. If the ratio is irrational, then the number of equal parts is infinite, but then the curve would not be algebraic. Hence the curve is contained in those without a diameter

which we have investigated.

362. If two similar and equal parts lie on opposite sides of the straight lines AO and BO, as in *figure 76,* in such a way that the part AOa is similar and equal to the part OBb, then we draw the two straight lines AR and BS in such a way that the angle $OAR = OBS = \tfrac{1}{2}AOB$. Now the straight lines AR and BS are parallel. We join A to B with the straight line AB and then through the midpoint C we draw CV parallel to AR and BS. Then the parts aA and bB are similar and equal with respect to the straight line CV. If $ba \neq 0$, then there is an arc Ee on the other side, corresponding to the arc bB, from b to a which is similar and equal to aA. Likewise corresponding to this part, there will be a point Dd on the other side from a to e where $ae = ba$. Hence this curve would have an infinite number of similar and equal parts on both sides of the straight line CV. However this cannot happen if the curve is algebraic.

363. But the curve would be just as described above if the straight line AB were oblique to the two parallel line AR and BS, or, which comes to the same thing, if in the triangle AOB the side $AO \neq BO$. On the other hand, if $AO = BO$, then the straight line AB is perpendicular to AR, BS and CV, which also passes through O. In this case the point b coincides with a. Since the parts aA and bB not only are equal and similar, but there are equally disposed on both sides of the straight line CV, so that this straight line CV is a diameter for the curve. This case falls under the classification already given for curves with diameters. It follows that absolutely all algebraic curves with two or more equal parts fall under the classification given in this chapter.

CHAPTER XVI

On Finding Curves from the
Properties of the Ordinate.

364. Let P and Q be any two non-irrational functions of the abscissa x. Suppose that the nature of the curve is expressed by the equation $y^2 - Py + Q = 0$. To any abscissa x there correspond either two or no value for the ordinate. The sum of the two values of the ordinate is equal to P and their product is equal to Q. If P is a constant, then the sum of the two ordinates is constant for all abscissas, so that the curve has a diameter. The same phenomenon occurs if $P = a + nx$. Then the straight line with equation $z = \frac{1}{2}a + \frac{1}{2}nx$ will be the diameter, in the wider sense of this name, where obliquity is not excluded. If Q is constant, then the product of the ordinates is constant, so that the curve never intersects the axis. If $Q = \alpha + \beta x + \gamma x^2$, and this expression has two real factors, then the curve passes through the axis at two points, and Q is the product of two intervals on the axis. It follows that the product of the ordinates has a constant ratio to the product of the intervals on the axis.

365. These properties, which we noted before when we discussed conic sections, belong to many other curves also. Thus, the constant product of the two values of the ordinates which correspond to a single

abscissa was seen when the asymptote of a hyperbola is taken as the axis. This same property is shared by any curve with the equation $y^2 - Py \pm a^2 = 0$. Let us take the straight line EF, which intersects the curve in the two points E and F, for the axis, as in *figure 19*. As for conic sections, the product $PM \cdot PN$ has a constant ratio to the product $PE \cdot PF$. This property of conic sections is shared by all curves with the equation $y^2 - Py + ax - nx^2 = 0$. That is $PM \cdot PN = PE \cdot PF$, or $pm \cdot pn = Ep \cdot pF$, if $y^2 - Py = ax - x^2$. The property of the circle, whereby it is made up of elements, is common not only to an infinite number of higher order curves, but also to the other conics. That is, let $P = b + nx$, then the equation is $y^2 - nxy + x^2 = ax + by$. We have a circle when $n = 0$, and the angle EPM is a right angle. This equation is for an ellipse when $n^2 < 4$; a hyperbola when $n^2 > 4$ and a parabola when $n^2 = 4$.

366. From this we conclude that for all conic sections $AEBF$, as in *figure 77*, with axes or principal diameters AB and EF, if we draw any two straight lines pq and mn with an angle of inclination with the principal axis equal to half of a right angle and intersecting each other at h, then $mh \cdot nh = ph \cdot qh$. The following must be included among the prize winning properties: if we draw the straight lines PQ and MN through the center C with angles of inclination equal to half of the right angle, then we have $MC \cdot NC = PC \cdot QC$. Since all straight lines parallel to these follow the same law, we have the result that $mh \cdot nh = ph \cdot qh$. We should note here also, that if the straight lines MN and PQ had been drawn in such a way that the angle of inclination with respect to one of the principal axes was the same, that is, $PCA = NCA$, since $CP = CN$, then all straight lines drawn parallel to these which intersect each other so that the product of their parts are equal, that is, $mh \cdot hn = ph \cdot hq$.

367. After these observations, let us consider other questions concerning the two ordinates corresponding to any abscissa in the equation $y^2 - Py + Q = 0$. Let $AP = x$ be the abscissa with PM and PQ the two corresponding ordinates in *figure 78*. First we look for all curves with the property that $PM^2 + PN^2 = a^2$, where a is constant. Since $PM + PN = P$ and $PM \cdot PN = Q$, we have $PM^2 + PN^2 = P^2 - 2Q$, so that the required condition is satisfied if $P^2 - 2Q = a^2$, that is, if $Q = \dfrac{P^2 - a^2}{2}$. Hence, in order to obtain the desired curves, the equation $y^2 - Py + \dfrac{P^2 - a^2}{2} = 0$ must be satisfied. If we let $P = 2nx$, we obtain the conic section with this property,

$$y^2 - 2nxy + 2n^2x^2 - \tfrac{1}{2}a^2 = 0,$$

which is the equation for an ellipse, with the center taken as origin.

368. From this we obtain a rather elegant property of an ellipse. We draw any two conjugate diameters AB and EF of an ellipse, as in *figure 79,* and the parallelogram $GHIK$ whose sides are the tangent lines at the points A, B, E, and F. We let MN be any chord parallel to either diameter EF which intersects the two diagonals GK and HI of the parallelogram in the points P and p. Then the sum of the squares, $PM^2 + PN^2 = pM^2 + pN^2 = 2CE^2$, which is constant. In a similar way, if we draw the chord RS parallel to the other diameter AB, then $PR^2 + PS^2 = \pi R^2 + \pi S^2 = 2CA^2$. In order to see this, we let $CA = CB = a$, $CE = CF = b$, $CQ = t$, and $QM = u$, then $a^2u^2 + b^2t^2 = a^2b^2$. We have $\dfrac{a}{b} = \dfrac{CQ}{PQ}$ and $\dfrac{CP}{CQ}$ is some ratio, say, $\dfrac{m}{1}$. Hence, when we let $CP = x$ and $PM = y$, then $x = mt$ and $y = u + \dfrac{bt}{a}$, that is, $t = \dfrac{x}{m}$, and $u = y - \dfrac{bx}{ma}$. When we substitute

these values in the given equation we obtain

$$a^2 y^2 - \frac{2abxy}{m} + \frac{2b^2 x^2}{m^2} = a^2 b^2.$$

We let $\dfrac{b}{ma} = n$, so that

$$y^2 - 2nxy + 2n^2 x^2 = b^2,$$

which is the equation found earlier, which proves that $PM^2 + PN^2$ is a constant.

369. Now we look for curves in which the sum of the cubes $PN^3 + PN^3$ is always constant. Since $Pm + PN = p$ we have $PN^3 + PN^3 = p^2 - 3PQ$. If we let $PM^3 + PN^3 = a^3$, we have $Q = \dfrac{p^3 - a^3}{3P}$. Hence, the general equation for these curves is

$y^2 - Py + \dfrac{1}{3}P^2 - \dfrac{a^3}{3P} = 0$, where for P we may substitute any non-irrational function of x. The simplest curve which has this property is a third order line, where we let $P = 3nx$ and $a = 3nb$, to obtain

$$xy^2 - 3nx^2 y + 3n^2 x^3 - 3n^2 b^3 = 0,$$

which belongs to the second species in the classification given above.

370. In a similar way, if we seek a curve such that $PM^4 + PN^4$ is constant, then since $PM^4 + PN^4 = P^4 - 4P^2 Q + 2Q^2$, the quantity Q must satisfy the equation $P^4 - 4P^2 Q + 2Q^2 = a^4$, that is, $Q = P^2 + \sqrt{\tfrac{1}{2}P^4 + \tfrac{1}{2}a^4}$. Since both P and Q have to be non- irrational functions of x and y can have no more than two values for abscissa x, the quantity $\sqrt{\tfrac{1}{2}P^4 + \tfrac{1}{2}a^4}$ must be non-irrational. However, this is impossible, so the function Q is always double-valued, and the ordinate y has four-values for each abscissa x. However, from the equation
tion

$$y^2 - Py + Q = 0,$$

we obtain $y = \frac{1}{2}P \pm \left(-\frac{3}{4}P^2 \pm \sqrt{\frac{1}{2}P^4 + \frac{1}{2}a^4} \right)^{\frac{1}{2}}$, hence, it is clear

that the value of y cannot be real unless $\sqrt{\frac{1}{2}P^4 + \frac{1}{2}a^4}$ is taken with the

positive sign. It follows, notwithstanding the fact that Q is a double-valued function, that y never has more than two values, the sum of the fourth powers of which is constant, as the problem requires.

371. If now we want a curve such that the sum of the fifth powers of the two ordinates is a constant, that is, $PM^5 + PN^5 = a^5$, then we must have $P^5 - 5P^3Q + 5PQ^2 = a^5$. Since from the equation for the curve $y^2 - Py + Q = 0$, we have $Q = -y^2 + Py$, so that

$$P^5 - 5P^4y + 10P^3y^2 - 10P^2y^3 + 5Py^4 = a^5,$$

or $(P - y)^5 + y^5 = a^5$. In the same way we find, when we require $PM^6 + PN^6 = a^6$, that the equation is $(P - y)^6 + y^6 = a^6$. In general, if we desire a curve in which $PM^n + PN^n = a^n$, then the equation will be $(P - y)^n + y^n = a^n$, where for P any single-valued function of x can be chosen. The idea behind this equation can be easily seen by noting that P is the sum of the two ordinates, so that if y is one of these, then $P - y$ is the other. Hence it is immediate that $(P - y)^n + y^n = a^n$.

372. Now if we eliminate P instead of Q by putting $P = \frac{y^2 + Q}{y}$ in the relation between P and Q, for the equation $PM^n + PN^n = a^n$, we obtain the equation $y^n + \frac{Q^n}{y^n} = a^n$. Since the product of the ordinates is equal to Q, if one of the ordinates is y, then the other is equal to $\frac{Q}{y}$, and the equation that was discovered is immedi-

ate. For curves in which $PM^n + PN^n = a^n$, we have found two general equations $(P - y)^n + y^n = a^n$ and $y^n + \dfrac{Q^n}{y^n} = a^n$. From the second of these we obtain $y^{2n} = a^n y^n - Q^n$, so that

$$y^n = \tfrac{1}{2}a^n \pm \left[\frac{1}{4}a^{2n} - Q^n\right]^{\frac{1}{2}},$$

that is,

$$y = \left[\tfrac{1}{2}a^n \pm \left[\frac{1}{4}a^{2n} - Q^n\right]^{\frac{1}{2}}\right]^{\frac{1}{n}}.$$

This is only a two valued function, and for any abscissa x there are two and no more than two values of the ordinate, provided Q^n is a non-irrational   or   single-valued   function   of x.   The   equations $y^n + (P - y)^n = a^n$ has the advantage of being an equation of lower degree.

373. These equations solve the problem not only for the cases when n is a positive integer, but also for negative integers and fractions. Thus

if we desire	we have the equation $aP = Py - y^2$
$\dfrac{1}{PM} + \dfrac{1}{PN} = \dfrac{1}{a}$	or $aQ + ay^2 = Qy$
$\dfrac{1}{PM^2} + \dfrac{1}{PN^2} = \dfrac{1}{a^2}$	$a^2y^2 + a^2(P - y)^2 = y^2(P - y)^2$ or $a^2Q^2 + a^2y^4 = Q^2y^2$
$\dfrac{1}{PM^3} + \dfrac{1}{PN^3} = \dfrac{1}{a^3}$	$a^3y^3 + a^3(P - y)^3 = y^3(P - y)^3$ or $a^3Q^3 + a^3y^6 = Q^3y^3$

etc.

For fractional exponents, the results are as follows:

if we desire	we have the equation
$\sqrt{PM} + \sqrt{PN} = \sqrt{a}$	$\sqrt{y} + \sqrt{P-y} = \sqrt{a}$

$$\text{or}$$

$$y = \sqrt{ay} - \sqrt{Q}$$

which can be reduced to rationality as

$$y^2 - Py + \frac{1}{4}(a - P)^2 = 0$$

$$\text{or}$$

$$y^2 - (a - 2\sqrt{Q})y + Q = 0$$

| $PM^{\frac{1}{3}} + PN^{\frac{1}{3}} = a^{\frac{1}{3}}$ | $y^{\frac{1}{3}} + (P-y)^{\frac{1}{3}} = a^{\frac{1}{3}}$ |

$$\text{that is}$$

$$y^2 - Py + \frac{1}{27a}(a - P)^3 = 0$$

$$\text{or}$$

$$y^{\frac{1}{3}} + (Q/y)^{\frac{1}{3}} = a^{\frac{1}{3}}$$

$$\text{that is}$$

$$y^2 - (a - 3(aQ)^{\frac{1}{3}})y + Q = 0$$

etc.

In this way, all algebraic curves in which we always have $PM^n + PN^n = a^n$, one general equation covers all possibilities, whether n is a positive or negative integer, or a fraction.

374. Until this time we have been considering curves with two values of the ordinate corresponding to each value of the abscissa. We

can use the same method to consider curves with three values of the ordinate for each value of the abscissa. The general equation for curves of this kind is

$$y^3 - Py^2 + Qy - R = 0,$$

where P, Q, and R are any non-irrational functions of x. Let p, q, and r be the three values of the ordinate corresponding to the abscissa x, one of which must always be real, but we are especially interested in those parts of the curve where all three are real. From the nature of the equation we have $P = p + q + r$, $Q = pq + pr + qr$, and $R = pqr$. Hence, if we want to find a curve in which either $p + q + r$ or $pq + pr + qr$, or pqr is constant, all that needs to be done is to make P or Q or R constant, with the other two remaining arbitrary.

375. We can also find curves in which $p^n + q^n + r^n$ remains constant everywhere. We have the following results from the first volume:

$$p + q + r = P$$

$$p^2 + q^2 + r^2 = P^2 - 2Q$$

$$p^3 + q^3 + r^3 = P^3 - 3PQ + 3R$$

$$p^4 + q^4 + r^4 = P^4 - 4P^2Q + 2Q^2 + 4PR$$

$$p^5 + q^5 + r^5 = p^5 - 5P^3Q + 5PQ^2 + 5P^2R - 5QR$$

etc. If n is a negative integer, we let $z = \dfrac{1}{y}$ so that $z^3 - \dfrac{Qz^2}{R} + \dfrac{Pz}{R} - \dfrac{1}{R} = 0$ has the three roots $\dfrac{1}{p}, \dfrac{1}{q}$, and $\dfrac{1}{r}$. Then we have

$$\frac{1}{p} + \frac{1}{q} + \frac{1}{r} = \frac{Q}{R}$$

$$\frac{1}{p^2} + \frac{1}{q^2} + \frac{1}{r^2} = \frac{Q^2 - 2PR}{R^2}$$

$$\frac{1}{p^3} + \frac{1}{q^3} + \frac{1}{r^3} = \frac{Q^3 - 3PQR + 3R^2}{R^3}$$

$$\frac{1}{p^4} + \frac{1}{q^4} + \frac{1}{r^4} = \frac{Q^4 - 4PQ^2R + 4QR^2 + 2P^2R^2}{R^4}$$

etc. In this way by setting the quantity equal to a constant, be obtain a suitable relationship between P, Q, and R. By using this relationship and the equation $y^3 - Py^2 + Qy - R = 0$, one of the functions P, Q, or R can be eliminated to obtain an equation for the desired curve. For example, if we want a curve for which $p^3 + q^3 + r^3 = a^3$, then $P^3 - 3PQ + 3R = a^3$. Since $R = y^3 - Py^2 + Q$, we have the equation $3y^3 - 3Py^2 + 3Qy + p^3 - 3PQ = a^3$ for the desired curve.

376. Whether n is a positive or a negative integer, the solution to the problem is easily found through the given formulas. There is a major difficulty if n is a fraction. Let us try to find a curve for which $\sqrt{p} + \sqrt{q} + \sqrt{r} = \sqrt{a}$. When we square both sides of this equation, since $p + q + r = P$, we have

$$P + 2\sqrt{pq} + 2\sqrt{pr} + 2\sqrt{qr} = a,$$

that is, $\dfrac{a - P}{2} = \sqrt{pq} + \sqrt{pr} + + \sqrt{qr}$. Once more, we square both sides, and since $pq + pr + qr = Q$, we have

$$\frac{(a - P)^2}{4} = Q + 2\sqrt{p^2 qr} + 2\sqrt{pq^2 r} + 2\sqrt{pqr^2}$$

$$= Q + 2\left(\sqrt{p} + \sqrt{q} + \sqrt{r}\right)\sqrt{pqr} = 2\sqrt{aR} + Q.$$

It   follows   that $(a - P)^2 = 4Q + 8\sqrt{aR}$,   that   is,

$Q = \dfrac{(a - P)^2}{4} - 2\sqrt{aR}$. Hence the curve we are looking for has the

equation

$$y^3 - Py^2 + \left[\frac{1}{4}(a - P)^2 - 2\sqrt{aR}\right]y - R = 0.$$

We can remove the irrationality, since $R = \dfrac{(a^2 - 2aP + P^2 - 4Q)^2}{64a}$,

the equation can be written

$$y^3 - Py^2 + Qy - \frac{(a^2 - 2aP + P^2 - 4Q)^2}{64a} = 0.$$

377. This sort of thing is rather tedious, especially if the radicals
are of higher powers. We will try some other method, which we will
illustrate with the following example. Suppose we are looking for a curve
such that $p^{\frac{1}{3}} + q^{\frac{1}{3}} + r^{\frac{1}{3}} = a^{\frac{1}{3}}$. We let $(pq)^{\frac{1}{3}} + (pr)^{\frac{1}{3}} + (qr)^{\frac{1}{3}} = v$.
Since $(pqr)^{\frac{1}{3}} = R^{\frac{1}{3}}$, we have $p^{\frac{2}{3}} + q^{\frac{2}{3}} + r^{\frac{2}{3}} = a^{\frac{2}{3}} - 2v$, and

$$p + q + r = a - 3va^{\frac{1}{3}} + 3R^{\frac{1}{3}} = P.$$

Hence

$$(pq)^{\frac{2}{3}} + (pr)^{\frac{2}{3}} + (qr)^{\frac{2}{3}} = v^2 - 2(aR)^{\frac{1}{3}}$$

and

$$pq + pr + qr = Q = v^3 - 3v(aR)^{\frac{1}{3}} + 3R^{\frac{2}{3}}.$$

Now that expressions have been found for P and Q, we take for v any function of x and write the equation for the desired curve,

$$y^3 - (a - 3v\sqrt{a} + 3R^{\frac{1}{3}})y^2 + (v^3 - 3v(aR)^{\frac{1}{3}} + 3R^{\frac{2}{3}})y - R = 0.$$

378. Notwithstanding these difficulties, it is still possible to find a general solution to the problem. From the equation $y^3 - Py^2 + Qy - R = 0$, we let y denote these three ordinates, p, q, and r. If we let $p = y$, then $P = y + q + r$ and $Q = qy + ry + qr$, that is, $q + r = P - y$ and $qr = Q - y(q + r) = Q - Py - y^2$. From this we obtain $q - r = \sqrt{P^2 + 2Py - 3y^2 - 4Q}$, so that

$$q = \tfrac{1}{2}(P - y) + \tfrac{1}{2}\sqrt{P^2 + 2Py - 3y^2 - 4Q}$$

and

$$r = \tfrac{1}{2}(P - y) - \tfrac{1}{2}\sqrt{P^2 + 2Py - 3y^2 - 4Q}.$$

Hence when we want a curve for which $p^n + q^n + r^n = a^n$, it must satisfy the equation

$$y^n + (\tfrac{1}{2}(P - y) + \tfrac{1}{2}\sqrt{P^2 + 2Py - 3y^2 - 4Q})^n +$$

$$(\tfrac{1}{2}(P - y) - \tfrac{1}{2}\sqrt{P^2 + 2Py - 3y^2 - 4Q})^n = a^n.$$

This equation solves the problem whether n is an integer or a fraction.

379. There are innumerable other questions about the conditions on the three ordinates which can be settled with this same method, for instance if for a^n we substitute any function of x. We can also ask about other functions of p, q, and r besides the sums of powers, provided these quantities enter into the function in an equal way, in the

sense that no change takes place when the variables are permuted. Thus, these three ordinates p, q, and r corresponding to the same abscissa x could be defined in such a way that the triangles formed from sides equal to these three numbers have a constant area. The area of such a triangle is equal to $\frac{1}{4}\sqrt{2p^2q^2 + 2p^2r^2 + 2q^2r^2 - p^4 - q^4 - r^4}$, which we set equal to a^2. Since $p^4 + q^4 + r^4 = P^4 - 4P^2Q + 4PR + 2Q^2$ and $p^2q^2 + p^2r^2 + q^2r^2 = Q^2 - 2PR$, we have

$$16a^4 = 4P^2Q - 8PR - P^4$$

and

$$R = \tfrac{1}{2}PQ - \frac{1}{8}P^3 - \frac{2a^4}{P}.$$

Hence, we have the equation

$$y^2 - Py^2 + Qy - \tfrac{1}{2}PQ + \frac{1}{8}P^3 + \frac{2a^4}{P} = 0.$$

If P is taken to be equal to constant $2b$, then the perimeters of all of those triangles is also constant. Hence, if we take $Q = mx^2 + nbx + ka^2$, we obtain a third order line expressed by the equation

$$y^2 + mx^2y - 2by^2 + nbxy - mbx^2 + ka^2y$$

$$- nb^2x + \frac{a^4}{b} - ka^2b + b^3 = 0$$

which has the property that the three ordinates p, q, and r corresponding to each abscissa, have a constant sum equal to $2b$, and also the area of the triangle with sides p, q, and r is always equal to a^2.

380. Similar questions about four or more ordinates corresponding to a single abscissa can be answered with the same methods. There is no

further difficulty which arises in such a discussion, so we proceed to other questions. We now consider relationships between ordinates when they do not correspond to the same abscissa, but to different abscissas. First consider some relationship between the ordinates PN and QN corresponding to the abscissas $AP = x$ and $AQ = -x$, as illustrated in *figure 80*. Let $y = X$ be the equation for the curve, where X is some function of x which gives the ordinate value PM. If instead of x we substitute $-x$, the corresponding ordinate is QN. If X is an even function of x, say P, then $QN = PM$; if X is an odd function of x, say Q, then $QN = -PM$. If both P and R are even functions, while Q and S are odd function of x, then for the curve whose equation is $y = \dfrac{P + Q}{R + S}$, we have $PM = \dfrac{P + Q}{R + S}$ and $QN = \dfrac{P - Q}{R - S}$.

381. We would like to find a curve with the property that $PM + QN$ is equal to a constant, namely, $2AB = 2a$. It is clear that the equation $y = a + Q$ has the property if Q is an odd function of X. Since $PM = a + Q$ and $QN = a - Q$,   it   follows   that $PM + QN = 2a$, as required. When we let $y - a = u$, where $u = Q$, we have an equation for the same curve. If we take the straight line Bp for the axis and the point B for the origin, then $Bp = x$ and $pM = u$. The equation $u = Q$ gives a curve such that equal parts lie on either side of the center B in opposite quadrants. After we have drawn any such curve MBN and taken the straight line PQ as an axis, the question is answered, since when the perpendicular BA is dropped from the center B to the axis, and when we take the two equal abscissas $AP = AQ$, we always have the sum $PM + QN = 2AB$.

382. Previously we have found two different equations for curves with equal parts alternately disposed with respect to the center B,

namely, the following equations in the coordinates x and u.

I.

$$0 = \alpha x + \beta u + \gamma x^3 + \delta x^2 u + \epsilon x u^2 + \zeta u^3 + \eta x^5 + \theta x^4 u + \cdots$$

II.

$$0 = \alpha + \beta x^2 + \gamma x u + \delta u^2 + \epsilon x^4 + \zeta x^3 u + \eta x^2 u^2 + \theta x u^3 + \cdots.$$

It follows that if in either of these equations we let $u = y - a$, then we have general equations in x and y for algebraic curves which satisfy the condition we set. In the first place, any straight line through the point B satisfies the condition. Next, any conic section which has a center at the point B solves the problem. In this last case, for both abscissas AP and AQ, there correspond two ordinates. If the curve is a hyperbola, the ordinates are taken parallel to the other asymptote. Hence there are two pairs of ordinates which have equal sums.

383. If we are looking for a curve MBN in which, not the sum of the two ordinates PM and QN, but the sum of any power of the ordinates is constant, then the solution is found in a similar way. We need $PM^n + QN^n = 2a^n$. It is clear that this condition will be satisfied by the equation $y^n = a^n + Q$, where Q is any odd function of x; since $PM^n = a^n + Q$, we have $QN^n = a^n - Q$ and so $PM^n + QN^n = 2a^n$. When we let $y^n - a^n = u$, and let $u = Q$, we have expressed the nature of the curve as one in the coordinates x and u, with equal parts in opposite quadrants about the center B. For this reason, if we substitute $y^n - a^n$ for u in the two equations given in the preceding section, we will have general equations satisfying the conditions of this problem.

384. Since questions of this kind offer no difficulty, we turn to the

following problem. We look for a curve MBN such that when we take a fixed point A on the axis and choose equal abscissas AP and AQ, then the product of the ordinates $PM \cdot QN$, is equal to a constant a^2. We can give several particular solutions to this question, of which we will give some of special interest, before we look for a general solution. Let P be an even function and Q be an odd function of the abscissa $AP = x$, and we let the ordinate $PM = y = P + Q$. When we take x to be negative, we have $QN = P - Q$. Since we want

$$PM \cdot QN = P^2 - Q^2 = a^2,$$

that is $P = \sqrt{a^2 + Q^2}$ and since, Q^2 is an even function of x, the expression $\sqrt{a^2 + Q^2}$ is a fitting value for the even function P. Hence for the desired curve we have the equation $y = Q + \sqrt{a^2 + Q^2}$, where Q is any odd function of x.

385. Since the sign of the radical is ambiguous, to each abscissa there correspond two ordinates, one positive and the other negative. Hence, to the abscissa AP there correspond the ordinates $Q + \sqrt{a^2 + Q^2}$ and $Q - \sqrt{a^2 + Q^2}$. It follows that the curve has equal parts alternately placed about the point A as a center. It is impossible to remove the ambiguity of sign, since if we take a function like $\dfrac{a^2}{4x} - x$ as the odd function Q, so that $a^2 + Q^2$ is a perfect square and $\sqrt{a^2 + Q^2} = \dfrac{a^2}{4x} + x$. This is an odd function, which cannot be substituted for the even function P. It follows that a Q should not be chosen such that $a^2 + Q^2$ is a perfect square.

386. In a similar way, if we let $y = (P + Q)^n$, then $QN = (P - Q)^n$ so that $(P^2 - Q^2)^n = a^2$. Hence we let $P^2 = a^{\frac{2}{n}} + Q^2$ and $P = \left(a^{\frac{2}{n}} + Q^2 \right)^{\frac{1}{2}}$. As long as this is an irrational

quantity, it can be assigned to P. It follows that we obtain a curve which satisfies the given condition from the equation

$$y = \left(Q + \left(a^{\frac{2}{n}} + Q^2 \right)^{\frac{1}{2}} \right)^n.$$ The construction of these curves is easy;

we draw any curve which has two parts which are equal and similar alternately about the center A. The ordinate of this curve corresponds to the abscissa $AP = x$, which we set equal to z. Then z is an odd function of x which we substitute for Q. From the equation which we

found, $y^{\frac{1}{n}} - Q = \left(a^{\frac{2}{n}} + Q^2 \right)^{\frac{1}{2}}$. Hence $Q = z = \dfrac{y^{\frac{2}{n}} - a^{\frac{2}{n}}}{2y^{\frac{1}{n}}}$. We let

$\dfrac{1}{n} = m$, then in the given equation in z and x, we substitute

$z = \dfrac{y^{2m} - a^{2m}}{2y^m}$ to obtain an equation in x and y for the desired

curve. Since we have found two such general equations in x and z, namely

$$0 = \alpha + \beta x^2 + \gamma xz + \delta z^2 + \epsilon x^4 + \zeta x^3 z + \eta x^2 z^2 + \theta xz^3 + \cdots$$

and

$$0 = \alpha x + \beta z + \gamma x^3 + \delta x^2 z + \epsilon xz^2 + \zeta z^3 + \eta x^5 + \theta x^4 + \cdots,$$

when we substitute $z = y^m - \dfrac{a^{2m}}{y^m}$. We have neglected the divisor 2,

since any multiple of z can be used for Q. Hence, we have two general equations for curves with the required condition.

387. Besides P, we let R also be even, and besides Q, S is also odd. We look for a curve satisfying $y = \dfrac{P + Q}{R + S} = PM$. Then

$QN = \dfrac{P - Q}{R - S}$, so that $\dfrac{P^2 - Q^2}{R^2 - S^2} = a^2$. This condition is easily satisfied by letting $y = \dfrac{P + Q}{P - Q}a$, or even $y = ((P + Q)/(P - Q))^n a$.

In this way we avoid the previous inconvenience that to each abscissa there were two or more ordinate values, since to each abscissa there corresponds only one ordinate. Hence, the simplest curve satisfying the condition is the second order line satisfying the equation $y = \dfrac{b + x}{b - x}a$, which is a hyperbola. In fact, the hyperbola also satisfies the equation found previously, $y = Q + \sqrt{a^2 + Q^2}$, when we let $Q = nx$. In this case $y^2 - 2nxy = a^2$. Hence, the hyperbola is a solution to this problem in two different ways.

388. With these results it becomes clear that an equation for the desired curve should be such that when $- x$ is substituted for x and when $\dfrac{a^2}{y}$ is substituted for y, the equation should not change. Formulas of this kind are $\left(y^n - \dfrac{a^{2n}}{y^n}\right)P$ and $\left(y^n - \dfrac{a^{2n}}{y^n}\right)Q$, where P is an even function and Q is an odd function of x. If we form an equation made up of such terms, then the curve will satisfy the desired condition. Let M, P, R, T, etc be any even functions of x, and let N, Q, S, V, etc be any odd functions. Then we have the following general equation:

$$0 = M + \left(\frac{y}{a} + \frac{a}{y}\right)P + \left(\frac{y^2}{a^2} + \frac{a^2}{y^2}\right)R + \left(\frac{y^3}{a^3} + \frac{a^3}{y^3}\right)T + \cdots$$

$$+ \left(\frac{y}{a} - \frac{a}{y}\right)Q + \left(\frac{y^2}{a^2} - \frac{a^2}{y^2}\right)S + \left(\frac{y^3}{a^3} - \frac{a^3}{y^3}\right)V + \cdots.$$

If this equation is multiplied by an odd function of x, then the even functions become odd and the odd functions become even, so that the following equation also satisfies the desired condition:

$$0 = N + (\frac{y}{a} + \frac{a}{y})Q + (\frac{y^2}{a^2} + \frac{a^2}{y^2})S + (\frac{y^3}{a^3} + \frac{a^3}{y^3})V + \cdots$$

$$+ (\frac{y}{a} - \frac{a}{y})P + (\frac{y^2}{a^2} - \frac{a^2}{y^2})R + (\frac{y^3}{a^3} - \frac{a^3}{y^3})T + \cdots .$$

When we free these equations of fractions we obtain the following polynomials of indefinite degree n:

I.

$$0 = a^n y^n M + a^{n-1}y^{n+1}(P + Q) + a^{n-2}y^{n+2}(R + S)$$

$$+ a^{n-3}y^{n+3}(T + V) + \cdots$$

$$+ a^{n+1}y^{n-1}(P - Q) + a^{n+2}y^{n-2}(R - S)$$

$$+ a^{n+3}y^{n-3}(T - V) + \cdots .$$

II.

$$0 = a^n y^n N + a^{n+1}(P + Q) + a^{n-2}y^{n+2}(R + S)$$

$$+ a^{n-3}y^{n+3}(T + V) + \cdots$$

$$- a^{n+1}y^{n-1}(P - Q) - a^{n+2}y^{n-2}(R - S)$$

$$- a^{n+3}y^{n-3}(T - V) \cdots .$$

389. In the formulas $(y^n + \frac{a^{2n}}{y^n})P$, and $(y^n - \frac{a^{2n}}{y^n})Q$, the exponent n can be a fraction as well as an integer. In particular, if we substitute for n the fraction $\frac{1}{2}, \frac{3}{2}, \frac{5}{2}, \frac{7}{2}$, etc. in the general equations, the irrationalities will be spontaneously eliminated in the equation

$$0 = \frac{y + a}{(ay)^{\frac{1}{2}}}P + \frac{y^3 + a^3}{ay(ay)^{\frac{1}{2}}}R + \frac{y^5 + a^5}{a^2y^2(ay)^{\frac{1}{2}}}T + \cdots$$

$$+ \frac{y - a}{(ay)^{\frac{1}{2}}} Q + \frac{y^3 - a^3}{ay(ay)^{\frac{1}{2}}} S + \frac{y^5 - a^5}{a^2 y^2 (ay)^{\frac{1}{2}}} V + \cdots$$

or the equation

$$0 = y \frac{+ a}{(ay)^{\frac{1}{2}}} Q + \frac{y^3 + a^3}{ay(ay)^{\frac{1}{2}}} S + \frac{y^5 + a^5}{a^2 y^2 (ay)^{\frac{1}{2}}} V + \cdots$$

$$+ \frac{y - a}{(ay)^{\frac{1}{2}}} P + \frac{y^3 - a^3}{ay(ay)^{\frac{1}{2}}} R + \frac{y^5 - a^5}{a^2 y^2 (ay)^{\frac{1}{2}}} T + \cdots .$$

But when these equations are freed of fractions, we have

$$0 = + a^n y^{n+1}(P + Q) + a^{n-1} y^{n+2}(R + S)$$

$$+ a^{n-2} y^{n+3}(T + V) + \cdots$$

$$+ a^{n+1} y^n (P - Q) + a^{n+2} y^{n-1}(R - S)$$

$$+ a^{n+3} y^{n-2}(T - V) \cdots$$

and

$$0 = + a^n y^{n+1}(P + Q) + a^{n-1} y^{n+2}(R + S)$$

$$+ a^{n-2} y^{n+3}(T + V) + \cdots$$

$$- a^{n+1} y^n (P - Q) - a^{n+2} y^{n-1}(R - S)$$

$$- a^{n+3} y^{n-2}(T - V) \cdots .$$

390. From these four equations it is easy to find the equations which solve the problem with lines from each order. In the first place, the straight line through the point B and parallel to the axis AP is the first order solution. From the second order, the two earlier equations, when $n = 1$, give us $\alpha a x y + y^2 - a^2 = 0$, which comes from the second

equation when we let $N = \alpha x$, $P = 1$, and $Q = 0$. The first equation gives no curves. The next two equations give

$$y(\alpha + \beta x) \pm a(\alpha - \beta x) = 0$$

when $n = 0$. For the third order, the two earlier equations give

$$0 = ay(\alpha + \beta x^2) + y^2(\gamma + \delta x) + a^2(\gamma - \delta x)$$

and

$$0 = \alpha ayx + y^2(\gamma + \delta x) - a^2(\gamma - \delta x)$$

when we let $n = 1$. The next two equations give us

$$0 = y(\alpha + \beta x + \gamma x^2) \pm a(\alpha - \beta x + \gamma x^2)$$

when $n = 0$, and

$$0 = ay^2(\alpha + \beta x) + y^3 \pm a^2y(\alpha - \beta x) \pm a^3$$

when $n = 1$. In a similar way, all the lines of all orders which solve the problem are found by this method.

CHAPTER XVII

On Finding Curves from Other Properties.

391. The questions which were treated in the preceding chapter were stated in such a way that they were easily answered by equations. Now we consider properties which are not immediately stated in terms of parallel ordinates; for instance, some property of straight lines from some given point to the curve. Let C be a point from which the straight lines CM and CN are drawn to the curve, as in *figure 81*. Some properties concerning these lines will be proposed. It will be convenient to change the method we have been using to express the coordinates of the curve, in order to bring these straight lines into the equation.

392. Since there are many different ways in which the equations for curves can be expressed in terms of two variables, in the present treatment, we will choose the length of the straight line CM from the given point C to the curve as one of the variables. Then we need another variable which will indicate the position of the line. In order to accomplish this, we take a straight line CA through the point C as the axis, and the angle ACM or some function of this angle will take the place of the other variable. We let the straight line $CM = z$, and we let the angle $ACM = \phi$, whose sine or tangent will enter into the equation. It is clear that if we have an equation in z and $\sin \phi$ or $\tan \phi$, then the

nature of the curve AMN is determined, since for any angle ACM we have the length of the straight line CM, so that a point on the curve is determined.

393. We will consider this method of expressing a curve a bit more carefully. In the first place, let us equate the distance z to some function of the sine of the angle ϕ. Since this function is single-valued, it seems that the straight line CM intersects the curve in a single point. However, if the angle ϕ is increased by two right angles, straight line drawn through C is in the same position, the only difference is that it is in the opposite direction. Hence, the same straight line can give a second intersection with the curve, even though z is set equal to a single-valued function of the sine of the angle ϕ. For example, let P be a function of $\sin \phi$, so that $z = P$ and we obtain the point M on the curve, as in *figure 82*. Now we increase the angle by two right angles, or we take the negative sine, so that the function P becomes Q, and $z = Q$. Hence, we have a second intersection m with the curve by the same straight line CM, taking $Cm = Q$.

394. Although P is a single-valued function of $\sin \phi$, still the straight line CM, drawn through the point C at the angle $ACM = \phi$, meets the curve in two points M and m unless $Q = -P$. Hence, if each straight line CM is to meet the curve in only one point, the function P must be an odd function of $\sin \phi$. This also takes place if P is an odd function of $\cos \phi$. For this reason, every curve which meets each straight line in only one point is of the form $z = P$, provided P is an odd function of both $\sin \phi$ and $\cos \phi$.

395. Since curves which are met in a single point by all straight lines drawn from the point C have an equation $z = P$ where P is an odd function of $\sin \phi$ and $\cos \phi$, or a function which takes a negative

value when both sin φ and cos φ are taken to be negative, it is easy to write the equations of such curves in rectangular coordinates. If we drop the perpendicular MP from the point M to the axis CA, and if we let $CP = x$ and $PM = y$, then $\frac{y}{z} = \sin \phi$, and $\frac{x}{z} = \cos \phi$. Hence, if P is an odd function of $\frac{x}{z}$ and $\frac{y}{z}$, all of these curves have an equation of the form $z = P$. We begin with the simplest cases. Let

$$z = \frac{\alpha x}{z} + \frac{\beta y}{z} + \frac{\gamma z}{x} + \frac{\delta z}{y}$$

and then proceeding to higher powers, we have

$$z = \frac{\alpha x}{z} + \frac{\beta y}{z} + \frac{\gamma z}{x} + \frac{\delta z}{y} + \frac{\epsilon x^3}{z^3} + \frac{\zeta x^2 y}{z^3}$$

$$+ \frac{\eta x y^2}{z^3} + \frac{\theta y^3}{z^3} + \frac{\iota x^2}{yz} + \frac{\kappa y^2}{xz} + \frac{\lambda yz}{x^2} + \cdots .$$

396. If this equation is divided by z, then only even powers of z appear, and since $z = \sqrt{x^2 + y^2}$, no irrationality will remain when z is eliminated. There remains a non-irrational equation in x and y. Hence, the general equation has the form of a function in x and y, with degree -1, equated to 1 or some constant. Suppose that P is such a function, so that $C = P$. Hence $\frac{1}{C} = \frac{1}{P}$, but $\frac{1}{P}$ is a first degree equation in x and y. Hence if any first degree equation in x and y is set equal to a constant, the equation will be for a curve which intersects straight lines through the point C in only one point.

397. Let P be a function of degree n in x and y and let Q be a function of degree $n + 1$. Then $\frac{Q}{P}$ is a first degree function, so that all curves with the form $\frac{Q}{P} = c$, that is, $Q = cP$, have this same property.

If we let n be any number, then the general equation for such a curve is

$$\alpha x^{n+1} + \beta x^n y + \gamma x^{n-1} y^2 + \delta x^{n-2} y^3 + \epsilon x^{n-3} y^4 + \cdots$$

$$= c(A x^n + B x^{n-1} y + C x^{n-2} y^2 + D x^{n-3} y^3 + \cdots)$$

From this we find that the lines of the various orders which intersect straight lines through the point C in a single point are as follows.

I.

$$\alpha x + \beta y = c$$

II.

$$\alpha x^2 + \beta xy + \gamma y^2 = c(Ax + By)$$

III.

$$\alpha x^3 + \beta x^2 y + \gamma xy^2 + \delta y^3 = c(Ax^2 + Bxy + Cy^2)$$

IV.

$$\alpha x^4 + \beta x^3 y + \gamma x^2 y^2 + \delta xy^3 + \epsilon y^4$$

$$= c(Ax^3 + Bx^2 y + Cxy^2 + Dy^3)$$

etc.

398. In the first equation a straight line has the property that it can be intersected in at most one point by all lines through a given point. In the second equation we have a general equation for a conic section. Provided the point C lies on the conic section, all lines through C have C in common, so this point it not counted. Since a conic section can be intersected in no more than two points by any straight line, it follows that any straight line through the point C on the curve intersects the curve in only one other point. Curves of higher orders which pass through the point C likewise have the common point of intersection with

all straight lines through C, so once more we no not count this point. Hence in all of the given equations we have those which contain the point C and have only one other point of intersection with any straight line through the point C. It follows that we have classified all of the algebraic curves which have no more than one point of intersection with all lines through the given point C.

399. Next we consider those curves which have either two points of intersection, or none, with every straight line through the point C. There will be no points of intersection if the roots of the equation which give the points of intersection happen to be complex. For any angle $ACM = \phi$, the straight line $CM = z$ will have two points of intersection, if its values are defined by a quadratic equation, so we let $z^2 - Pz + Q = 0$, where P and Q are functions of the angle ϕ or of its sine or cosine. Since the straight line CM is to intersect the curve in no more than the two points M and N, not only must P and Q be single-valued functions of the angle ϕ, but the increase of ϕ by two right angles should not introduce any new intersections. This can be provided for by demanding that P be an odd function of the sine and cosine of the angle ϕ, so that it takes a negative value if the sine and cosine are taken to be negative. The function Q, on the other hand, must be an even function of the same sine and cosine.

400. When we change to rectangular coordinates, $CP = x$ and $PM = y$, we have $\dfrac{y}{z} = \sin \phi$ and $\dfrac{x}{z} = \cos \phi$. Hence, P must be an odd function of $\dfrac{x}{z}$ and $\dfrac{y}{z}$, Q must be an even function of $\dfrac{x}{z}$ and $\dfrac{y}{z}$. From this we conclude that $\dfrac{P}{z}$ will be a non-irrational function of x and y which is also a homogeneous function of degree -1. In a similar way we see that $\dfrac{Q}{z^2}$ is a non-irrational homogeneous function of degree -2 in

x and y. Hence, if L is a homogeneous function of degree $n + 2$, if M is a homogeneous function of degree $n + 1$, and if N is a homogeneous function of degree n, all in x and y, then $\dfrac{M}{L}$ is a convenient function for $\dfrac{P}{z}$, and $\dfrac{N}{L}$ is convenient for $\dfrac{Q}{z^2}$. Since $z^2 - Pz + Q = 0$, we have $1 - \dfrac{P}{z} + \dfrac{Q}{z^2} = 0$, so that the general equation for a curve which is intersected in two points by straight lines through the point C is $1 - \dfrac{M}{L} + \dfrac{N}{L} = 0$, that is $L - M + N = 0$, where $P = \dfrac{Mz}{L}$ and $Q = \dfrac{Nz^2}{L} = \dfrac{N(x^2 + y^2)}{L}$. It follows that P is an irrational function of x and y since $z = \sqrt{x^2 + y^2}$ and Q is a non-irrational function of 0-degree.

401. Hence, now it is easy to find a line of any order which is cut in two points or not at all by any straight line through the given point C. For the second order, we let $n = 0$ to obtain the most general equation for a conic section:

$$\alpha x^2 + \beta xy + \gamma y^2 - \delta x - \epsilon y + \zeta = 0.$$

We take the point C to be anywhere, and any line through C either intersects the conic section in two points or it does not intersect it at all. Nevertheless, it can happen that a certain straight line intersects the curve in only one point, but this occurs in only one or at most two of the infinite number of straight lines through C, so this exception is of no importance. If there is a desire to resolve this paradox, it can be said that the second intersection occurs at infinity, so that this exception can be considered to have no force against our assertion.

402. In order that it may become clear when, precisely, this exception occurs, we transform the equation in x and y into an equation in z

and the angle $ACM = \phi$. Because $y = z \sin \phi$ and $x = z \cos \phi$, the equation becomes

$$z^2(\alpha(\cos \phi)^2 + \beta \sin \phi \cos \phi + \gamma(\sin \phi)^2)$$

$$- z(\delta \cos \phi + \epsilon \sin \phi) + \zeta = 0.$$

From this it becomes clear that if the coefficient of z^2 becomes equal to zero, there will be a single intersection. This occurs when $\alpha + \beta \tan \phi + \gamma(\tan \phi)^2 = 0$. If this equation has two real roots, there will be two cases in which a straight line through C intersects the curve in only one point. Since the roots of this same equation give the asymptotes of the curve, it becomes clear that a straight line through C which is parallel to one of the two asymptotes will intersect the curve in only one point, but there are only two such lines. In the case of a parabola, there is only one straight line, parallel to the axis of the curve, which is the exception. If the conic section is an ellipse, no matter where the point C may be located, every straight line through this point will intersect the curve in two points or not at all.

403. When we let $n = 1$ we obtain the general equation for a line of the third order with this property:

$$\alpha x^3 + \beta x^2 y + \gamma x y^2 + \delta y^3 - \epsilon x^2 - \zeta x y - \eta y^2 + \theta x + \iota y = 0.$$

This equation contains all third order lines, so that every one of these has the property, provided that the point C lies on the curve. If we let $x = 0$, then we also have $y = 0$. In a similar way, for a curve of the fourth order with the desired property, not only must the point C be on the curve, but it must also be a double point. Hence, every fourth order line with a double point satisfies the desired property, provided C is placed at the double point. If the point C is a triple point of the curve, then every straight line through it intersects the curve in a single point, so it belongs to the class of curves we considered at the beginning. In the same way a fifth order line has the property if the point C is placed at a triple point, and so forth. In all of this we should keep in mind that if the straight line through C is parallel to any straight line asymptote or to the axis of a parabolic asymptote, then there will be but a single intersection, since the other has gone to infinity.

404. There is a remarkable correspondence between the order of a line and the number of intersections with a straight line, since from the nature of a line of any order, the maximum number of possible intersections with a straight line is equal to the order of the line. The number of actual intersections is equal to the order if we include the complex intersects and those which have gone to infinity. Of all of these intersections, whether they are real or complex or at infinity, we exclude only those which take place at the point C. Then it becomes clear that since a line of order n is intersected by any straight line in n points, the multiplicity of the point C should be equal to $n - 2$ if the number of intersections is to be equal to 2.

405. After all of this has been noted, we can now easily consider some problems concerning the relationship between the two values of z, CM and CN. We will indicate the usual problems and give either their solution, or show the difficulties in finding the solution. Since the two values of z, namely, CM and CN are roots of the equation $z^2 - Pz + Q = 0$, their sum is equal to P and their product $CM \cdot CN = Q$. To begin, if we require that the sum $CM + CN$ should be constant, then the function P must be a constant. From the nature of the question about all straight lines through C intersecting the curve in two points, it is necessary that $P = \dfrac{Mz}{L} = \dfrac{M\sqrt{x^2 + y^2}}{L}$ (see section 399). But the quantity involving an irrationality cannot be a constant.

It follows that there is no curve which properly satisfies the proposed condition.

406. If we drop the condition that there are only two intersections with any straight line through C, and we seek the curve which has more than two intersections, but among them we have the two at M and N, such that $CM + CN$ is constant, then there are innumerable such curves. We let P be equal to the constant $CM + CN = a$. Then $z^2 - az + Q = 0$, where Q is a function $\dfrac{Nz^2}{L}$. Since this equation contains an irrationality, when we remove it we have $a^2z^2 = (z^2 + Q)^2$, that is, $a^2 = z^2\left(1 + \dfrac{N}{L}\right)^2$, or $a^2L^2 = (x^2 + y^2)(L^2 + 2LN + N^2)$, in which L is homogeneous of degree $n + 2$, while N is homogeneous of degree n in x and y. Hence, the simplest curve which answers this question is obtained by letting $L = x^2 + y^2$ and $N = \pm b^2$, so that $a^2(x^2 + y^2) = (x^2 + y^2 \pm b^2)^2$. This is a complex fourth order line. It is made up of two concentric circles with C as the common center. The next most simple curve which satisfies the condition is of order six, obtained by letting $L = \alpha x^2 + \beta xy + \gamma y^2$ and $N = \pm b^2$. We obtain the equation

$$a^2(\alpha x^2 + \beta xy + \gamma y^2)^2$$

$$= (x^2 + y^2)(\alpha x^2 + \beta xy + \gamma y^2 \pm b^2)^2.$$

We let $\alpha = 1$, $\beta = 0$, and $\gamma = 0$, so that $y^2 + x^2 = \dfrac{a^2x^4}{x^4 \pm 2b^2x^2 + b^4}$,

that is, $y = \dfrac{x\sqrt{a^2x^2 - x^4 - (\pm 2b^2x^4) - b^4}}{x^2 \pm b^2}$.

407. If we exclude, as is quite proper, such solutions in which the straight line through C intersects the curve in more than two points,

there are no curves which satisfy the given condition, and so there is no continuous line such that a straight line through C intersects the line in only the points M and N and such that $CM + CN$ is constant. However, if we ask that the product $CM \cdot CN$ should be constant, then one obvious solution is the circle, no matter where the point C may be placed. There are an infinite number of other curves which have the same property. The function Q must be a constant, which is equal to the product $CM \cdot CN = a^2$. Since $Q = \dfrac{Nz^2}{L}$, which is a non-irrational function in x and y, so there is no difficulty.

408. Let $\dfrac{Nz^2}{L} = a^2$, that is $L = \dfrac{Nz^2}{a^2} = \dfrac{N(x^2 + y^2)}{a^2}$, then all of the curves satisfying the desired conditions are contained in the general equation $\dfrac{N(x^2 + y^2)}{a^2} - M + N = 0$. This equation can also be written as $Ma^2 = N(x^2 + y^2 + a^2)$, where M is a homogeneous function of degree $n + 1$ and N is a homogeneous function of degree n, both in x and y, so that $\dfrac{M}{N} = \dfrac{x^2 + y^2 + a^2}{a^2}$ is a first degree function in x and y. This is a general equation for all curves which are intersected in only two points M and N by straight lines through the point C and the product $CM \cdot CN$ is always equal to the constant a^2.

409. Since $\dfrac{M}{N}$ is a first degree homogeneous function of x and y, the simplest case will be $\dfrac{M}{N} = \dfrac{\alpha x + \beta y}{a}$, from which we obtain the equation $x^2 + y^2 - a(\alpha x + \beta y) + a^2 = 0$, which is always the equation for a circle. Since this is the general equation for a circle in rectangular coordinates, it is clear that a circle satisfies the desired conditions, wherever the point C is located, but this was already known from Euclid's *Elements*. Besides the circle, there is no other conic section

which satisfies the given conditions. From all of the succeeding orders of lines we obtain curves which satisfy the conditions, so that there are an infinite number of such curves. Indeed, in any order, a curve which satisfies these conditions has an equation of the given form. Thus all third order curves with the given conditions have the general equation

$$\frac{\alpha x^2 + \beta xy + \gamma y^2}{a(\delta x + \epsilon y)} = \frac{x^2 + y^2 + a^2}{a^2},$$

that is

$$(\delta x + \epsilon y)(x^2 + y^2) - a(\alpha x^2 + \beta xy + \gamma y^2) + a^2(\delta x + \epsilon y) = 0.$$

In a similar way, from each of the succeeding orders there are lines satisfying the condition.

410. Now we propose to find all curves which are intersected in two points by line through the point C with the further property that the sum of the squares of the two distances is equal to a constant, that is $CM^2 + CN^2 = 2a^2$. Since $CM + CN = P$ and $CM \cdot CN = Q$, we have $CM^2 + CN^2 = P^2 - 2Q$. It follows that $P^2 - 2Q = 2a^2$, or $Q = \dfrac{P^2 - 2a^2}{2}$. Since $P = \dfrac{Mz}{L}$ and $Q = \dfrac{Nz^2}{L}$, we have $\dfrac{2Nz^2}{L} = \dfrac{M^2z^2}{L^2} - 2a^2$. Hence, $N = \dfrac{M^2}{2L} - \dfrac{a^2L}{z^2}$. Since L is a function of degree $n + 2$, M is a function of degree $n + 1$, and N is a function of degree n in x and y, there is no problem with the equation. When we choose L and M in this way, we have the following general equation for curves with the desired properties.

$$L - M + \frac{M^2}{2L} - \frac{a^2L}{z^2} = 0,$$

that is,

$$2L^2(x^2 + y^2) - 2LM(x^2 + y^2) + M^2(x^2 + y^2) - 2a^2L^2 = 0.$$

If we let $M = 0$, we have the equation for a circle with C at the center, and it is obvious that such a circle has the desired properties.

411. We let $n + 1 = 0$, so that M is a constant quantity equal to $2b$, and we let $L = \alpha x + \beta y$. In this way we obtain the fourth order line with the following equation

$$(\alpha x + \beta y)^2(x^2 + y^2 - a^2) - 2b(\alpha x + \beta y)(x^2 + y^2)$$

$$+ 2b^2(x^2 + y^2) = 0.$$

We obtain an equation for a different fourth order line, if we let $L = x^2 + y^2$ and $M = 2(\alpha x + \beta y)a$, then we can divide the equation by $2x^2 + 2y^2$ to obtain

$$(x^2 + y^2)^2 - 2a(\alpha x + \beta y)(x^2 + y^2)$$

$$+ 2a^2(\alpha x + \beta y)^2 - a^2(x^2 + y^2) = 0.$$

Unless the equation is divisible by $x^2 + y^2$, after we substitute $2M$ for M, we have

$$L^2(x^2 + y^2) - 2LM(x^2 + y^2) + 2M^2(x^2 + y^2) - a^2L^2 = 0,$$

which is for a curve of order $2n + 6$. It follows that we have such an equation for curves of every even order. Furthermore, if L is divisible by $x^2 + y^2$, for example $L = (x^2 + y^2)N$, where N is homogeneous of degree n, then we have the other general equation.

$$N^2(x^2 + y^2)^2 - 2MN(x^2 + y^2) + 2M^2 - a^2N^2(x^2 + y^2) = 0,$$

which is for lines of order $2n + 4$. Hence we have two general equation in each even order for curves with the proposed property. We have from order six the two general equation for such curves

$$(\alpha x^2 + \beta xy + \gamma y^2)^2(x^2 + y^2 - a^2)$$

$$- 2a(\delta x + \epsilon y)(x^2 + y^2)(\alpha x^2 + \beta xy + \gamma y^2 - a(\delta x + \epsilon y)) = 0,$$

and

$$(\delta x + \epsilon y)^2(x^2 + y^2)(x^2 + y^2 - a^2) =$$

$$2a(\alpha x^2 + \beta xy + \gamma y^2)((\delta x + \epsilon y)(x^2 + y^2) - a(\alpha x^2 + \beta y^2 + \gamma y^2)).$$

There are no lines of odd order which satisfy the given conditions.

412. If we now want curves such that $CM^2 + CM \cdot CN + CN^2$ is constant, rather than $CM^2 + CN^2$ being constant, or more generally, such that $CM^2 + n \cdot CM \cdot CN + CN^2$, the problem has a similar solution. Since

$$CM^2 + n \cdot CM \cdot CN + CN^2 = P^2 + (n - 2)Q,$$

we let $P^2 + (n - 2)Q = a^2$, so that $Q = \dfrac{a^2 - P^2}{n - 2}$, and this equation offers no difficulties. Since $P = \dfrac{Mz}{L}$ and $Q = \dfrac{Nz^2}{L}$, we have

$$\frac{M^2z^2}{L^2} + \frac{(n - 2)Nz^2}{L} = a^2.$$

Hence, $N = \dfrac{a^2 L}{(n - 2)z^2} - \dfrac{M^2}{(n - 2)L}$. It follows that the curve whose equation is $L - M + N = 0$ will have the property that $CM^2 + n \cdot CM \cdot CN + CN^2 = a^2$. The equation takes the form

$$(n - 2)L^2z^2 - (n - 2)LMz^2 + a^2L^2 - M^2z^2 = 0$$

or, since $z^2 = x^2 + y^2$,

$$a^2L^2 + (x^2 + y^2)((n - 2)L^2 - (n - 2)LM - M^2) = 0,$$

where L is a function of degree $m + 2$, and M of degree $m + 1$ in x

and y. Let N be any homogeneous equation of degree m and let $L = (x^2 + y^2)N$ to obtain a second general equation

$$a^2(x^2 + y^2)N^2 + (n - 2)(x^2 + y^2)^2N^2$$

$$- (n - 2)(x^2 + y^2)MN - M^2 = 0.$$

413. If we let $n = 2$, so that $(CM + CN)^2 = a^2$, then $a^2L^2 = (x^2 + y^2)M^2$, or $M^2 = a^2(x^2 + y^2)N^2$. Since both of these equations are homogeneous, there are at least two equations with the form $\alpha y = \beta x$, so that the conditions cannot be satisfied unless there are at least two straight lines through the point C. The fact that this cannot happen, in the sense of the proposed condition, makes it clear that the problem has no solution. But we have already seen that there is no solution to the problem when $CM + CN = a$. Suppose now that $n = -2$, so that the square of the difference, $(CN - CM)^2$ is constant, which means that the distance MN must be constant. In this case we have the two equation

$$a^2L^2 = (x^2 + y^2)(2L - M)^2$$

and

$$a^2(x^2 + y^2)N^2 = (2(x^2 + y^2)N - M)^2,$$

so that we have the simplest solution when we let $N = 1$ and $M = 2bx$. Thus we obtain

$$a^2(x^2 + y^2) = 4(x^2 + y^2 - bx)^2,$$

or, when we let $a^2 = 8c^2$,

$$(x^2 + y^2)^2 = 2(c^2 + bx)(x^2 + y^2) - b^2x^2.$$

Hence, $x^2 + y^2 = c^2 + bx \pm c\sqrt{c^2 + 2bx}$, and

$$y = \left(c^2 + bx - x^2 \pm c \sqrt{c^2 + 2bx} \right)^{\frac{1}{2}}.$$

414. It follows that there are innumerable curves which, when cut by a straight line through C in the two points M and N, have the interval MN always constant. This condition is clearly satisfied by a circle with its center at C, since in that case the interval MN is always a diameter of the circle. We obtain a circle from the general equations when we let $M = 0$. Then, after the circle, there are fourth order lines with the equations $a^2(x^2 + y^2) = 4(x^2 + y^2 - bx)^2$ and also $a^2x^2 = (x^2 + y^2)(2x - 2b)^2$. In order that we may more easily recognize these curves, we transform the equations into equations in z and the angle ϕ. Since $x^2 + y^2 = z^2$, $x = z \cos \phi$, and $y = z \sin \phi$, when we let $a = 2c$, the first equation takes the form

$$c^2z^2 = (z^2 - bz \cos \phi)^2,$$

that is,

$$b \cos \phi \pm c = z.$$

From the second equation we have

$$c^2(\cos \phi)^2 = (z \cos \phi - b)^2,$$

that is,

$$z = \frac{b}{\cos \phi}.$$

From these equation, the construction of the curves is easily seen.

415. In order to construct the curve from the equation $z = b \cos \phi \pm c$, we draw the straight line ABC through C, on which we take $CD = b$, then from D we measure the distance c in both directions in order to find the point A and B such that $DA = DB = c$. The

points A and B are the first points on the desired curve. Then we draw any straight line NCM through C and drop the perpendicular DL from D to that line. From the point L we measure the distance c in both directions to find the points M and N on the curve, where $LM = LN = c$. It follows that we always have the length of the interval $MN = 2c$, as is required. If $CD = b < c$, then the curve has a conjugate point at C as in *figure 83*. If $b = c$, then the curve has a cusp at C, when the interval AC vanishes, as in *figure 84*. Finally, if $C < b$, the point A lies between C and B, the curve has a node, that is a double point, at C, as in *figure 85*. Furthermore, the diameter of all these curves is the straight line ACB, and the normal line ECF has a length equal to $2c$.

416. Besides these fourth order curves which remain in a bounded region, there are others of the same order which go to infinity. These have the equation $z = \dfrac{b}{\cos \phi} \pm c$. The construction, as depicted in *figure 86*, is as follows. We draw the principal straight line CAB through C and take $CD = b$ and $DA = DB = c$. The points A and B lie on the curve. Then through D we draw the normal line EDF, and take any straight line CL, where L lies on the normal line, so that $\dfrac{b}{\cos \phi} = CL$ where the angle $DCL = \phi$. Finally we measure the distance c in either direction from L to obtain the points M and N on the desired curve, where $LM = LN = c$. From this construction it is clear that the curve is the ancient *conchoid* of Nicomedes, with its pole at C and the straight line EF is an asymptote to which the four branches of the curve converge at infinity. The part hBh is called the *exterior* of the conchoid and gAg is the *interior*. Besides these parts, there is also a conjugate point at C.

417. These curves are fourth order lines which satisfy the condition. It is easy to find as many other curves as may be desired from higher orders. Indeed, if P is an odd function of the sine and cosine of the angle ϕ, then the equation $z = bP \pm c$ presents a continuous curve which is intersected in the two points M and N by every straight line through C, and the length of the interval MN is equal to the constant $2c$. All of these curves can be referred to as belonging to the conchoidal genus, where instead of the straight line directrix EF, we substitute any curve with the equation $z = bP$. We previously saw that a curve defined by this equation is intersected by straight lines through C in only one point. Hence, due to the arbitrary length c, from a single curve $z = bP$ there are innumerable curves which can be drawn.

418. For instance, let $CEDLF$ be an arbitrary curve, shown in *figure 87*, which is intersected in only one point D or L by any straight line through C. Then on each such straight line CL we take on either side of L the equal intervals $LM = LN = c$. The continuous curve, $AMPCQBNRC$, is described by the points M and N on all the straight lines through C, and the length of the interval MN is always equal to $2c$. We note that if the curve $CEDF$ is a circle through the point C, then the curve described will be the same fourth order curve which we found before in section 414.

419. We have answered the question about such curves AMN which are intersected in two points M and N by straight lines through C and have the property that $CN - CM$ or $CM^2 - 2CM \cdot CN + CN^2$ is always a constant. There has been little consideration given to the case where $CM^2 + CM \cdot CN + CN^2$ is constant. Now we let $n = 1$ in the equation obtained in section 412, so that

$$a^2 L^2 = (x^2 + y^2)(L^2 - LM + M^2),$$

where L is of degree $m + 1$ and M is of degree m in x and y. We also have the following equation

$$a^2(x^2 + y^2)N^2 = (x^2 + y^2)^2 N^2 - (x^2 + y^2)MN + M^2,$$

in which M is homogeneous and one degree higher than N in x and y.

420. First we note that if we let $M = 0$, then we have a circle whose center is at C. Since every straight line from the center to the curve is equal, the circle satisfies all conditions of this kind. For the present case, after the circle, the simplest case occurs when we let $M = b$ and $L = x$, so that

$$a^2 x^2 = (x^2 + y^2)(x^2 - bx + b^2)$$

that is,

$$y^2 = \frac{x^2(a^2 - b^2 + bx - x^2)}{b^2 - bx + x^2}.$$

If we let $N = 1$ and $M = bx$ in the other equation we obtain an equation for another fourth order line.

$$a^2(x^2 + y^2) = (x^2 + y^2)^2 - bx(x^2 + y^2) + b^2 x^2,$$

that is

$$x^2 + y^2 = \tfrac{1}{2}bx + \tfrac{1}{2}a^2 \pm \sqrt{\frac{1}{4}a^4 + \frac{1}{2}a^2 bx - \frac{3}{4}b^2 x^2}$$

which also satisfies the given condition.

421. Now that we have solved these problem, we move on to a consideration of higher powers of the two values of z in the equation $z^2 - Pz + Q = 0$, where $P = \dfrac{Mz}{L}$ and $Q = \dfrac{Nz^2}{L}$, where L is a homogeneous function of degree $n + 2$, M is of degree $n + 1$, and N of

degree n in x and y. We let the abscissa $CP = x$, and the ordinate $PM = y$. The proposed question is concerned with the two intersections M and N with the required property $CM^3 + CN^3 = a^3$. From the nature of the equation $z^2 - Pz + Q = 0$, we have $CM^3 + CN^3 = P^3 - 3PQ$, so that $P^3 - 3PQ = a^3$. Since P^3 and PQ are irrational quantities, there is no solution. There is no solution as long as we demand only two intersections. If we allow two or more intersections, then there are infinitely many different solutions; we let $Q = \dfrac{P^3 - a^3}{3P}$ and choose P to be any function of the sine and cosine of the angle $ACM = \phi$.

422. If we desire a curve such that $CM^4 + CN^4 = a^4$, then we let $P^4 - 4P^2Q + 2Q^2 = a^4$. Since there is no irrationality in this equation, there is no problem. We let $Q = P^2 + \sqrt{\frac{1}{2}P^4 + \frac{1}{2}a^4}$ where we note the sign of the radical. We can consider this function to be single-valued, since if $\sqrt{\frac{1}{2}P^4 + \frac{1}{2}a^4}$ were taken with the negative sign, the resulting values of z would be complex. Hence we have $\dfrac{Nz^2}{L} = M^2z^2 + \sqrt{\dfrac{M^4z^4}{2L^4} + \frac{1}{2}a^4}$. since for the curve we have $L - M + N = 0$ or $z^2 - \dfrac{Mz^2}{L} + \dfrac{Nz^2}{L} = 0$, it follows that $z^2 - \dfrac{Mz^2}{L} + \dfrac{M^2z^2}{L^2} + \sqrt{\dfrac{M^4z^4}{2L^4} + \frac{1}{2}a^4} = 0$. When the irrationality is removed we have

$$\frac{z^4}{L^4}(L^2 - LM + M^2)^2 = \frac{M^4z^4}{2L^4} + \frac{1}{2}a^4,$$

that is

$$(x^2 + y^2)^2(2(L^2 - LM + M^2)^2 - M^4) = a^4L^4,$$

which is the desired general equation for the curves satisfying the given condition.

423. There is a different, easier method, which was given in section 372, and it can be used to answer this and similar questions. Since $CM \cdot CN = Q$, if one of these ordinates is called z, then the other is

equal   to $\dfrac{Q}{z} = \dfrac{Nz}{L}$,   since $Q = \dfrac{Nz^2}{L}$.   Hence,   if   we   want

$CM^n + CN^n = a^n$, then $z^n + \dfrac{N^n z^n}{L^n} = a^n$, so that $z^n = \dfrac{a^n L^n}{L^n + N^n}$.

If n is even, then the equation involves no irrationality and it satisfies the desired conditions. If n is odd, we have to take squares to remove the irrationality, but this doubles the number of intersections, so that in the strict sense of only two intersections, there is no solution. Thus, if

we   want $CM^2 + CN^2 = a^2$,   then $z^2 = x^2 + y^2 = \dfrac{a^2 L^2}{L^2 + N^2}$.   This

agrees   with   what   we   found   above   in   section   410,   that   is,

$x^2 + y^2 = \dfrac{a^2 L^2}{(L - M)^2 + L^2}$, since $L - M + N = 0$. In general, if

$CM^n + CN^n = a^n$, with n even, we obtain the equation

$$z^n = \left(x^2 + y^2\right)^{\frac{n}{2}} = \frac{a^n L^n}{L^n + N^n} = \frac{a^n L^n}{L^n + (L - M)^n},$$

where L has degree $m + 2$, M has degree $m + 1$, and N has degree m, all function in x and y.

424. The same solution is arrived at if we begin with the sum $CM + CN = P$. If we let either one of CM and CN be equal to z, then the other is equal to $P - z$. Hence, if $CM^n + CN^n$ is to be equal

to a constant, then $z^n + (P - z)^n = a^n$. We have seen that $P = \dfrac{Mz}{L}$

and $Q = \dfrac{Nz^2}{L}$ are necessary conditions just as is $L - M + N = 0$. It

follows   then,   that $z^n + \dfrac{z^n (M - L)^n}{L^n} = a^n$,   that   is

$$z^n = \frac{a^n L^n}{L^n + (M - L)^n} \quad \text{or} \quad z^n = \frac{a^n L^n}{L^n + N^n}.$$ If we eliminate L, then

we have $z^n = \dfrac{a^n (M - N)^n}{(M - N)^n + N^n}$. When n is even, these equations give

curves which satisfy the desired conditions. However, if n is odd, there

are indeed two intersections M and N such that $CM^n + CN^n = a^n$,

but in addition, there are two other intersections with the same property,

so that any straight line through C, satisfies the conditions in two

different ways simultaneously.

425. With these discussions behind us, we can now answer some

rather difficult questions with relative ease. Suppose we want to find a

curve such that any straight line through C intersects it in two points M

and N and

$$CM^n + CN^n + \alpha\, CM \cdot CN (CM^{n-2} + CN^{n-2})$$

$$+ \beta\, CM^2 \cdot CN^2 (CM^{n-4} + CN^{n-4}) + \cdots = a^n.$$

We let one of the ordinates $CM = z$, and the other $CN = \dfrac{Q}{z} = \dfrac{Nz}{L}$.

When we substitute these values we obtain the general equation for the

desired curve,

$$z^n (L^n + N^n + \alpha LN(L^{n-2} + N^{n-2}) + \beta L^2 N^2 (L^{n-4} + N^{n-4}) + \cdots)$$

$$= a^n L^n.$$

Since $L - M + N = 0$, and L, M, and N are homogeneous functions of

x and y with respective degrees $m + 2$, $m + 1$, and m, it follows that

when we let $L = M - N$ or $N = M - L$, we have an infinite number

of solutions for the given problem.

426. Now we move on to the investigation of curves which are

intersected in three points by straight lines through a fixed point C. We express the general equation for such a curve as follows,

$$z^3 - Pz^2 + Qz - R = 0,$$

where z designates the distance of any point of the curve from c, and P, Q, and R are functions of the angle $ACM = \phi$ or of its sine and cosine. For the reasons given before, in order to insure no more than three intersections, P and R must be odd functions of $\sin \phi$ and $\cos \phi$, while Q must be an even function. When we change to rectangular coordinates $CP = x$ and $PM = y$, so that $x^2 + y^2 = z^2$. We let K, L, M, and N denote homogeneous functions in x and y of degrees $n + 3$, $n + 2$, $n + 1$, and n respectively. Then $P = \dfrac{Lz}{K}$, $Q = \dfrac{Mz^2}{K}$, and $R = \dfrac{Nz^3}{K}$, so that we have the following general equation for such curves in rectangular coordinates,

$$K - L + M - N = 0.$$

From this equation it becomes clear that the point C will be a multiple point with multiplicity equal to n.

427. First we note that all third order lines satisfy this equation if we take the point C distinct from the curve. Furthermore this equation contains all fourth order lines provided the point C lies on the curve. In the third place, all fifth order lines in which there is a double point satisfying this equation provided C is chosen as one of the double points. In a similar way, lines of higher order satisfy this condition provided the curve has a multiple point with multiplicity equal to n when the order of the line is $n + 3$.

428. Let p, q, and r be the three values of z which we obtain from the equation $z^3 - Pz^2 + Qz - R = 0$ when we assign any value to the angle $CAM = \phi$. From the nature of the equation we have $P = p + q + r$, $Q = pq + pr + qr$, and $R = pqr$. Since P and R

cannot be represented rationally by x and y, it is clear that there are no curves of the kind we are considering in which $p + q + r$ or pqr are constant. Nor is there such a curve when an odd function of p, q, and r is set equal to a constant. There is no problem when we set even function of p, q, and r equal to a constant. Hence, if we require that

$$pq + pr + qr = a^2, \qquad \text{then} \qquad Q = \frac{Mz^2}{K} = a^2, \qquad \text{so} \qquad \text{that}$$

$M(x^2 + y^2) = a^2 K$. When we substitute this value for K in the equation $K - L + M - N = 0$, we obtain the general equation for all curves with this property,

$$M(x^2 + y^2) - a^2 L + a^2 M - a^2 N = 0.$$

When we eliminate M, the equation has the form

$$(x^2 + y^2)K - (x^2 + y^2)L + a^2 K - (x^2 + y^2)N = 0.$$

429. In the same way, similar questions can easily be answered. Suppose we want to find a curve which the straight lines through C intersect in the three points and $p^2 + q^2 + r^2 = a^2$. Since

$$p^2 + q^2 + r^2 = P^2 - 2Q \quad \text{and} \quad P = \frac{Lz}{K} \quad \text{and} \quad Q = \frac{Mz^2}{K}, \quad \text{we have}$$

$$\frac{L^2 z^2}{K^2} - \frac{2Mz^2}{K} = a^2, \quad \text{that is} \quad (x^2 + y^2)L^2 - 2(x^2 + y^2)KM = a^2 K^2.$$

Since for all curves with three intersections we have the general equation $K - L + M - N = 0$, where the essential fact is that the degree in x and y of K is three more than the degree of N. Hence, for a simultaneous solution to these two equation, we first multiply the equation $K - L + M - N = 0$ by $2(x^2 + y^2)K$, so that M can be eliminated. The result is

$$2(x^2 + y^2)K^2 - 2(x^2 + y^2)KL + (x^2 + y^2)L^2$$

$$- a^2K^2 - 2(x^2 + y^2)KN = 0.$$

The member of this equation with the highest degree is $2(x^2 + y^2)K^2$, and that degree is $2n + 8$ in x and y. The member with the lowest degree is $2(x^2 + y^2)KN$, which has degree $2n + 5$.

430. Since neither the member with highest degree nor that with lowest degree can vanish, in order to find the simplest such curve, we let $n = 0$, $N = b^3$, $K = x(x^2 + y^2)$, and $L = 0$, which gives the equation

$$2(x^2 + y^2)^3x^2 - a^2x^2(x^2 + y^2)^2 - 2b^3x(x^2 + y^2)^2 = 0.$$

If we divide this equation by $2x(x^2 + y^2)^2$, we obtain

$$x(x^2 + y^2) - \tfrac{1}{2}a^2x - b^3 = 0,$$

which is of order three. If we had not made $L = 0$, but rather $L = 2c(x^2 + y^2)$ we should have had the following equation for a fourth order curve;

$$x^2(x^2 + y^2) - 2cx(x^2 + y^2) + 2c^2(x^2 + y^2) - \tfrac{1}{2}a^2x^2 - b^3x = 0,$$

that is,

$$x^2(x^2 + y^2) + (2c - x)^2(x^2 + y^2) = a^2x^2 + 2b^3x.$$

In a similar way we can obtain many other curves of higher order which satisfy the desired conditions.

431. Next, we can find curves in which $p^4 + q^4 + r^4$ is a constant. Since $p^4 + q^4 + r^4 = P^4 - 4P^2Q + 2Q^2 + 4PR$, we let $P^4 - 4P^2Q + 2Q^2 + 4PR = c^4$. Then

$$z^4(L^4 - 4KL^2M + 2K^2M^2 + 4K^2LN) = c^4K^4,$$

so that

$$4K^2LNz^4 = c^4K^4 - z^4(L^4 - 4KL^2M + 2K^2M^2).$$

From this equation we find the value of N which is then substituted into the equation $K - L + M - N = 0$, to obtain the general equation for curves which satisfy this condition.

432. It is possible simultaneously to satisfy the conditions $p^4 + q^4 + r^4 = c^4$ and $p^2 + q^2 + r^2 = a^2$. For this to be true we must have $z^2L^2 - 2z^2KM = a^2K^2$, so that $2z^2KM = z^2L^2 - a^2K^2$. Since

$$4K^2LNz^4 = c^4K^4 - L^4z^4 + 4KL^2Mz^4 - 2K^2M^2z^4$$

we have

$$4K^2LNz^2 = c^4K^4 + L^4z^4 - 2a^2K^2L^2z^2 - 2K^2M^2z^4,$$

and we also have

$$4K^2LMz^4 = 2KL^3z^4 - 2a^2K^3Lz^2.$$

When we substitute these values for M and N in the equation $K - L + M - N = 0$, or the equation

$$4K^3Lz^4 - 4K^2L^2z^4 + 4K^2LMz^4 - 4K^2LNz^4 = 0,$$

we obtain the following equation for the curve:

$$4K^3Lz^4 - 4K^2L^2z^4 + 2KL^3z^4 - 2a^2K^3Lz^2$$

$$- c^4K^4 - L^4z^4 + 2a^2K^2L^2z^2 + 2K^2M^2z^4 = 0.$$

Since

$$KMz^2 = \tfrac{1}{2}L^2z^2 - \tfrac{1}{2}a^2K^2$$

we have

$$2K^2M^2z^4 = \tfrac{1}{2}L^4z^4 - a^2K^2L^2z^2 + \tfrac{1}{2}a^4K^4.$$

It follows that the general equation for the desired curves is

$$8K^3Lz^4 - 8K^2L^2z^4 + 4KL^3z^4 - 4a^2K^3Lz^2$$

$$- 2c^4K^4 - L^4z^4 + 2a^2K^2L^2z^2 + a^4K^4 = 0.$$

433.   Since K is a homogeneous function of x and y of one degree greater than the degree of L, the simplest curve with three intersections and the simultaneous conditions $p^2 + q^2 + r^2 = a^2$ and $p^4 + q^4 + r^4 = c^4$ can be found by letting $K = z^2$ and $L = bx$. The resulting equation is

$$8bxz^6 - 8b^2x^2z^4 + 4b^3x^3z^2 - 4a^2bxz^4$$

$$- 2c^4z^4 - b^4x^4 + 2a^2b^2x^2z^2 + a^4z^4 = 0.$$

Since $z^2 = x^2 + y^2$, there is no irrationality in the equation, and we have a seventh order line, whose point C is a quadruple point. Another seventh order line can be obtained by letting $K = x$ and $L = b$. In this case we have

$$8bx^3z^4 - 8b^2x^2z^4 + 4b^3xz^4 - 4a^2bx^3z^2$$

$$- 2c^4x^4 - b^4z^4 + 2a^2b^2x^2z^2 + a^4x^4 = 0,$$

that is,

$$z^4 = \frac{4a^2bx^3z^2 - 2a^2b^2x^2z^2 + 2c^4x^4 - a^4x^4}{8bx^3 - 8b^2x^2 + 4b^3x - b^4}.$$

It follows that

$$z^2 = \frac{2a^2bx^3 - a^2b^2x^2}{b(2x - b)(4x^2 - 2bx + b^2)}$$

$$\frac{\pm \, x^2\sqrt{(2bx - b^2)(2c^4(b^2 - 2bx + 4x^2) - 2a^4(b^2 - 2bx + 2x^2))}}{b(2x - b)(4x^2 - 2bx + b^2)}$$

434. We could proceed to the equation for curves which are intersected in four points by straight lines through the point C, and then to find particular such curves which satisfy certain conditions. However, if careful attention has been paid to the preceding arguments, there are no further difficulties to be encountered, and everything of this sort which could be desired can be found with very little labor. Except for the question about the existence of genuine solutions, the answers are immediate. For this reason we delay no longer on this material, but proceed to other ways of obtaining knowledge concerning curves.

CHAPTER XVIII

On the Similarities and Affinities of Curves.

435. In every equation for a curve, besides the rectangular coordinates x and y, there must be included constants, either one or more, such as a, b, c, etc. We call these the thread constants, which with the variables x and y will everywhere have the same thread degree. If in one term we have the product of n threads, there must be the same number of threads multiplied in each term otherwise we would have to compare heterogeneous quantities, which cannot be done. Hence, in every equation for a curve the thread constants a, b, c, etc. with the variables x and y everywhere have the same degree, unless by chance one of the thread constants is equal to 1 or is expressed by some other absolute number. After having made this remark, if no thread constants appear in an equation, then in each term the variables x and y have the same degree, so that they constitute a homogeneous function. We noted before that this kind of equation does not give a curve, but rather several straight lines which all intersect each other in a single point.

436. First we consider an equation with only one thread constant a besides the two variables x and y. Then there are three threads, a, x, and y with the same degree in each term of the equation. Hence, this kind of equation, according to the different values given to the thread variable, can give an infinite number of different curves. But these curves differ from one another only in their size, otherwise they are completely similar. Hence, all curves which can be included in such an equation are properly said to be similar to each other, and there is no other real difference between them than the difference we understand in circles with different radii.

437. In order that we may understand this similarity more clearly, let us consider a specific equation which has, besides the variables x and y, with a single thread constant a which we can call a *parameter:*

$$y^3 - 2x^3 + ay^2 - a^2x + 2a^2y = 0.$$

Let AC be the value of the parameter a in *figure 88,* and let AMB be a curve given by this equation when we take the straight line AB for the axis. We let $AP = x$ and $PM = y$ be the coordinates. When we give the parameter a any other value $ac = a$, as in *figure 89,* the curve which the equation gives becomes amb. We say that these two curves AMB and amb are similar to each other. If we keep $AC = a$, $AP = x$, and $PM = y$, and let $ac = \frac{1}{n}AC = \frac{a}{n}$, then it follows that $ap = \frac{1}{n}AP = \frac{x}{n}$ and $pm = \frac{1}{n}PM = \frac{y}{n}$. If we now substitute $\frac{a}{n}$, $\frac{x}{n}$, and $\frac{y}{n}$ for a, x, and y respectively, we then have each term divided by n^3, so that the same equation results.

438. Similar curves have the following property, from which the nature of similarity is all the more apparent, that when the abscissas AP and ap are taken in the same ratio as the parameters AC and ac, then the ordinate PM and pm are also in the same ratio. That is if we take $AP/ap = AC/ac$, then also $PM/pm = AC/ac$. Since then $AP/PM = ap/pm$, the curves are similar to each other in the geometric

sense, and except for the size, have the same properties. Indeed, if we take the abscissas AP and ap homologous, that is in the same ratio as the parameters AC and ac, not only will the ordinates PM and pm preserve the same ratio of the parameters, but also all other lines which are similarly drawn, so that also the arcs of the curves AM and am are also in the same ratio as AC and ac. It also follows that the similar areas APM and apm will be as the square of the ratio of the parameters. That is as AC^2 to ac^2. Furthermore, if we take any two homologous points O and o, so that $AO/ao = AC/ac$, and the angles $AOM = aom$, with the straight lines OM and om drawn to the respective curves, then we also have $OM/om = AC/ac$. Finally, because of the similarity, the tangents at the homologous points M and m will make equal angles with the respective axes, and the osculating radii will be in the same ratio as the parameters AC and ac.

439. It follows that all circles with the equation $y^2 = 2ax - x^2$ are similar figures. Likewise all curves with the equation $y^2 = ax$, that is all parabolas, are similar figures. From equations of this kind by which we see that curves are similar, since the coordinates x and y and the parameter a have the same degree in each term, when y is evaluated, it is equal to a first degree homogeneous function in x and a. Conversely, if P is a first degree homogeneous function in x and a, then $y = P$ is the equation for innumerable different but similar curves, each of which is obtained by assigning a value to the parameter a. In a similar way, from this kind of equation for similar curves, x is equal to a first degree homogeneous function in a and y, and the parameter a is equal to a first degree homogeneous function in x and y.

440. Given any curve AMB, there are infinitely many other curves amb which are similar and they can easily be found in the following way. We take any ratio which the homogeneous parts are to have; we take

this ratio to be $1/n$. If the given curve AMB is referred to the axis AB by rectangular coordinates AP and PM, then on the similar axis ab we choose the abscissa ap so that $AP/ap = 1/n$ and at p we erect the perpendicular pm so that $PM/pm = 1/n$. Then the point m is on the similar curve amb, so that the points M and m are homologous. An alternate method is to choose any fixed point O and then find a similar fixed point o, then with the angle $aom = AOM$, we cut off the distance om so that $OM/om = 1/n$, then the points m is on the similar curve amb. In this way, for any ratio $1/n$ chosen arbitrarily, a similar curve can be drawn. This is the usual way of constructing mechanical instruments for drawing similar figures of any desired size.

441. If the nature of a given curve AM is expressed by some equation in the coordinates $AP = x$ and $PM = y$, then it is easy to find the equation for the curve am. Let the homologous abscissa $ap = X$ and the ordinate $pm = Y$, then from the construction $x/X = 1/n$ and $y/Y = 1/n$, so that $x = \dfrac{X}{n}$ and $y = \dfrac{Y}{n}$. When these values are substituted into the original equation in x and y, we obtain an equation in X and Y for similar curves. If in the new equation we consider only the coordinates X and Y and the letter n, then the degree of the equation is zero for each term. If we remove the fractions from the equation by multiplying by the proper power of n, then we obtain an equation in which each term has the same degree in X, Y, and n. We saw before that in every equation for similar curves the two coordinates and the constant, which varies to give the different similar curves, appear everywhere in the same degree, and this is the criterion for equations of similar curves.

442. Just as in similar curves the homologous abscissas and ordinates increase and decrease in the same ratio, so if the abscissas follow one ratio and the ordinates follow a different ratio, the curves are no

longer similar. Nevertheless curves with this property do have some relationship to each other, and we say that the two curves are *affine*. In some way affinity includes similarity as a kind of species, since affine curves become similar when the two ratios, which the abscissas and ordinates follow respectively, become equal to each other. From any given curve AMB there are infinitely many different affine curves amb which we can find as follows. We choose the abscissa ap in such a way that $AP/ap = 1/m$, then the ordinate pm is constructed so that $PM/pm = 1/n$. By changing the ratios $1/m$ and $1/n$, either one, the other, or both, we obtain innumerable other curves which are affine to the original curve AMB.

443. Let the nature of the given curve AMB be expressed by some equation in the rectangular coordinates $AP = x$ and $PM = y$. According to the description already given the affine curve amb has an equation in the coordinates $ap = X$ and $pm = Y$. Since $x/X = 1/m$ and $y/Y = 1/n$, we have $x = \dfrac{X}{m}$ and $y = \dfrac{Y}{n}$. If these values are substituted into the equation in x and y, we obtain a general equation in X and Y for all curves affine to the original curve. In order that we may investigate more carefully the nature of this equation, let us suppose that the equation for the given curve AMB is such that the ordinate y is equal to some function P of x, so that $y = P$. Now if we substitute $\dfrac{X}{m}$ for x in P, then P becomes a zero degree function in X and m, so that the general equation for the affine curves will be such that $\dfrac{Y}{n}$ is equal to a zero degree function in X and m. That is, a zero degree function of Y and n is equal to a zero degree function of X and m.

444. There is a distinction between similar and affine curves which it is quite important to note. Curves which are similar with respect to

one axis or a fixed point, are also similar with respect to any other axis or homologous point. On the other hand, affine curves are such only with respect to their axis to which they are referred, not with respect to any arbitrary axis or homologous points to which their affinity might be referred. We must also note that just as all similar curves belong to the same genus of lines, so also all curves which are affine to each other belong to the same order and genus. In order that these remarks may be clarified it will be useful to give some examples of same better known curves to illustrate what has been noted.

445. Let the given curve be a circle related to its diameter, whose nature is given by the equation $y^2 = 2cx - x^2$. We let $x = \dfrac{X}{n}$ and $y = \dfrac{Y}{n}$, so that the resulting equation in X and Y is the general equation for all similar curves. This equation is $\dfrac{Y^2}{n^2} = \dfrac{2cX}{n} - \dfrac{X^2}{n^2}$, that is, $Y^2 = 2ncX - X^2$. From this equation it is clear that all curves similar to the circle are circles with diameters equal to $2nc$, which distinguishes one from the other. In order to find the curves which are affine to the circle, we substitute $x = \dfrac{X}{m}$ and $y = \dfrac{Y}{n}$, to obtain $\dfrac{Y^2}{n^2} = \dfrac{2cX}{m} - \dfrac{X^2}{m^2}$, that is, $m^2 Y^2 = 2mn^2cX - n^2X^2$, which is the general equation for an ellipse related to one of its principal axis. In this way we see that all ellipsis are curves which are affine to a circle. It follows that all ellipses are affine to each other. In a similar way it can be seen that all hyperbolas are affine to each other. Ellipses in which the ratio of the two principal axis are equal are similar to each other; the same is true for hyperbolas.

446. As to a parabola expressed by the equation $y^2 = cx$, it is clear that all similar curves are also parabolas, so that all parabolas are

similar to each other. However, if we consider curves affine to the parabola, we substitute $y = \dfrac{Y}{n}$ and $x = \dfrac{X}{m}$ to obtain $Y^2 = \dfrac{n^2 c}{m} X$, which is also a parabola, so that it is clear that the curves which are affine to the parabola are also similar to the parabola. It follows that in this case that similarity is the same as affinity. The same phenomenon occurs in all curves whose equations have only two terms, such as $y^3 = c^2 x$, $y^3 = cx^2$, $y^2 x = c^3$, etc. For these curves, both the parabolic and hyperbolic, affinity and similarity are the same concept, which is not true for other genera, as we have already noted with regard to circles and ellipses.

447. Insofar as we include the constants a, b, c, etc. in an equation in x and y, if we assign particular values to each constant, then a single curve is determined. If one of the constants, for example a, is allowed to change, and successive different values are assigned to a, we obtain different curves for each different value, all together we have an infinite number of different curves, which will be similar if none of the other constants besides a are in the equation, otherwise the curves will not be similar. If besides a, we allow another constant b to change, then because of the mutability of b with each fixed value of a we have an infinite number of curves. It follows that when we allow both a and b to vary, we have an infinity of infinities of different curves. If in addition we take a third variable constant c, then we have another infinity of curves for a fixed a and b. Thus, the more constants we have which are allowed to change, the higher the power of infinities of curves we have.

448. Let us consider a bit more carefully the infinity of curves which arise from a single equation when we let only one of the thread constants vary. An equation of this kind, provided the axis and origin remain the same, not only gives an infinity of curves, but also gives the

position of each curve. It follows that the infinite number of curves fills up the whole space, so that for each point there is a curve from the infinity of curves which passes through that point. Depending on how the equation is defined, the infinity of curves will be similar or dissimilar, which can be judged by what was said previously. It can also happen that all the curves are not only similar, but also equal, differing only in position. For instance, the equation $y = a + \sqrt{2cx - x^2}$ gives an infinite number of circles with equal radii equal to c and centers all lying on the perpendicular to the axis, when a is allowed to change.

449. Conversely, if the same curve is drawn in infinitely many different positions in the plane according to some law, we can find an equation whereby the infinity of equal curves can be given by allowing a single constant to change. Let the curve shown in *figure 90* which is to be situated in an infinite number of positions be the circle with radius equal to c. The infinite number of curves are to be drawn in such a way that the vertices A, and a lie on the given curve AaL which is called the *directrix*. The diameter ab is to remain parallel to the axis AB. In order to find the equation for the infinity of circles, let us take any point a on the directrix. From this point we drop the perpendicular aK to the principal axis. We let $AK = a$, and because of the given directrix, we have Ka through the point a. We let $Ka = A$ so that A is some function of the given a. Then from the point a we draw ab parallel to the principal axis, and this will be the diameter of the circle which has its vertex at the point a. We choose any point m of this new circle and draw the ordinate $mP = y$ corresponding to the abscissa $AP = x$. Then we have $ap = x - a$ and $pm = y - A$. When we let $ap = t$ and $pm = u$, from the nature of a circle we have $u^2 = 2ct - t^2$. Now since $t = x - a$ and $u = y - A$ we have

$$(y - A)^2 = 2c(x - a) - (x - a)^2$$

which is the general equation for all the circles along the directrix AaL in the sense already described. All of these circles come from this equation if the tread a, upon which A also depends, is allowed to vary.

450. In a similar way, if instead of the circle, some other curve amb is moved along the directrix AaL so that its vertex, or origin, a is on the directrix and the axis ab always remains parallel, then the same curve is drawn an infinite number of times and the equation can be found whereby these curves are all represented simultaneously. Let the nature of the curve be represented by an equation in the coordinates $ap = t$ and $pm = u$, and choose the straight line AB parallel to ab as the principal axis to which all curves are to be referred. The principal axis is also the axis for the directrix AaL. Now, as before, we let $AK = a$, and $Ka = A$, so that A is a function of a. We let the abscissa $AP = x$ and the ordinate $Pm = y$, so that $t = x - a$ and $u = y - A$. If these values are substituted for t and u in the given equation, we obtain the general equation for all the curves amb taken together. For each different value of a we have a single one of the infinite number of curves amb, and as a moves along the directrix we obtain all of the curves. For instance, if the curve amb is a parabola with the equation $u^2 = ct$, then the infinite number of equal parabolas whose vertices lie along the directrix AaL with axes parallel to the straight line AB are represented by the equation $(y - A)^2 = c(x - a)$.

451. Just as we have supposed that the vertex A of the curve has been moved along the given directrix curve in such a way that its axis always remains parallel, so a much more general equation can be found for the same curve drawn in the given plane an infinite number of times according to some law. In order that this move become clearer, we

suppose first that the vertex A of the curve travels along the circumference of a circle Aa, as in *figure 91*, in such a way that the axis of the
curve ab is always directed through the center of the circle O. The
rotary motion of the curve AMB with the axis BAO around the point O
gives all of the infinite number of the same curve AMB and they all
should be represented by a single equation with some variable constant.

452. We require that the radius be fixed, so that $AO = aO = c$
and we let the angle $AOa = \alpha$ which is allowed to vary. From any one
of these curves amb we choose some point m and drop the perpendicular
mP to the principal axis. We let $OP = x$ and $Pm = y$. Then we also
drop the perpendicular mp to the axis ab of the curve amb, then we let
$ap = t$ and $pm = u$. We suppose that we have an equation in t and u
which expresses the nature of the curve amb. From P we draw Ps
parallel to Ob which intersects the extension of the ordinate mp in the
point s, so that $ps = x \sin \alpha$ and $Op - Ps = x \cos \alpha$. Since the angle
$Pms = AOa = \alpha$, we have $Ps = y \sin \alpha$ and

$$mp = u = y \cos \alpha - x \sin \alpha.$$

We substitute into the equation in t and u the values
$t = x \cos \alpha + y \sin \alpha - c$ and $u = y \cos \alpha - x \sin \alpha$ to obtain a
general equation in the coordinates x and y which gives all of the curves
when we let the angle α vary.

453. Now we let the vertex of the curve AMB move along any
directrix AaL, while the positions of the axis ab changes, so that the
angle AOa depends in some way on the point a, as illustrated in *figure
92*. For instance, as the vertex a changes, we let $AK = a$, $Ka = A$,
and the angle $AOa = \alpha$. Since the directrix is given, A is some known
function of a and the sine and cosine of α are also function of a. With
this granted, we have $KO = \dfrac{A}{\tan \alpha}$ and $Oa = \dfrac{A}{\sin \alpha}$. From an

arbitrary point m on the curve amb we first drop the perpendicular mP to the principal axis AO, then we drop the perpendicular mp to its own axis, so that $AP = x$, $Pm = y$, $ap = t$, and $pm = u$. We suppose that an invariant equation for the curve has been given in the coordinates t and u, from which we find the variable equation in x and y which is the general equation for all of the curves.

454. In order to show this in detail we draw the normal line Ps from P to the extension of the line mp, which will be parallel to the axis abO of the curve. Since the angle $Pms = AOa = \alpha$, we have $Ps = y \sin \alpha$ and $ms = y \cos \alpha$. Then since

$$OP = a + \frac{A}{\tan \alpha} - x,$$

we have

$$ps = a \sin \alpha + A \cos \alpha - x \sin \alpha$$

and

$$Op - Ps = a \cos \alpha + \frac{A \cos \alpha}{\tan \alpha} - x \cos \alpha.$$

Hence

$$Op = a \cos \alpha + \frac{A \cos \alpha}{\tan \alpha} - x \cos \alpha + y \sin \alpha$$

$$= \frac{A}{\sin \alpha} - t.$$

It follows that

$$t = A \sin \alpha - a \cos \alpha + x \cos \alpha - y \sin \alpha$$

and

$$u = - a \sin \alpha - A \cos \alpha + x \sin \alpha + y \cos \alpha.$$

Now if we substitute into the equation in t and u

$$t = (x - a) \cos \alpha - (y - A) \sin \alpha$$

$$u = (x - a) \sin \alpha + (y - A) \cos \alpha,$$

we obtain the desired equation in x and y. By whatever rule the curve amb is drawn in the plane an infinite number of times, in this way we have found a general equation for those curves.

455. In this way we have included in a single equation an infinite number of copies of the same curve, which differ from one another only as to their location, provided the original equation in t and u is invariant and contains no changeable constant a. However, if the equation in t and u does contain one or more constants which we will assume also depend on a, then we obtain an infinite number of different curves, either similar, or dissimilar contained in the same general equation. They will all be similar curves if the equation in t and u is so constituted that u is equal to some first degree homogeneous function in t and f, where f is some quantity which depends on a. If this is not the case, then the curves will be dissimilar.

456. In order that we may illustrate this argument with an example of different curves, we suppose that an infinite number of circles AB, αB, amB, all passing through the point B and all having their centers on the straight line AE, as shown in *figure 93*. Circles of this kind are frequently used to represent meridians on geographical maps. We drop the perpendicular BC to the straight line AC and let $BC = c$, which is a constant distance. Now we consider amB, one of the infinite number of circles. We drop the ordinate mP and let $CP = x$ and $Pm = y$. The radius of this circle is constant, but different circles have different radii,

so we let $aE = BE = a$, so that $CE = \sqrt{a^2 - c^2}$ and $PE = x + \sqrt{a^2 - c^2}$. Since $PE^2 + Pm^2 = a^2$, we have

$$y^2 + x^2 + 2x\sqrt{a^2 - c^2} - a^2 - c^2 = a^2,$$

that is,

$$y^2 = c^2 - 2x\sqrt{a^2 - c^2} - x^2.$$

If we now let the interval CE be a variable instead of a constant, we obtain all of the circles which pass through B and have centers on the straight line AE. In order to simplify the equation we let $CE = a$, so that the equation becomes $y^2 = c^2 - 2ax - x^2$. In a similar way any infinite set of curves which follow some rule can be included in a single equation provided we carefully distinguish between variable and invariable constants.

CHAPTER XIX

On the Intersection of Curves.

457. In the previous chapters we have frequently considered curves intersected by a straight line, we have shown that second order lines cannot be intersected by a straight line in more than two points, that third order lines have no more than three intersections, and fourth order lines no more than four intersections, and so forth. Since we will investigate in this chapter the intersections of any two curves, we should begin this account with straight lines and investigate those points at which any given straight line intersects a given curve. In this way we will prepare the way for the mutual intersections of given curves, which information is very useful for the construction of equations of higher degree. We will pursue this topic more fully in the next chapter.

458. Let AMm be any given curve whose nature is expressed by an equation in the rectangular coordinates $AP = x$ and $PM = y$. We draw any straight line BMm and try to find how many and which points are intersection points, as shown in *figure 94*. For this we need the equation of the straight line in the rectangular coordinates x and y with respect to the same axis AP and the same origin A. The equation for the straight line will have the form $\alpha x + \beta y = \gamma$. If we let $x = 0$, then $y = AD = \dfrac{\gamma}{\beta}$, while if we let $y = 0$, then $x = -AB = \dfrac{\gamma}{\alpha}$. Hence B

is the intersection of the straight line with the axis and we note the angle at B has its tangent equal to $\dfrac{AD}{AB} = \dfrac{-\alpha}{\beta}$. Thus, both the curve and the straight line are expressed in the common coordinates x and y.

459. If we take the same value for the abscissa x in both equations, then if the ordinates y are different, then we see how far apart the two curves are at the point corresponding to the abscissa. If the values of the ordinates y for both equations are equal, then at that point the curve and the straight line have a common point, so that we have a point of intersection. In order to find the intersections we need not only the abscissas to be equal in both equations, but also both ordinates. Hence we have two equations in two variables x and y and the intersections can be found by finding either the abscissa x or the ordinate y. That is, if we eliminate y from both equations, we have an equation in only x, whose solutions give the abscissas AP and ap. From these we obtain the corresponding ordinates PM and pm so that we have the points of intersection M and m.

460. Since the equation for the straight line BMm is $\alpha x + \beta y = \gamma$, we have $y = \dfrac{\gamma - \alpha x}{\beta}$. When this value is substituted into the equation for the curve, we have an equation in x alone, whose real roots give all of the abscissas which correspond to the intersections. It follows that the number of intersections is known from the number of real roots of the equation we found. Since in $y = \dfrac{\gamma - \alpha x}{\beta}$ the variable x appears only to the first power, after the substitution the new equation has no higher degree than the equation for the curve in x and y. Hence the new equation has the same degree, or a lower degree if through a summation the highest power of x is eliminated.

461. When we have found the abscissas AP and Ap which correspond to intersections, the points M and m are easily found, since the ordinates which are erected at the points P and p pass through the intersections, and we note these points where the straight line BMm intersects the ordinates. We can also note the points where the ordinates intersect the curve AMm, but it frequently happens that one ordinate intersects the curve in several points, so that there can be some doubt which point on the curve is the desired intersection. This inconvenience can be avoided if the intersections are taken with the straight line, since it can intersect any ordinate in only one point. If it should happen that two of the values for x are equal, then the two points of intersection M and m become a single point. In this case either the straight line BM is tangent to the curve, or it cuts the curve in a double point.

462. If the variable y is eliminated and the resulting equation by which x is found has no real roots, then we know that the straight line never intersects the curve, nor is it tangent. The real roots of that equation, however many there may be, give all of the intersections. For any real abscissa there is only one corresponding real ordinate for the straight line BMm, which is equal to the corresponding ordinate for the curve, and this would not occur if there were not an intersection. This should here be carefully noted, since when two curves intersect, it is not always true that all of the roots give all of the intersections. The reason for this will soon be seen when we consider two curves and investigate their intersections.

463. Now let any two curves MEm and MFm be given which intersect each other, as in *figure 95*. In order to find the points of intersection we express each of the curves by an equation in the rectangular coordinates x and y with respect to the same axis AB and origin A. If we take the same abscissa x for both curves where an intersection occurs,

then the corresponding ordinates for both curves will also be equal. Hence, if we eliminate y from both equations, we obtain a new equation in x alone whose real roots will indicate all of the intersections M, m, and m, however many there may be. That is, the abscissas AP, Ap, Ap, etc. which correspond to the intersections M, m, m etc., are values of x which are the roots of that equation.

464. Once we have found the abscissas AP, Ap, etc. which correspond to intersections, it is not so easy to find the points of intersection. If to each abscissa AP there correspond several ordinates, which will be the case when for both curves y is a multiple-valued function of x, then from this double multitude of ordinate values we must choose those which are equal to each other. This investigation will be all the more difficult, the more values of y we have from each curve. It is easy to get around this difficulty. If we take the equation in which y is expressed in terms of x when y is eliminated from both equations, then for any value of x we can find the distance between the point P and the point of intersection from the value of the ordinate. In order to accomplish this we do not have to consider the nature of both of the curves, or even of one of the curves.

465. Let one of the curves be a parabola expressed by the equation $y^2 - 2xy + x^2 - 2ax = 0$ and let the other curve be the circle expressed by the equation $y^2 + x^2 - c^2 = 0$. In order to eliminate y we subtract the first equation from the second. The remaining equation is $2xy + 2ax - c^2 = 0$ so that $y = \dfrac{c^2 - 2ax}{2x}$, from which it is clear that for each value of x there will always be a corresponding real value of y. When we substitute this value for y in the second equation we have

$$c^4 - 4ac^2x + 4(a^2 - c^2)x^2 + 4x^4 = 0,$$

and for each of the real roots of this equation we have a true intersection. If we let $c = 2a$ so that

$$4a^4 - 4a^3x - 3a^2x^2 + x^4 = 0,$$

then one root is $x = 2a$. When we remove this root, there remains the equation

$$x^3 + 2ax^2 + a^2x - 2a^3 = 0,$$

which still has one real root. Either of the ordinates can be found conveniently from the equation $y = \dfrac{2a^2 - ax}{x}$. In the first case, when $x = 2a$, the corresponding value is $y = 0$, so that the intersection lies on the axis.

466. From this we should understand that having compared the two equations in x and y, in the process of eliminating y, whenever we can find a non-irrational function of x to express y, then for each real root x of the final equation, after y has been eliminated, there will be a corresponding real intersection. However, if every function of x which expresses y, from either equation, is irrational, then it can happen that not all real roots of the final equation will correspond to true intersections. From time to time such a value of x can occur which corresponds to no real ordinate for either curve, and in this case the result cannot be blamed on an error in calculating. Since both ordinate values which correspond to this kind of abscissa are complex, we recall that two complex values can be either equal or unequal, just as with real values. If it happens that the two complex values are equal, then there will be no true intersection.

467. In order to see this more clearly, we construct the parabola *EM* with the parameter equal to $2a$ on the same axis *BAE* with the circle *AmB* with radius equal to c, with no intersection, as shown in *figure*

96. Let the interval $AE = b$ so that there is certainly no intersection. We take A as the origin with positive values toward E and negative values toward B. The equation for the parabola is $y^2 = -2ax - 2ab$ and for the circle we have $y^2 = -2cx - x^2$. Now if we want to look for what might be an intersection, we eliminate y, and immediately get the equation $x^2 + 2(a + c)x - 2ab = 0$. From this equation be obtain two real values for x, namely,

$$x = -a - c \pm \sqrt{(a + c)^2 + 2ab}\,,$$

one of which is positive and the other is negative, while as a matter of fact there is no real intersection. For these two abscissas both the parabola and the circle have complex ordinate values, which are equal to each other. When we substitute this value for x, we obtain

$$y = \left(-2a^2 - 2ac - 2ab \pm 2a\sqrt{a^2 + 2ac + c^2 + 2ab} \right)^{\frac{1}{2}},$$

and both of these expressions are complex.

468. From this example we see that there is such a thing as a complex intersection of curves, which are not real intersections, but still are found by the same kind of calculations as those which give the real intersections. For this reason, the number of real roots of the final equation in x is not always the number of intersections. It can happen that there may be more roots than there are intersections, and there may be no intersections at all, even though there are two or more real roots x. Nevertheless, each intersection always provides a real root of x in the final equation, and for this reason, there are always at least as many real roots as there are intersections. It can happen that there are more real roots than intersections. It is easy to see whether each real root x corresponds to an intersection; if the value of the corresponding y is real, then the intersection is real, while if the value of y is complex, then there

is no intersection, or it is a complex intersection.

469. This exceptional difference in the number of real roots x and the number of intersections occurs only when either in both equations y appears only with even powers and the principal axis is a diameter of both curves, or if both equations are such that when y^2 is eliminated, y also is eliminated, so that y can only be expressed as an irrational function of x. For example, suppose one equation is $y^2 - xy = a^2$ and the other is $y^4 - 2xy^3 + x^3y = b^2x^2$. From the first equation we have $(y^2 - xy)^2 = a^4$, that is $y^4 - 2xy^3 = a^4 - x^2y^2$. When this value is substituted into the second equation, we have $a^4 - x^2y^2 + x^3y = b^2x^2$, that is $y^2 - xy = \dfrac{a^4 - b^2x^2}{x^2} = a^2$. Hence $x^2 = \dfrac{a^4}{a^2 + b^2}$, so that $x = \dfrac{\pm\, a^2}{(a^2 + b^2)^{\frac{1}{2}}}$. It would seem that there should be two intersections, but we see that for each of these roots we have two real values for y from the equation $y^2 - xy = a^2$. Indeed, all of the roots of $y^2 = \dfrac{\pm\, a^2 y}{(a^2 + b^2)^{\frac{1}{2}}} + a^2$ are real, so that there are four intersections. That is, for each abscissa $x = \dfrac{\pm\, a^2}{(a^2 + b^2)^{\frac{1}{2}}}$ there are two corresponding real intersections.

470. When it is not true that the axis is the diameter of both curves, nor that when the higher powers of y have been eliminated, then y itself has also been eliminated, then since y can be expressed as a non-irrational function of x, each real root of the final equation corresponds to a true intersection, so that in these cases there is no need for special caution. As we have already seen, this happens when one of the curves is a straight line, or if its ordinates are expressed by a single-valued

function of x. In that case no abscissa corresponds to a complex ordinate, so that each root x gives a true intersection. It frequently happens that although y has higher degrees in both equations, still it is possible to eliminate the higher powers of y to arrive at an equation by means of which y can be expressed as a non-irrational function, that is, a single-valued function of x.

471. Whenever we find by calculation only complex intersections, this happens not only when neither curve has real ordinates corresponding to the abscissas of the discovered roots, as happened in the previous example of the parabola and circle; it can also happen when one of the curves has real ordinates corresponding to the abscissas of all of the discovered roots, however, all of the real roots do not correspond to intersections. Here is an example of a third order line:

$$y^3 - 3ay^2 + 2a^2y - 6ax^2 = 0,$$

which has real ordinates for all abscissas. Indeed, for abscissas $x < \dfrac{a}{3}(1/3)^{\frac{1}{4}}$, the ordinate has three values. If this curve is combined with the parabola having the equation $y^2 - 2ax = 0$, which has no real ordinate if the abscissa is negative, then no negative abscissa can provide an intersection.

472. Now we eliminate y. From the second equation we have $y^2 = 2ax$, so that the first equation becomes $2axy - 6a^2x + 2a^2y - 6ax^2 = 0.$ From this we have $y = \dfrac{6a^2x + 6ax^2}{2a^2 + 2ax} = 3x.$ Since the previous equation in divisible by $y - 3x$, if we preform this division we have an equation free of y, $2a^2 + 2ax = 0.$ Hence $x = -a$, and there should be an intersection of the curves corresponding to the abscissa $x = -a$, to which there is no

corresponding real ordinate for the parabola. If we substitute $x = -a$ in the equation for the third order line, we have $y^3 - 3ay^2 + 2a^2y - 6a^3 = 0$, from which we have the real ordinate $y = 3a$. The other two values of y in the equation $y^2 + 2a^2 = 0$ are complex. Indeed, these complex values are equal to the complex values for the parabola corresponding to $x = -a$. There are two real intersections which arise from the factor $y - 3x = 0$. From this value, we have $9x^2 - 2ax = 0$. The first of the intersections corresponds to the origin where $x = 0$ and $y = 0$ the other corresponds to the abscissa $x = \dfrac{2a}{9}$, where $y = 3x = \dfrac{2a}{3}$.

473. In this case we found complex intersections even though in the process of eliminating y we obtained the equation $2axy - 6a^2x + 2a^2y - 6ax^2 = 0$, in which y appears only to the first power, so that y can be expressed by a function without irrationality. This was one of the criteria we noted previously for having no complex intersections. Indeed, if this equation had had no divisions, then there would have been no complex intersections. In this case we found by division an equation which no longer involved the ordinate y, which was as if y could not be expressed by a non-irrational function of x. Whenever a function of this kind can be expressed as a product of factors, we have to judge each factor separately. One factor may completely disallow complex intersections, while another may admit them.

474. Now that we have considered these problems, we would like to show more clearly how the intersections of two given curves should be found. Since this investigation depends on the elimination of one of the coordinates y, we will consider only the powers to which y is raised in either equation. Elimination is carried out in the same way no matter

what the power of x may be. Let P, Q, R, S, T, etc. and p, q, r, s, t etc. be any non-irrational functions of x. First we suppose that the two curves whose intersections we investigate have the form

I.

$$P + Qy = 0$$

II.

$$p + qy = 0.$$

If we multiply the first equation by p and the second by P, then subtract one from the other, we obtain an equation from which y has been eliminated

$$pQ - Pq = 0.$$

The roots of this equation, which contains only constants and the variable x, give the points on the axis corresponding to the points of intersection, and the ordinate values of these points can be found from either equation $y = \dfrac{-P}{Q} = \dfrac{-p}{q}$. Hence, if the ordinate of either curve is expressed by a single-valued function of x, there can be no complex intersections.

475. Now let us suppose that for one of the curves the ordinate y is expressed by a single-valued function of x, as before, but that the ordinate of the other is expressed by a double-valued function of x, so that

I.

$$P + Qy = 0$$

II.

$$p + qy + ry^2 = 0.$$

If we multiply the first equation by p and the second by P and then subtract one from the other we have

III.

$$pQ - Pq - Pry = 0,$$

that is,

$$(Pq - pQ) + Pry = 0.$$

If we multiply the first by Pr, the third by Q, and then take the difference, we obtain an equation from which y has been eliminated,

$$P^2r - PQq - pQ^2 = 0.$$

The roots of this equation give the abscissas of the intersections, and the corresponding real ordinates $y = \dfrac{-P}{Q} = \dfrac{pQ - Pq}{Pr}$, so that the intersections are all real.

476. Now we suppose, as before, that one curve has its ordinate equal to a single-valued function of x, and the ordinate of the other curve is expressed by a cubical equation, that is by a triple-valued function of x, so that the two given equations are

I.

$$P + Qy = 0$$

II.

$$p + qy + ry^2 + sy^3 = 0.$$

We multiply the first equation by p, the second by P, and then the difference is divided by y to obtain

III.

$$(Pq - pQ) + Pry + Psy^2 = 0.$$

If we substitute into this equation $y = \dfrac{-P}{Q}$ from the first equation, and then remove the fractions, we obtain the equation

$$PQ^2q - pQ^3 - P^2Qr + P^3s = 0,$$

that is,

$$Q^3p - PQ^2q + P^2Qr - P^3s = 0,$$

which is the same result we would have obtained if we had substituted $y = \dfrac{-P}{Q}$ from the first equation immediately into the second equation. All of these are real intersections.

477. In a similar way, if the second curve is expressed by an equation of the fourth or higher degree in y, provided the first has an ordinate given by a single-valued function or a non-irrational function of x, it is easy to eliminate the variable y. Let the given equation be

I.

$$P + Qy = 0$$

II.

$$p + qy + ry^2 + sy^3 + ty^4 = 0.$$

From the first we have $y = \dfrac{-P}{Q}$, which can be substituted into the second to obtain an equation in only x and known constants,

$$Q^4p - PQ^3q + P^2Q^2r - P^3Qs + P^4t = 0.$$

All of the real roots of this equation will supply all of the true intersections, since each abscissa x can be assigned a real ordinate from the first equation, namely, $y = \dfrac{-P}{Q}$.

478. Now we suppose that both curves have an ordinate y expressed by a quadratic equation. First we consider the case where both are pure quadratics represented as follows:

I.

$$P + Ry^2 = 0$$

II.

$$p + ry^2 = 0.$$

When y is eliminated we immediately have

$$Pr - Rp = 0,$$

whose real roots give true intersections only if the discovered values of x make $\dfrac{-P}{R}$ or $\dfrac{-p}{r}$ positive. Since $y^2 = \dfrac{-P}{R} = \dfrac{-p}{r}$, there are two corresponding values of y, one positive and one negative. Hence each such root of $Pr - Rp = 0$ provides two intersections, at equal distances above and below the axis. This would not happen if the axis were not a diameter for both curves. If some root of this final equation $Pr - Rp = 0$ makes $\dfrac{-P}{R} = \dfrac{-p}{r}$ negative, then, since y is complex, the intersections will also be complex.

479. Suppose now that in both proposed quadratic equations there is also a term containing y, so that both equations have the form

I.

$$P + Qy + Ry^2 = 0$$

II.

$$p + qy + ry^2 = 0.$$

In order to eliminate the variable y from the equations, we multiply the first by p and second by P, then we take the difference and divide by y. Then we have

III.

$$(Pq - Qp) + (Pr - Rp)y = 0.$$

Now we multiply the first equation by r, the second by R, and take the difference to obtain

IV.

$$(Pr - Rp) + (Qr - Rq)y = 0.$$

From the third and fourth equations we have

$$y = \frac{Qp - Pq}{Pr - Rp} = \frac{Rp - Pr}{Qr - Rq},$$

so that

$$(Qp - Pq)(Qr - Rq) + (Pr - Rp)^2 = 0,$$

that is,

$$P^2r^2 - 2PRpr + R^2p^2 + Q^2pr - PQqr - QRpq + PRq^2 = 0.$$

Each of the real roots of this equation indicates a true intersection, since for each value of x we have a real value of y from equations III or IV. It can happen, however, that the same intersections may be imaginary, if the equations III and IV have factors, so that by division we could obtain an equation with no y. Then we would solve this equation for its roots, and look for corresponding values of y from the first equations. If the values of y are complex, then this indicates that the intersections are complex.

480. Now suppose that for one curve the ordinate y is defined by a double-valued function of x and for the other curve, by a triple-valued function of x. That is, the given equations for the curves are

I.

$$P + Qy + Ry^2 = 0$$

II.

$$p + qy + ry^2 + sy^3 = 0.$$

We multiply the first by p, the second by P, take the difference and divide by y to obtain

III.

$$(Pq - Qp) + (Pr - Rp)y + Psy^2 = 0.$$

When we take this equation with the first, we have reduced the problem to that of the preceding section. Hence, where before we had p, q, and r, now we have $Pq - Qp$, $Pr - Rp$, and Ps, so that we have

$$y = \frac{PQq - Q^2p - P^2r + PRp}{P^2s - PRq + QRp},$$

$$y = \frac{Prq - QPp - P^2s}{PQs - PRr + R^2p},$$

so that

$$0 = (PRq - QRp - P^2s)^2$$
$$+ (PQs - PRr + R^2p)(PQq - Q^2p - P^2r + PRp).$$

When this equation is expanded we have

$$3P^2QRps - PQR^2pq - 2P^3Rqs + P^2R^2q^2 - PQ^3ps$$

$$+ Q^2R^2p^2 + P^4s^2 - P^3Qrs + P^2Q^2qs$$

$$+ P^3Rr^2 - P^2QRqr + PQ^2Rpr - Q^2R^2p^2$$

$$- 2P^2R^2pr + PR^3p^2 = 0.$$

Since

$$Q^2R^2p^2 - Q^2R^2p^2 = 0,$$

the whole equation is divisible by P, so that we have the equation

$$P^3s^2 - 2P^2Rqs - P^2Qrs + 3PQRps + PQ^2qs - Q^3ps + R^3p^3$$

$$+ P^2Rr^2 - PQRqr - 2PR^2pr + Q^2Rpr + PR^2q^2 - QR^2pq = 0.$$

From this equation we take the real roots to obtain the intersections, since real values of y correspond to the values of x.

481. Now we express each curve by a cubical equation in the following form.

I.

$$P + Qy + Ry^2 + Sy^3 = 0$$

II.

$$p + qy + ry^2 + sy^3 = 0.$$

We multiply the first equation by p, the second by P, then after subtraction we have

III.

$$(Pq - Qp) + (Pr - Rp)y + (Ps - Sp)y^2 = 0.$$

Next we multiply the first by s, the second by S, and then take the difference to obtain

IV.

$$(Sp - Ps) + (Sq - Qs)y + (Sr - Rs)y^2 = 0.$$

If we compare equations III and IV with the two in section 479, we obtain the following

$$P = Pq - Qp \quad\Big|\quad p = Sp - Ps$$
$$Q = Pr - Rp \quad\Big|\quad q = Sq - Qs$$
$$R = Ps - Sp \quad\Big|\quad r = Sr - Rs$$

When we substitute these values into the final equation in that section, we obtain

$$(Pq - Qp)^2(Sr - Rs)^2 - 2(Pq - Qp)(Ps - Sp)(Sp - Ps)(Sr - Rs)$$

$$+ (Ps - Sp)^2(Sp - Ps)^2 + (Pr - Rp)^2(Sp - Ps)(Sr - Rs)$$

$$- (Pq - Pq)(Pr - Rp)(Sq - Qs)(Sr - Rs)$$

$$- (Pr - Rp)(Ps - Sp)(Sp - Ps)(Sq - Qs)$$

$$+ (Pq - Qp)(Ps - Sp)(Sq - Qs)^2 = 0.$$

There are seven terms in this equation, each of which is divisible by $(Sp - Ps)$ except for the first and the fifth. If we join these two exceptional terms we obtain two factors, the first is

$$(Pq - Qp)(Sr - Rs)$$

and the second is

$$(Pq - Qp)(Sr - Rs) - (Pr - Rp)(Sq - QS).$$

This second factor can be expressed as

$$PQrs + RSpq - PRqs - QSpr = (Sp - Ps)(Rq - Qr),$$

so that the sum of the first and fifth terms has the form

$$(Pq - Qp)(Sr - Rs)(Sp - Ps)(Rq - Qr).$$

It follows that $Sp - Ps$ is a divisor of the whole expression. After division we have

$$0 = (Pq - Qp)(Sr - Rs)(Rq - Qr)$$

$$+ 2(Pq - Qp)(Sp - Ps)(Sr - Rs)$$

$$+ (Sp - Ps)^3 + (Pr - Rp)^2(Sr - Rs)$$

$$+ (Pr - Rp)(Sp - Ps)(Sq - Qs)$$

$$- (Pq - Qp)(Sq - Qs)^2,$$

and this can be expressed as

$$S^3p^3 - 3PS^2p^2s + P^2Sr^3 + 2PR^2prs - P^2Rr^2s$$

$$+ P^2Qrs^2 + PRSq^2r - P^3s^3 + 3P^2Sps^2 - R^3p^2s$$

$$- 2PRSpr^2 + R^2Sp^2r - RS^2p^2q - Q^2Rprs - PR^2q^2s$$

$$- PQSqr^2 + PQRqrs + 3PS^2pqr - 3P^2qrs + PQSprs$$

$$+ Q^2Spr^2 + QR^2pqs - QRSpqr - 3PQRps^2 + 3QRSp^2s$$

$$- PRSpqs + 2P^2Rqs^2 + 2PQSq^2s - PS^2q^3 - PQ^2qs^2$$

$$- 2QS^2p^2r - 2Q^2Spqs + Q^3ps^2 + QS^2pq^2 = 0.$$

482. In order that the method for eliminating y from two equations with higher powers may become clearer, we suppose that both equations are of the fourth degree in y as follows.

I.

$$P + Qy + Ry^2 + Sy^3 + Ty^4 = 0$$

II.

$$p + qy + ry^2 + sy^3 + ty^4 = 0.$$

We multiply the first equation by p, the second by P, and the difference gives

III.

$$(Pq - Qp) + (Pr - Rp)y + (Ps - Sp)y^2 + (Pt - Tp)y^3 = 0.$$

Then we multiply the first by t, the second by T, and after subtracting we have

IV.

$$(Pt - Tp) + (Qt - Tq)y + (Rt - Tr)y^2 + (St - Ts)y^3 = 0.$$

For the sake of brevity we make the following identifications:

$Pq - Qp = A$	$Pt - Tp = a$	$Sq - Qs = \alpha$
$Pr - Rp = B$	$Qt - Tq = b$	$Rq - Qr = \beta$
$Ps - Sp = C$	$Rt - Tr = c$	
$Pt - Tp = D$	$St - Ts = d$	

We note that not only is $a = D$ but also

$$Ad - Cb = (Pt - Tp)(Sp - Qs) = D\alpha$$

$$Ac - Bb = (Pt - Tp)(Rq - Qr) = D\beta.$$

When we make these substitutions into III and IV, we have the following:

III.

$$A + By + Cy^2 + Dy^3 = 0$$

IV.

$$a + by + cy^2 + dy^3 = 0.$$

Next we multiply these equations respectively by d and D, take the difference, and obtain

V.

$$(Ad - Da) + (Bd - Db)y + (Cd - Dc)y^2 = 0.$$

Then we multiply equations III and IV by a and A, take the difference, and obtain

VI.

$$(Ab - Ba) + (Ac - Ca)y + (Ad - Da)y^2 = 0.$$

Again, for the sake of brevity, we make the identifications

$$
\begin{array}{c|c|c}
Ab - Ba = E & Ad - Da = e & \\
Ac - Ca = F & Bd - Db = f & Cb - Bc = \zeta \\
Ad - Da = G & Cd - Dc = g &
\end{array}
$$

Furthermore, $G = e$, and $Eg - Ff = G\zeta$, so that $Eg - Ff$ is divisible by G. Hence we have the following equations:

V.

$$E + Fy + Gy^2 = 0$$

VI.

$$e + fy + gy^2 = 0.$$

By similar operations we obtain

VII.

$$(Ef - Fe) + (Eg - Ge)y = 0$$

VIII.

$$(Eg - Ge) + (Fg - Gf)y = 0.$$

Finally, for the sake of brevity we let

$$
\begin{array}{c|c}
Ef - Fe = H & EG - Ge = h \\
Eg - Ge = I & Fg - Gf = i
\end{array}
$$

so that $I = h$ and we have

VII.

$$H + Iy = 0$$

VIII.

$$h + iy = 0.$$

From these two equations we obtain an equation without y.

$$Hi - Ih = 0.$$

If we successively restore the values in the identification made for the sake of brevity we eventually obtain an equation in only P, Q, R, p, q, r, etc. The equation in E, F, g, e, f, and g is divisible by $G = e$. If we proceed to the letters A, B, C, D, a, b, c, and d, the resulting equation can be divided by $D^2 = a^2$, so that the terms in the final equation will each have only eight letters, four upper case and four lower case. In general, by this method, whatever the power of y, both equations may contain, the variable y can always be eliminated to find an equation which contains only x.

483. Although this method for eliminating one unknown from two equations has wide applications, still we will add another method which does not require so many repeated substitutions. Suppose we have the following two equations of any degrees.

I.

$$Py^m + Qy^{m-1} + Ry^{m-2} + Sy^{m-3} + \cdots = 0$$

II.

$$py^n + qy^{n-1} + ry^{n-2} + sy^{n-3} + \cdots = 0.$$

We want to find a single equation without y from these two equations. In order to achieve this we multiply the second equation by

$$Py^{k-n} + Ay^{k-n-1} + By^{k-n-2} + Cy^{k-n-3} + \cdots$$

where there are $k - n$ arbitrary letters A, B, C, etc. The first equation is multiplied by

$$py^{k-m} + ay^{k-m-1} + by^{k-m-2} + cy^{k-m-3} + \cdots$$

where there are $k - m$ arbitrary letters a, b, c, etc. Then both products are set equal to each other so that all terms with the same power of y will cancel each other. The final terms, which contain no y will provide the desired equation. The sum of all terms with the same power of y cancel each other, but the term with the highest power of y in both products is Ppy^k. There are $k - 1$ remaining terms, which gives $k - 1$ equations for determining the arbitrary letters. The number of arbitrary letters is $2k - m - n$, which must be equal to $k - 1$. Hence we let $k = m + n - 1$.

484. It follows that the first equation is to be multiplied by the inderminate quantity

$$py^{n-1} + ay^{n-2} + by^{n-3} + cy^{n-4} \cdots.$$

The second equation is multiplied by

$$Py^{m-1} + Ay^{m-2} + By^{m-3} + Cy^{m-4} + \cdots.$$

When we equate the coefficients of the same power of y in the two products we obtain

$$Pp = Pp$$

$$Pa + Qp = pA + qP$$

$$Pb + Qa + Rp = pB + qA + rP$$

$$Pc + Qb + Ra + Sp = pC + qB + rA + sP$$

etc. If we include the first equation, $Pp = Pp$, we have all together $m + n$ equations. With these equations we determine the arbitrary letters A, B, C, a, b, c, etc. The final equation contains only P, Q, R, p, q, r etc., so that we have the desired equation.

485. The determination of the arbitrary letters is facilitated by the introduction of other indeterminate quantities α, β, γ, etc. This will become clear from the following example. Let the two given equations be

I.

$$Py^2 + Qy + R = 0$$

II.

$$py^3 + qy^2 + ry + s = 0.$$

The first equation is multiplied by $py^2 + ay + b$, and the second by $Py + A$. Thus we obtain these equations

$$Pp = Pp$$

$$Pa + Qp = pA + qP = \alpha$$

$$Pb + Qa + Rp = qA + rP = \beta$$

$$Qb + Ra = rA + sP$$

$$Rb = sA.$$

We omit the first equation which is an identitiy. From the second we have

$$a = \frac{\alpha - Qp}{P}$$

$$A = \frac{\alpha - qP}{p}.$$

From the third we have

$$b = \frac{\beta}{P} - \frac{Qa}{P} - \frac{Rp}{P} = \frac{\beta}{P} - \frac{\alpha Q}{P^2} + \frac{Q^2 p}{P^2} - \frac{Rp}{P}$$

and

$$\beta = \frac{\alpha q}{p} - \frac{q^2 P}{p} + rP.$$

When we substitute this value for β, we have

$$b = \frac{\alpha q}{Pp} - \frac{q^2}{p} + r - \frac{\alpha Q}{P^2} + \frac{Q^2 p}{P^2} - \frac{Rp}{P},$$

that is

$$b = \frac{\alpha(Pq - Qp)}{P^2 p} + \frac{Q^2 p^2 - P^2 q^2}{P^2 p} + \frac{Pr - Rp}{P}.$$

When we substitute this value for b into the fourth equation, we have

$$\frac{\alpha Q(Pq - Qp)}{P^2 p} - \frac{Q(Pq - Qp)(Qp - Pq)}{P^2 p} + \frac{Q(Pr - Rp)}{P}$$

$$+ \frac{\alpha R}{P} - \frac{RQp}{P} = \frac{\alpha r}{p} - \frac{Prq}{p} + Ps.$$

When this equation is multiplied by $P^2 p$, we obtain

$$\alpha Q(Pq - Qp) + \alpha P(Rp - Pr) - Q(Pq - Qp)(Pq + Qp)$$

$$+ PQp(Pr - 2Rp) + P^3 qr - P^3 ps = 0.$$

From this it follows that

$$\alpha = \frac{P^2 Qq^2 - Q^3 p^2 - P^2 Qpr + 2PQRp^2 - P^3 qr + P^3 ps}{PQq - Q^2 p + PRp - P^2 r}.$$

The final equation gives

$$\frac{\alpha R(Pq - Qp)}{P^2 p} - \frac{R(P^2 q^2 - Q^2 p^2)}{P^2 p} + \frac{R(Pr - Rp)}{P}$$

$$= \frac{\alpha S}{p} - \frac{Pqs}{p}.$$

From this it follows that

$$\alpha = \frac{P^2 R q^2 - Q^2 R P^2 - P^2 Rpr + PR^2 p^2 - P^3 qs}{PRq - QRp - P^2 s}.$$

From the two expressions for α we obtain the desired expression, which has the same form as that found in section 480 for the same equations.

CHAPTER XX

On the Construction of Equations.

486. The material on the intersection of curves which was discussed in the previous chapter is generally treated in order to construct higher degree equations. Just as we found an equation whose roots provided the location of points of intersection for two given equations, so conversely the intersection of two curves can be used to find the roots of equations. This method is most useful if the roots of some equation are to be expressed by lines. Once the curves suitable for this purpose have been found, it is easy to note their intersections, so that if the ordinates are dropped to the axis, the abscissas provide the true roots of the equation. If the inconvenience mentioned in the previous chapter occurs in which there are more roots than required by the construction, still all of the abscissas found in this way will be roots.

487. We suppose that an algebraic equation in the variable x is given by assigning its roots, and that we are to find two curves, that is, two equations in the two variable x and y such that if the variable y is eliminated, we obtain the given equation. When this is accomplished, the two curves are referred to the same axis with the same origin, and the points of intersection are noted. Then from these points of intersection we drop the perpendiculars to the axis, which give the abscissas which are equal to all of the roots of the given equation. In this way the true values of all of the desired roots will be assigned, unless it happens that the equation has more roots than there were intersections.

488. Before we give the method by means of which the two curves, which are used to construct the given equation, can be found, we consider *a posteriori* those equations which are obtained from the two given curves. First we suppose that the two curves are the straight lines EM and FM which intersect in the point M, as shown in *figure 97*. We take the straight line EF as axis and choose the point A as origin. We construct the perpendicular ABC at the point A, where B is the intersection with the first straight line and C is the intersection with the second straight line. We let $AE = a$, $AF = b$, $AB = c$, and $AC = d$, then we let the abscissa $AP = x$, the ordinate $PM = y$, so that for the first straight line we have $a/c = (a + x)/y$, that is , $ay = c(a + x)$. For the second line we have $b/d = (b - x)/y$, that is $by = d(b - x)$. From the elimination of y from these equations, we obtain $bc(a + x) = ad(b - x)$, that is , $x = \dfrac{abd - abc}{bc + ad} = \dfrac{ab(d - c)}{bc + ad}$. Hence the linear equation $x = \dfrac{ab(d - c)}{bc + ad}$ can be interpreted as the intersection of two straight lines. We can always put any linear equation into the form.

489. The circle comes after the straight line in the ease with which it is described, and for this reason let us see how the intersection of a straight line and a circle can be interpreted. We take AP for the axis with A as the origin, and draw the straight line EM, as in *figure 98*. Let $AE = a$, $AB = b$, and the coordinates $AP = x$ and $PM = y$. We have $a/b = (a + x)/y$, so that $ay = b(a + x)$, which is an equation for the straight line. Let the radius of the circle $CM = c$; we drop the

perpendicular CD from the center to the axis and let $AD = f$, $CD = g$, so that $DP = x - f$, and $PM - CD = y - g$. From the nature of a circle we have $CM^2 = DP^2 + (PM - CD)^2$, so that the equation for the circle is

$$c^2 = x^2 - 2fx + f^2 + y^2 - 2gy + g^2 = (x - f)^2 + (y - g)^2.$$

From the equation for the straight line we have $y = \dfrac{ab + bx}{a}$ so that

$$y - g = \frac{a(b - g) + bx}{a} = b - g + \frac{bx}{a}.$$

When we substitute this in the other equation we have

$$c^2 = x^2 - 2fx + f^2 + (b - g)^2 + \frac{2b(b - g)x}{a} + \frac{b^2x^2}{a^2},$$

that is

$$(a^2 + b^2)x^2 + 2(ab(b - g) - a^2f)x + a^2(b - g)^2 + a^2f^2 - a^2c^2 = 0.$$

The roots of this equation are found from the intersections M and m of the straight line and the circle, so that when we drop the perpendiculars MP and mp from the points of intersection to the axis, the values of x will be AP and Ap.

490. Since all quadratic equations are included in the equation given above, this provides the construction for quadratic equations. Suppose that the given quadratic equation is

$$Ax^2 + Bx + C = 0.$$

This can be put in a form so that the first term agrees with the first term of that given above by multiplying by $\dfrac{a^2 + b^2}{A}$, to obtain

$$(a^2 + b^2)x^2 + \frac{B(a^2 + b^2)x}{A} + \frac{C(a^2 + b^2)}{A} = 0$$

When we equate corresponding terms we have

$$2Aab(b - g) - 2Aa^2f = B(a^2 + b^2)$$

so that

$$af = b(b - g) - \frac{B(a^2 + b^2)}{2Aa}.$$

Since we also have

$$a^2(b - g)^2 + a^2f^2 - a^2c^2 = \frac{C(a^2 + b^2)}{a},$$

it follows that

$$(a^2 + b^2)(b - g)^2 - \frac{Bb(b - g)(a^2 + b^2)}{Aa}$$

$$+ \frac{B^2(a^2 + b^2)^2}{4A^2a^2} - a^2c^2 = \frac{C(a^2 + b^2)}{A}.$$

Hence

$$(b - g)^2 = \frac{Bb(b - g)}{Aa} - \frac{B(a^2 + b^2)}{4A^2a^2} + \frac{a^2c^2}{a^2 + b^2} + \frac{C}{A},$$

so that

$$b - g = \frac{Bb}{2Aa} \pm \left(\frac{a^2c^2}{a^2 + b^2} + \frac{C}{A} - \frac{B^2}{4A^2} \right)^{\frac{1}{2}}.$$

There are still three quantities which have not been determined, a, b, and c, but they should be chosen in such a way that $\frac{a^2c^2}{a^2 + b^2} + \frac{C}{A} - \frac{B^2}{4A^2} > 0$ since otherwise $b - g = AB - CD$ would be complex, and CD would also be complex.

491. There is no problem if we let $b = 0$. In this case we have

$$g = \left(c^2 - \frac{B^2 + 4AC}{4A^2} \right)^{\frac{1}{2}} \quad \text{and} \quad f = \frac{-B}{2A}. \quad \text{Since the equation}$$

$Ax^2 + Bx + C = 0$ has no real roots unless $B^2 > 4AC$, it follows that

$\dfrac{B^2 - 4AC}{4A^2} > 0$. If we let c^2 be equal to this positive quantity, so that

$c = \dfrac{\sqrt{B^2 - 4AC}}{2A}$ then $g = 0$ and a no longer needs to be calculated.

The straight line EM coincides with the axis AP and the center of the

circle C coincides with the point D. Since $AD = \dfrac{-B}{2A}$ we have the

center, and the radius $c = \dfrac{\sqrt{B^2 - 4AC}}{2A}$, so that the intersection of

this circle with the axis gives the roots of the proposed equation. In

order to avoid the irrationality in this construction, we let

$g = c - \dfrac{k}{2A}$, so that

$$c^2 - \frac{2ck}{2A} + \frac{k^2}{4A^2} = c^2 - \frac{B^2 + 4AC}{4A^2}.$$

Then $c = \dfrac{k^2 + B^2 - 4AC}{4kA}$ and $g = \dfrac{B^2 - 4AC - k^2}{4kA}$. We still have

a choice as to the value of k, and no matter what value we assign to it,

since CM coincides with the axis, we should describe the circle as follows.

Let $AD = \dfrac{-B}{2A}$, take the perpendicular $CD = \dfrac{B^2 - 4AC - k^2}{4Ak}$.

With C as a center we draw the circle with radius equal to

$\dfrac{B^2 - 4AC + k^2}{4Ak}$. The intersection of this circle with the axis gives the

roots of the proposed equation. If we let $k = -B$, and take

$AD = \dfrac{-B}{2A}$, then $CD = \dfrac{C}{B}$ and the circle with center at C has radius

equal to $\dfrac{-B^2 + 2AC}{2AB} = \dfrac{-B}{2A} + \dfrac{C}{B}$. It follows that the radius of the circle is equal to $AD + CD$. This construction is the most convenient to use.

492. Now we consider two mutually intersecting circles, as in *figure 99*. To begin we let $AD = a$, $CD = b$, and we let the radius $CM = c$. If $AP = x$, $PM = y$, then $DP = a - x$ and $CD - PM = b - y$. From the nature of a circle we have

$$x^2 - 2ax + a^2 + y^2 - 2by + b^2 = c^2.$$

In a similar way for the other circle we let $Ad = f$, $dc = g$, and let the radius $cM = h$, so that

$$x^2 - 2fx + f^2 + y^2 + 2gy + g^2 = h^2.$$

When we take the difference between these two circles we have

$$2(f - a)x + a^2 - f^2 - 2(b + g)y + b^2 - g^2 = c^2 - h^2$$

so that

$$y = \frac{a^2 + b^2 - f^2 - g^2 - c^2 + h^2 - 2(a - f)x}{2(b + g)},$$

then

$$b - y = \frac{b^2 + 2bg - a^2 + f^2 + g^2 + c^2 - h^2 + 2(a - f)x}{2(b + g)},$$

and

$$a - x = \frac{2a(b + g) - 2(b + g)x}{2(b + g)}.$$

Since $(a - x)^2 + (b - y)^2 = c^2$, when we make the substitutions, we obtain

$$(4(a - f)^2 + 4(b + g)^2)x^2 + (4(a - f)(c^2 - h^2) - 4(a + f)(b + g)^2$$

$$- 4(a - f)(a^2 - f^2))x + (b + g)^4 + 2(a^2 - c^2)(b + g)^2$$

$$+ 2(f^2 - h^2)(b + g)^2 + (a^2 - c^2 - f^2 + h^2)^2 = 0.$$

By means of this equation there are an infinite number of ways in which the equation $Ax^2 + Bx + C$ can be interpreted. At the same time we keep in mind that an equation with degree higher than a quadratic cannot be interpreted by the intersection of two circles, since two circles cannot mutually intersect in more than two points. Since the same quadratic equation can be interpreted as the intersection of a straight line and a circle, this is the preferred interpretation, rather than two circles, unless in some very special cases when the values of a, b, f, g, c, and h are obviously determined.

493. Now we consider the intersection of a circle and a parabola, as in *figure 100*. We drop the perpendicular CD from the center of the circle C to the axis AP, then let $AD = a$, $CD = b$, and we let the radius of the circle $CM = c$. In the rectangular coordinates $AP = x$ and $PM = y$ the equation for the circle is $(x - a)^2 + (y - b)^2 = c^2$. We take the axis for the parabola FB normal to the axis AP. We let $AE = f$, $EF = g$ and we take the parameter of the parabola to be equal to $2h$. Then from the nature of a parabola we have $EP^2 = 2h(EF + PM)$, that is $(x - f)^2 = 2h(g + y)$ so that $y = \dfrac{(x - f)^2}{2h} - g$ and $y - b = \dfrac{(x - f)^2}{2h} - (b + g)$. If this value is substituted into the first equation, we can eliminate y, to obtain

$$\frac{(x - f)^4}{4h^2} - \frac{(b + g)(x - f)^2}{h} + (b + g)^2 + (x - a)^2 = c^2,$$

that is,

$$x^4 - 4fx^3 + (6f^2 - 4hb - 4hg + 4h^2)x^2$$

$$+ (4fhb + 4fhg - 4f^3 - 8ah^2)x + f^4$$

$$- 4f^2h(b + g) + 4h^2(b + g)^2 + 4a^2h^2 - 4c^2h^2 = 0.$$

The roots of this equation are the abscissas AP, Ap, Ap, and Ap, from which points the ordinates reach the points of intersection M, m, m, and m.

494. There are six constants a, b, c, f, g, and h in this equation, of which two, b and g, can be counted as the one $b + g$. If we let $b + g = k$, then there are five constants and the equation takes the form

$$x^4 - 4fx^3 + (6f^2 - 4hk + 4h^2)x^2 + (4fhk - 4f^3 - 8ah^2)x$$

$$+ f^4 - 4f^2hk + 4h^2k^2 + 4a^2h^2 - 4c^2h^2 = 0.$$

Since every biquadratic equation can be put into the form

$$x^4 - Ax^3 + Bx^2 - Cx + D = 0,$$

when we compare coefficients, we see that $4f = A$ or

$$f = \frac{1}{4}A, 6f^2 - 4hk + 4h^2 = B$$

or

$$\frac{3}{8}A^2 - 4hk + 4h^2 = B,$$

so that

$$k = \frac{3A^2}{32h} + h - \frac{B}{4h},$$

$$4f^3 - 4fhk + 8ah^2 = C,$$

that is,

$$\frac{1}{16}A^3 - \frac{3}{32}A^3 - Ah^2 + \frac{1}{4}AB + 8ah^2 = C.$$

It follows that

$$a = \frac{A^3}{256h^2} + \frac{A}{8} - \frac{AB}{32h^2} + \frac{C}{8h^2}.$$

Finally, we also have

$$(f^2 - 2hk)^2 + 4a^2h^2 - 4c^2h^2 = D.$$

But

$$f^2 - 2hk = \frac{B}{2} - 2h^2 - \frac{A^2}{8}$$

and

$$2ah = \frac{A^3}{128h} + \frac{Ah}{4} - \frac{AB}{16h} + \frac{C}{4h}.$$

When these values are substituted, we obtain an equation involving c and h. It will be very convenient to define them in such a way that both values will be real.

495. Since it is easy to remove the second term in any biquadratic, we suppose that it has already been removed, and we interpret the following equation.

$$x^4 + Bx^2 - Cx + D = 0.$$

It follows first that $f = 0$, secondly $k = h - \dfrac{B}{4h}$, thirdly $a = \dfrac{C}{8h^2}$.

Since $2hk - f^2 = 2h^2 - \dfrac{B}{2}$ and $2ah = \dfrac{C}{4h}$, we have in the fourth place

$$4h^4 - 2Bh^2 + \frac{1}{4}B^2 + \frac{c^2}{16h^2} - 4c^2h^2 = D.$$

It follows that

$$64c^2h^4 = C^2 + 4B^2h^2 - 32Bh^4 + 64h^6 - 16Dh^2,$$

so that

$$8ch^2 = \sqrt{4h^2(B - 4h^2)^2 + C^2 - 16Dh^2}.$$

Since we want to be sure that both c and h have real values, we let $c = h - \dfrac{B + q}{4h}$, so that

$$C^2 - 16Dh^2 + 8Bh^2q - 32h^4q - 4h^2q^4 = 0.$$

In order to obtain what we want we have to consider two cases, when D is negative and when D is positive.

I.

Suppose $D = E^2$ is positive, so that we have to interpret the equation

$$x^4 + Bx^2 - Cx + E^2 = 0$$

Here we let $q = 0$, so that $c = \dfrac{4h^2 - B}{4h}$, $h^2 = \dfrac{C^2}{16E^2}$, $h = \dfrac{C}{4E}$, and $c = \dfrac{C^2 - 4BE^2}{4CE}$. Furthermore, $k = c = \dfrac{C^2 - 4BE^2}{4CE}$, $a = \dfrac{2E^2}{C}$, and $f = 0$.

II.

If D is negative, we let $D = - E^2$ and the equation to be interpreted is

$$x^4 + Bx^2 - Cx - E^2 = 0$$

and

$$64c^2h^4 = C^2 + 4h^2(4h^2 - B)^2 + 16E^2h^2.$$

In this case c will be real, whatever value h may take, since

$$c = \frac{\sqrt{c^2 + 4h^2(4h^2 - B)^2 + 16E^2h^2}}{8h^2},$$

and h can be arbitrarily chosen. In either case we choose it in such a way that the value of c can be easily found. When this is done we have, as before, $AE = f = 0$, $CD + EF = k = \dfrac{4h^2 - B}{4h}$, and $AD = a = \dfrac{C}{8h^2}$. If we let $E = 0$, we obtain the cubic equation

$$x^3 + Bx - C = 0$$

The interpretation of this equation depends on Backer's rule, which is sufficiently well known.

496. If we take any two second order lines, that is, conic sections, whose equations are taken with respect to the same axis and origin, they can be written as follows.

$$ay^2 + byx + cx^2 + dy + ex + f = 0$$

$$\alpha y^2 + \beta yx + \gamma x^2 + \delta y + \epsilon x + \zeta = 0.$$

We eliminate y from these two equations by the method given previously. This can be accomplished by comparing these equations with those given in section 479, namely,

$$P + Qy + Ry^2 = 0$$

$$p + qy + ry^2 = 0.$$

We see that P and p are quadratic in x, Q and q are linear in x, while R and r are constant. It follows that the resulting equation will be biquadratic. It follows that it is impossible to obtain a higher degree equation than a biquadratic from the intersection of two conic sections. We have already seen that a biquadratic equation is obtained from the intersection of a circle and a parabola. We could also see this from the nature of second order lines, which can be intersected in two points by a straight line. Hence two straight lines can intersect a second order line in four points. But two straight lines together form a species of second order line. It follows that two second order lines can have four points of mutual intersection.

497. We now consider the intersection of curves, one of which is a second order line and the other is a third order line, represented by the following equations.

$$P + Qy + Ry^2 = 0$$

$$p + qy + ry^2 + sy^3 = 0,$$

where P is quadratic in x, Q is linear, and R is constant, while p is cubic, q is quadratic, r is linear, and s is constant. From the discussion in section 480, it is clear that the final result, after the elimination of y, will be a sixth degree equation. It follows that we cannot obtain an equation of degree higher than six from the intersection of a conic section with a third order line. This should be clear also from the nature of the two different orders, since a third order line can be intersected by a straight line in three points. This same curve can be intersected by two straight lines, which together constitute a second order curve, in six points.

498. If we use the method of eliminating y discussed above, or if we consider the argument from the intersection by straight lines, and apply these considerations to higher order lines, it becomes clear that the intersection of two third order lines gives ninth degree equations. The intersection of two fourth order lines gives equations with degree no higher than sixteen. In general, the intersection of two curves, of which

one is of order m and the other of order n, give equations of degree no more than mn. Thus in order to obtain an equation of degree 100, we need two curves, each of order 10, or one of order 20 and the other of the fifth order, and so forth, by expressing the number 100 as the product of two factors. If the degree of the equation to be constructed is a prime number or if it has no convenient factors, then we choose a convenient larger number, so that we can construct equations for two curves with a larger number of intersections. With these equations we can find other equations with fewer intersections. For example, for a 39th degree equation, we can use two curves with sixth and seventh order respectively. With these two equations we obtain one of degree 42, and this is a simpler construction than if one curve was third order and the other was thirteenth order.

499. From this discussion it is clear that given any equation we can find two curves whose intersections give the real roots of that equation in many, indeed infinitely many, different ways. From among these infinitely many ways it is best to choose that which is both as simple as possible and easy to describe. It is especially important that all of the intersections give real roots. This can be accomplished by choosing curves such that there are no complex intersections. We have already seen that there will be no complex intersections if in the equation for one curve the ordinate y is equal to a linear function of x. In this case, since the curve has no complex ordinates, there can be no complex intersections, no matter how the other curve may be besmirched with complex ordinates. In this business of finding one of the curves we always suppose that its equation has the form $P + Qy = 0$, where P and Q are functions of x.

500. No matter what equation is given we will always choose one convenient curve in the form $P + Qy = 0$. The equation for the other

curve should be such that when the value $\dfrac{-P}{Q}$ is substituted for y, the resulting equation is the original equation. Conversely, we can find the equation for second curve from the general equation by substituting y for $\dfrac{-P}{Q}$. For example if the given equation is

$$x^4 + Ax^3 + Bx^2 + Cx + D = 0,$$

we take the parabola $ay = x^2 + bx$ as the equation for one of the curves. From this we have $x^2 = ay - bx$, which we substitute into the given equation to obtain

$$x^4 = a^2y^2 - 2abxy + b^2x^2$$

$$Ax^3 = Aaxy - Abx^2$$

so that we obtain the following equation for a second order curve

$$a^2y^2 + a(A - 2b)xy + (B - Ab + b^2)x^2 + Cx + D = 0,$$

whose intersections with the curve $ay = x^2 + bx$ give the roots of the given equation.

501. Since both of these curves are determined by the arbitrary constants a and b, there is an infinite variety in the ways they can be chosen. It is possible to introduced even greater variety, since for the previous equation $x^2 - ay + bx = 0$ we could write $acx^2 - a^2cy + abcx = 0$, and this will give more variety to the second equation, although the intersections still give the roots of the given equation. The two new equations are as follows.

I.

$$ay = x^2 + bx$$

II.

$$a^2 y^2 + a(A - 2b)xy + (B - Ab + b^2 + ac)x^2$$

$$- a^2 cy + (C + abc)x + D = 0.$$

This second equation is such that it can represent any conic section. Particular attention should be paid to the quantity $A^2 - 4B - 4ac$, since if this is positive, then the curve is a hyperbola; if it is equal to zero, then the curve is a parabola; if it is negative, then the curve is an ellipse. This second curve will be a circle if $b = \dfrac{1}{2}A$ and $a^2 = B - \dfrac{1}{4}A^2 + ac$, that is, if $c = a + \dfrac{A^2}{4a} - \dfrac{B}{a}$. In this case the equation will be

$$a^2 y^2 + a^2 x^2 - (a^3 + \frac{A^2 a}{4} - Ba)y$$

$$+ (C + \frac{Aa^2}{2} + \frac{A^3}{8} - \frac{AB}{2})x + D = 0.$$

That is

$$(y - \frac{a}{2} - \frac{A^2}{8a} + \frac{B}{2a})^2 + (x + \frac{C}{2a^2} + \frac{A}{4} + \frac{A^3}{16a^2} - \frac{AB}{4a^2})^2 =$$

$$(\frac{a}{2} + \frac{A^2}{8a} + \frac{B}{2a})^2 + (\frac{C}{2a^2} + \frac{A}{4} + \frac{A^3}{16a^2} - \frac{AB}{4a^2})^2 - \frac{D}{a^2},$$

where the right member is the square of the radius of the circle.

502. Hence we have an infinite number of conic sections whose intersections with the parabola $ay = x^2 + bx$ give the roots of the given equation. However the curves may be chosen, they always intersect the parabola in the same points, so that all of these curves intersect each other in the same points. It follows that we can choose any two of these infinite number of curves (ignoring the parabola originally chosen), which

are expressed with respect to the same axis, and their intersections will give the roots of the original equation. In this way we can accomplish with a circle and a parabola, as we have already seen, or with two parabolas, or with a parabola and an ellipse or a hyperbola, or with two ellipses, or with two hyperbolas, or with an ellipse and a hyperbola. An even greater variety is possible if we use this method with a higher degree equation.

503. Similarly, we can find equations of higher degree by taking for the first curve the line with parabolic genus given by the equation $y = P$. For example, if the given equation is

$$x^{12} - f^{10}x^2 + f^9gx - g^{12} = 0,$$

we choose the fourth order parabolic equation $x^4 = a^3y$. Since $x^{12} = a^9y^3$, after we make our substitution we obtain the equation for a third order line.

$$a^9y^3 - f^{10}x^2 + f^9gx - g^{12} = 0$$

If we use any multiple of the first equation, we can obtain an infinite number of different fourth order lines any two of which together give the roots of the original equation.

504. If it should happen that for the given equation the method we have been discussing does not give suitably convenient equations, then the given equation can be multiplied by x, x^2, x^3, or some suitable higher power of x. In this way we add some new vanishing roots, but in the intersections these are easily recognized and distinguished from the other roots. Thus, although the equation is of a higher degree, it frequently happens that the problem is simplified. For example, if the given equation is a cubic,

$$x^3 + Ax^2 + Bx + C = 0,$$

when we let $x^2 = ay$ in order to find the other curve, it will always be a hyperbola. When we substitute ay for x^2, we obtain the equation

$$axy + Aay + Bx + C = 0.$$

If we add the equation $cx^2 - acy = 0$, we have the more general

$$axy + cx^2 + a(A - c)y + Bx + C = 0,$$

but this is also always the equation of a hyperbola. If we find it more convenient to have a circle or a parabola, then we multiply the given equation by x, to obtain the equation

$$x^4 + Ax^3 + Bx^2 + Cx = 0.$$

If we compare this with the biquadratic discussed in section 500, with $D = 0$, we can always make the second equation be a circle or a parabola.

505. Since for any equation of any degree we can construct two algebraic curves whose intersections give the roots of the original equation, and this can be done in an infinite number of ways, we can substitute one of these curves for another. From this fact there arises the question, to what extent, by means of one given curve and a given equation, the other curve can be found. First we should note that the curve should be such that its ordinate is expressed by a single-valued function of x, so that no complex intersections disturb the construction. Nor does it suffice that the curve, or a given part of the curve, has its abscissas all equal to a single root of the equation. This is the condition which is usually added when we are interested in only one root of the given equation. However, it can happen that this very arc of the curve has no intersection, even though the abscissa corresponding to some point of the arc is a

true root. This can occur since this root corresponds to a complex inter-section, or to an intersection of some other branch of the curve. For this reason we will delay no more on this question, which is more a curiosity, than something useful, since we have discussed in sufficient detail all of the truly fundamental facts of all the constructions of this kind.

CHAPTER XXI

On Transcendental Curves.

506. Until now we have been concerned with algebraic curves. These are characterized by the following: for any abscissa the corresponding ordinate can be expressed by an algebraic function of the abscissa. We can also say that the relationship between the abscissa and the ordinate can be expressed by an algebraic equation. It follows immediately that if the value of the ordinate cannot be expressed by an algebraic function, then the curve cannot be algebraic. A curve which is not algebraic is called *transcendental*. Hence a transcendental line, which is what such a curve is called, is defined to be one such that the relationship between the abscissa and the ordinate cannot be expressed by an algebraic equation. It follows that whenever the ordinate is equal to a transcendental function of the abscissa, then the genus of the curve is transcendental.

507. In previous sections we have given special consideration to two species of transcendental functions, namely, logarithms and trigonometric functions. If the ordinate y is equal to the logarithm of the abscissa x or to the arc of a circle whose sine, cosine, or tangent is expressed by the abscissa x, so that $y = \log x$, $y = \arcsin x$, $y = \arccos x$, or $y = \arctan x$, or if these values enter into an equation in x and y, then the curve is transcendental. These are only a few

species of transcendental curves. Besides these there are innumerable other transcendental expressions whose origins are treated at length in analysis of the infinite. The number of transcendental curves is much larger than the number of algebraic curves.

508. Any function which is not algebraic is transcendental, so that a curve is transcendental if its equation involves such a function. An algebraic equation either is non-irrational, with only integral exponents, or is irrational, with fractional exponents. In the latter case the equation can always be expressed in non-irrational form. Any curve whose equation expressing the relation between the coordinates x and y is not non-irrational, or cannot be put into such a form, is always transcendental. Suppose that in the equation there occur exponents which are neither integers nor fractions, then the equation cannot possibly be non-irrational. In this case the curve is transcendental. In this way we obtain the first species, indeed the simplest, of the transcendental curves. This is the case in which there are irrational exponents. Since these curves involve neither logarithms nor circular arcs, but simply the irrationality of exponents, they seem to pertain more to geometry. For this reason LEIBNITZ called such functions *intercendental*, as if they were situated between the algebraic and the transcendental.

509. Such an intercendental curve is defined by the equation $y = x^{\sqrt{2}}$. No matter how we operate on the exponent, the equation cannot be brought to non-irrational form. Such a curve can in no way be constructed geometrically. Only rational exponents can be expressed geometrically, for this reason such curves are most different from algebraic. If we wish to express the exponent $\sqrt{2}$ only approximately, we can substitute some fraction from the sequence $\dfrac{3}{2}, \dfrac{7}{5}, \dfrac{17}{12}, \dfrac{41}{29}, \dfrac{99}{70},$ which approximates the value of $\sqrt{2}$. These algebraic curves will

provide the approximations, but the order of the curve will be third, seventh, seventeenth, or forty-first, etc. Since $\sqrt{2}$ can be expressed rationally only by a fraction with infinitely large numerator and denominator, this curve should be considered to have infinite order, so that it cannot be thought to be algebraic. In addition, since $\sqrt{2}$ involves two values, one positive and one negative, y has two values, with the result that there are a pair of curves.

510. If we wish to construct this curve exactly, it is impossible without the aid of logarithms. Suppose $y = x^{\sqrt{2}}$. By taking logarithms, we obtain $\log y = \sqrt{2} \log x$. For any abscissa we multiply its logarithm by $\sqrt{2}$ to obtain the logarithm of the ordinate. Hence for any abscissa x, we obtain the corresponding ordinate from the table of logarithms. For example, if $x = 0$, then $y = 0$; if $x = 1$, then $y = 1$ and these values are easily obtained from the equation. However, if $x = 2$, then $\log y = \sqrt{2} \log 2 = \sqrt{2}(0.3010300)$. Since $\sqrt{2} = 1.41421356$, we have $\log y = 0.4257274$ and the approximate value of y is 2.665186. If $x = 10$, then $\log y = 1.4142356$, so that $y = 25.955870$. In this way, for each abscissa we can obtain the corresponding ordinate, so that the curve can be constructed, provided positive values are assigned to the ordinate. However, if the abscissa takes negative values, then it is difficult to say whether the values of y will be real or imaginary. For example, if $x = -1$, then $(-1)^{\sqrt{2}}$ cannot be defined, since approximations to the value $\sqrt{2}$ give no help.

511. There is even less doubt that equations in which there appear complex exponents should be given a transcendental genus. It is quite possible that an expression containing complex exponents could give a definite real value. We have already seen examples of this, so that a single example should now suffice. Suppose

$$2y = x^i + x^{-i},$$

where, although the expressions x^i and x^{-i} are both complex quantities, still the sum of the two has a real value. Indeed, if log $x = v$ and we take e to be the number whose natural logarithm is equal to 1, then $x = e^v$. When we substitute this value for x in the original equation, we obtain

$$2y = e^{vi} + e^{-vi}.$$

In Section 138 (of Book I) we saw that

$$\frac{e^{vi} + e^{-vi}}{2} = \cos v.$$

It follows that $y = \cos v = \cos \log x$. If we choose any positive value for x and take its natural logarithm, then on a unit circle we take an arc equal in length to that logarithm, then the cosine of this arc will give the value of the ordinate y. For example, if we take $x = 2$ so that $2y = 2^i + 2^{-i}$, then $y = \cos \log 2 = \cos 0.6931471805599$. Since an arc with length equal to $3.1415926535 \cdots$ contains $180°$, by the golden rule we calculate the arc with length equal to log 2 to contain $39°, 42', 51'', 52''', 9''''$, whose cosine is equal to 0. 76923890135408. This number gives the value of the ordinate corresponding to the abscissa $x = 2$. Since expressions of this kind involve both logarithms and circular arcs, they are correctly referred to as transcendental.

512. After the algebraic curves, the first of the transcendental curves are those which involve logarithms. The simplest of these are given by the equation $\log\frac{y}{a} = \frac{x}{b}$, or $x = b \log\frac{y}{a}$, where logarithms of any kind may be used, since multiplication by the constant b can be used to bring any system to some fixed system. For this reason we suppose that the symbol log refers to the natural logarithms. The curves given

by the equation $x = b \log \frac{y}{a}$ are commonly called LOGARITHMIC. Let

e be   the   number   such   that $\log e = 1$,   so   that

$e = 2.71828182845904523536028$ and $e^{\frac{x}{b}} = \frac{y}{a}$ or $y = ae^{\frac{x}{b}}$. From this

equation we easily recognize the nature of a logarithmic curve: if for x

we substitute successively the terms of an arithmetic progression, then

the values of the corresponding ordinates form a geometric progression.

In order to facilitate our construction we let $e = m^n$ and $b = nc$, so

that $y = am^{\frac{x}{c}}$, where m is any positive number greater than 1. It fol-

lows that if

$$x = 0,\ c,\ 2c,\ 3c,\ 4c,\ 5c,\ 6c,\ \cdots$$

then

$$y = a,\ am,\ am^2,\ am^3,\ am^4,\ am^5,\ am^6,\ \cdots$$

and if x is given negative values and we let

$$x = -c,\ -2c,\ -3c,\ -4c,\ -5c,\ \cdots,$$

then

$$y = \frac{a}{m},\ \frac{a}{m^2},\ \frac{a}{m^3},\ \frac{a}{m^4},\ \frac{a}{m^5},\ \cdots.$$

513. In *figure 101* it is clear that the values of the ordinate y are

always positive and increase to infinity when the abscissa x is positive

and increases to infinity. When the abscissa goes to infinity in the other

direction the ordinate decreases, so that the axis becomes an asymptote

Ap to the curve. That is, if we take A to be the origin of the abscissas,

then at this point the ordinate $AB = a$, and if we take the abscissa

$AP = x$, then the ordinate $PM = y = am^{\frac{x}{c}} = ae^{\frac{x}{b}}$. It follows that $\log \frac{y}{a} = \frac{x}{b}$. Hence, the abscissa AP divided by the constant b expresses the logarithms of the ratio $\frac{PM}{AB}$. If we take any other point a on the axis as the origin, the equation remain similar. Indeed, if $Aa = f$ and we let $aP = t$, since $x = t - f$, we have $y = ae^{\frac{t-f}{b}} = \frac{ae^{\frac{t}{b}}}{e^{\frac{f}{b}}}$. We let the constant $\frac{a}{e^{\frac{f}{b}}} = g$, so that $y = ge^{\frac{t}{b}}$. Since $ab = g$, it follows that $\frac{aP}{b} = \log \frac{PM}{ab}$, and so when any two ordinates PM and pm are drawn with the distance Pp separating them, we have $\frac{Pp}{b} = \log \frac{PM}{m}$. The constant b, upon which this relationship depends, acts as a logarithmic parameter.

514. It is easy to define the tangent to this logarithmic curve at any point M. Since $PM = ae^{\frac{x}{b}}$ if $AP = x$, when we draw any other ordinate QN at a distance $PQ = u$ from the previous ordinate we have $QN = ae^{\frac{x+u}{b}} = ae^{\frac{x}{b}}e^{\frac{u}{b}}$. We draw ML parallel to the axis, so that $LN = QN - PM = ae^{\frac{x}{b}}(e^{\frac{u}{b}} - 1)$. We now draw the straight line NMT through the points M and N, intersecting the axis at the point T, so that $\frac{LN}{ML} = \frac{PM}{PT}$. It follows that $PT = \frac{u}{e^{\frac{u}{b}} - 1}$. We have shown in a previous section that

$$e^{\frac{u}{b}} = 1 + \frac{u}{b} + \frac{u^2}{2b^2} + \frac{u^3}{6b^3} + \cdots,$$

so that

$$PT = \frac{1}{\dfrac{1}{b} + \dfrac{u}{2b^2} + \dfrac{u^2}{6b^3} + \cdots}.$$

When the interval $PQ = u$ vanishes and the points M and N coincide, the straight line NMT becomes the tangent to the curve and the subtangent becomes $PT = b$, which is constant. This is the principal property of the logarithmic curve: the logarithmic parameter b is everywhere equal to the subtangent which thus has constant length.

515. The question now arises whether the whole logarithmic curve is described in this way, and whether besides this branch MBm, which goes to infinity in both directions, there might be some other branches. We have already seen that there is no asymptote to which two branches do not converge. For this reason some say that the logarithmic curve consists of two parts symmetric with respect to the axis, so that the asymptote becomes a diameter. However, the equation $y = ae^{\frac{x}{b}}$ in no way justifies this property. Whenever $\frac{x}{b}$ takes an integral value or the value of a fraction with an odd denominator, then y has a single real value, and this is always positive. If the fraction $\frac{x}{b}$ has an even denominator, then the ordinate y does indeed have two values, one positive and the other negative. Hence the curve does have points on the other side of the asymptote; indeed an infinite number of such discrete points are below the asymptote, but they do not form a continuous branch of the curve, although infinitely small intervals constitute a continuous curve. This is a paradox which never occurs in algebraic curves. From this

there arises another paradox even more to be wondered at. Since the logarithms of negative numbers are complex, (this is both self-evident, and also can be understood from the fact that the ratio of $\log(-1)$ to i is finite,) it follows that $\log(-n)$ is a complex quantity, say equal to j. However, since the logarithms of the square of a quantity is equal to two lines the logarithm of the quantity, we have $\log(-n)^2 = \log n^2 = 2j$. Now $\log n^2$ is a real quantity, so that both the real quantity $\log n$ and the complex quantity j are both half of the real quantity $\log n^2$. Furthermore, any number has two different numbers equal to one half of it, one of which is real and other complex. In a similar way any number has three different numbers equal to one third of itself; it has four different numbers equal to one fourth of itself; and so forth. Note, however, that only one of each of these parts is real. It is not clear how this paradox can be reconciled with the usual idea of quantity.

516. If one concedes what we have assumed, then it follows that given a number a, half of a is both $\frac{a}{2} + \log(-1)$ and $\frac{a}{2}$. We also see that two times a is

$$a + 2\log(-1) = a + \log(-1)^2 = a + \log 1 = a.$$

Here we must note that $\log(-1) = -\log(-)$, although $\log(-1) \neq 0$, since from $-1 = \frac{1}{-1}$, it follows that

$$\log(-1) = \log 1 - \log(-1) = -\log(-1).$$

In a similar way, since $1^{\frac{1}{3}}$ is not only 1, but also $\frac{-1 \pm \sqrt{-3}}{2}$, we have $3\log\frac{-1 \pm \sqrt{-3}}{2} = \log 1 = 0$. It follows that one third of any quantity a is $\frac{a}{3}$, $\frac{a}{3} + \log\frac{-1 + \sqrt{-3}}{2}$,

$\dfrac{a}{3} + \log \dfrac{-1 - \sqrt{-3}}{2}$. Three times each of these quantities gives
the same result, namely a. In order to solve this conundrum, and it
seems that we should not admit it to be true, we must state one more
paradox: every number has an infinite number of logarithms, among
which no more than one is real. Hence, although $\log 1 = 0$, there are also
an infinite number of other complex logarithms of unity, among which
are $2 \log (- 1)$, $3 \log \dfrac{-1 + \sqrt{-3}}{2}$, $4 \log (- 1)$, $4 \log \pm i$, and an
infinite number of others may be obtained by the extraction of roots of
unity. This proposition becomes more intuitive when we consider the following. Let $x = \log a$, so that $a = e^x$. Then

$$a = 1 + x + \frac{x^2}{2} + \frac{x^3}{6} + \frac{x^4}{24} + \cdots .$$

Since this is an equation of infinite degree, it is not surprising that there
should be an infinite number of roots. Although we have explained the
second paradox, the first one still retains its vigor, insofar as we have
shown that the logarithmic curve has an infinite number of discrete
points below the axis.

517. It is much more obvious that there are these infinite number
of discrete points from a consideration of the equation $y = (- 1)^x$.
Whenever x is an even integer or a fraction with an even numerator,
then $y = 1$. However, if x is an odd integer or a fraction with both
numerator and denominator odd, then $y = - 1$. In all other cases, that
is, when x is either a fraction with even denominator, or an irrational
number, then the value of y is complex. It follows that $y = (- 1)^x$ has
an infinite number of discrete points on both sides of the axis located at
unit distance form the axis. No two of these points are contiguous, and
in spite of this, on the same side of the axis there are two such points

separated by a distance less than any assignable quantity. Indeed, between any two different values for the abscissa, no matter how close they may be, there is not only one, but an infinite number of different fractions with odd denominators. To each of these factions there corresponds a point belonging to the proposed equation. These points are located on two straight lines parallel to the axis, with each line unit distant from the axis. On these lines, there is no interval without one, indeed an infinite number of these discrete points of the equation $y = (- 1)^x$. This same anomaly holds for the equation $y = (- a)^x$ and others similar to it, where a negative quantity is raised to a variable power. It was necessary at this time to discuss the paradox which arises only with transcendental curves.

518. Any equation in which not only logarithms, but also any variable exponent occur should be assigned to the genus of logarithmic curve, indeed any equation which can be put into a form containing a logarithm is to be so assigned. It follows that this includes those curves which are usually called *exponentials*. For example, such a curve is given to be the equation $y = x^x$, or in logarithmic form $m \log y = x \log x$. If we let $x = 0$, then $y = 1$; if $x = 1$, then $y = 1$; if $x = 2$, then $y = 4$; if $x = 3$, then $y = 27$, etc. In *figure 102* we let BDM express the form of the curve with respect to the axis AP, so that if we take $AC = 1$, then $AB = CD = 1$. Between A and C the ordinates are less than one, since if $x = \frac{1}{2}$, then $y = \sqrt{\tfrac{1}{2}} = 0.7071068$. The minimum value of the ordinate occurs when the abscissa $x = \frac{1}{e} = 0.36787944$ and then the ordinate $y = 0.6922005$, as we shall show in what follows. With regard to the curve beyond B, as we shall see, when the abscissa is negative, then $y = \frac{1}{(- x)^x}$, so that this part consists only of discrete points converg-

ing to the axis as if to an asymptote. These discrete points lie on both sides of the axis depending on whether x is odd or even. Indeed there are an infinite number of these points below the axis when we take x to be a fraction with even denominator. If we let $x = \dfrac{1}{2}$, then $y = \sqrt{\tfrac{1}{2}}$, and $y = -\sqrt{\tfrac{1}{2}}$. The continuous curve MDB terminates abruptly at B, as opposed to the nature of algebraic curves. Instead of a continuation there are those discrete points; the reality of these points may be more clearly seen when they are considered as pairs. If this were not conceded, we would have to state that the whole curve ceases abruptly at the point B. But this is contrary to the law of continuity, and so it is absurd.

519. Among the infinite number of other curves of this genus, whose construction can be effected by means of logarithms, there are some for which the construction is not so easily seen. These constructions may be found with the aid of a suitable substitution. Such a curve has the equation $x^y = y^x$. It is immediately clear that when the ordinate is equal to the abscissa then the equation is satisfied. That is the straight line which makes half a right angle with the axis. Still, it is clear that this is not the whole curve since the equation contains more than the equation for the straight line $y = x$, nor does the equation for this straight line exhaust the content of the equation $x^y = y^x$, since it can be satisfied even if $x \neq y$. Indeed, if $x = 2$ and $y = 4$ the equation is still satisfied. Besides the straight line EAF in *figure 103*, the proposed equation has other parts. In order to find them and show that total curve, we make the substitution $y = tx$, so that $x^{tx} = t^x x^x$. If we now extract the xth root, we obtain $x^t = tx$, so that $x^{t-1} = t$. It follows that $x = t^{\frac{1}{t-1}}$ and $y = t^{\frac{t}{t-1}}$. If we let $t - 1 = \dfrac{1}{u}$, then $x = \left(1 + \dfrac{1}{u}\right)^u$ and $y = \left(1 + \dfrac{1}{u}\right)^{u+1}$. It follows that besides the straight line EAF,

there is a branch RS asymptotic to the straight lines AG and AH and the straight line AF is a diameter of the branch RS. The curve intersects the straight line AF in the point C so that $AB = BC = e$, where e is the number whose logarithm is equal to 1. In addition, the equation gives an infinite number of discrete points which along with the straight line EF and the curve RCS exhaust the equation. These infinite number of pairs of numbers x and y can be found to satisfy $x^y = y^x$. Such pairs can be found in the rational numbers as follows:

$$x = 2 \qquad\qquad y = 4$$

$$x = \frac{3^2}{2^2} = \frac{9}{4} \qquad\qquad y = \frac{3^3}{2^3} = \frac{27}{8}$$

$$x = \frac{4^3}{3^3} = \frac{64}{27} \qquad\qquad y = \frac{4^4}{3^4} = \frac{256}{81}$$

$$x = \frac{5^4}{4^4} = \frac{625}{256} \qquad\qquad y = \frac{5^5}{4^5} = \frac{3125}{1024}$$

$$\text{etc.}$$

In the first of any of these pairs is raised to the second power, this quantity is equal to the second raised to the first power. For example,

$$2^4 = 4^2 = 16$$

$$\left(\frac{9}{4}\right)^{\frac{27}{8}} = \left(\frac{27}{8}\right)^{\frac{9}{4}} = \left(\frac{3}{2}\right)^{\frac{27}{4}}$$

$$\left(\frac{64}{27}\right)^{\frac{256}{81}} = \left(\frac{256}{81}\right)^{\frac{64}{27}} = \left(\frac{4}{3}\right)^{\frac{256}{27}},$$

$$\text{etc.}$$

520. Although in this curve and in other similar curves we can determine an infinite number of points by algebraic methods, these curves are in no way algebraic. There are an infinite number of other points which cannot be found algebraically. We now move on to another genus of transcendental curves, which require circular arcs. In what follows we will always consider arcs of a unit circle, so that we introduce no more complication than is necessary. We will consider only the simplest of curves of this genus, such as $\frac{y}{a} = \arcsin\frac{x}{c}$ where the ordinate y is proportional to the arc of a circle whose sine is $\frac{x}{c}$. Since there are an infinite number of arcs with the same sine $\frac{x}{c}$, the ordinate y is an infinite valued function. Not only the ordinate, but other straight lines intersect the curve in an infinite number of points, and this property very clearly distinguishes this curve from algebraic curves. Let s be the shortest arc with sine $\frac{x}{c}$, and let π denote a semicircular arc, then the values of $\frac{y}{a}$ are the following:

$$s, \ \pi - s, \ 2\pi + s, \ 3\pi - s, \ 4\pi + s, \ 5\pi - s,$$

etc.

$$-\pi - s, \ -2\pi + s, \ -3\pi - s, \ -4\pi + s, \ -5\pi - s,$$

etc. In *figure 104* we take the straight line CAB as the axis with A as the origin. First we let the abscissa $x = 0$, then the ordinate $AA^1 = \pi a$, $AA^2 = 2\pi a$, $AA^3 = 3\pi a$, etc. In the other direction $AA^{-1} = \pi a$, $AA^{-2} = 2\pi a$, $AA^{-3} = 3\pi a$, etc., and the curve passes through each of these points. When we let the abscissa $x = AP$, the ordinate intersects the curve in an infinite number of points M, so that $PM^1 = as$, $PM^2 = a(\pi - s)$, $PM^3 = a(2\pi + s)$, etc. The whole curve is made up of an infinite number of parts AE^1A^1, $A^1F^1A^2$, $A^2E^2A^3$, $A^3F^2A^4$, etc. Each straight line parallel to the axis BC which passes through E or F will be a diameter of the curve. We note that $AC = AB = c$, and that the intervals E^1E^2, E^2E^3, E^1E^{-1}, $E^{-1}E^{-2}$ as well as F^1F^2, F^1F^{-1}, $F^{-1}F^{-2}$, are all equal to $2a\pi$. LEIBNIZ called this curve the *sine line*, because it can be easily used to find the sine of any arc. Since $\dfrac{y}{a} = \arcsin\dfrac{x}{c}$ we have $\dfrac{x}{c} = \sin\dfrac{y}{a}$. If we let $\dfrac{y}{a} = \dfrac{1}{2}\pi - \dfrac{z}{a}$, then $\dfrac{x}{c} = \arccos\dfrac{z}{a}$, so that this is also the *cosine line*.

521. In a similar way, from this consideration, we obtain the *tangent line*, whose equation is $y = \arctan x$ where, for the sake of brevity, we have let $a = 1$ and $c = 1$. When we write the equation in the form $x = \tan y = \dfrac{\sin y}{\cos y}$, the curve is easily obtained from the nature of tangents. The curve has an infinite number of asymptotes parallel to each other. Likewise we can draw the *secant line* from the equation $y = \operatorname{arcsec} x$, or $x = \sec y = \dfrac{1}{\cos y}$, which has an infinite number of branches, each of which goes to infinity. We note especially a curve of this genus called a CYCLOID or *trochoid,* which is defined to be the path of a point on the circumference of a circle which is rolling along a straight line. The equation of this curve expressed in rectangular coordinates is $y = \sqrt{1 - x^2} + \arccos x$. This curve is worthy of note both because of the ease with which it is described, and also because of the many important properties which it possesses. Since most of these properties require analysis of the infinite for their explanation, we will now consider only a few which can be derived immediately from its description.

522. In *figure 105* we let the circle ACB roll on the straight line

EA. In order that our investigation might be more general, we choose a point D which lies on the extended diameter rather than the point B on the circumference. The point moves along the curve Dd. Let the radius of this circle $CA = CB = a$, and the distance $CD = b$. We note that at the point D the curve reaches its maximum distance from the straight line. We let the circle reach the position $aQbR$, and we let the distance $AQ = z$, so that the arc $aQ = z$. When we divide this distance by the radius a we obtain the angle $acQ = \dfrac{z}{a}$. The point which is describing the curve is at d, so that $cd = b$ and the angle $dcQ = \pi - \dfrac{z}{a}$. Now we drop the perpendicular dp from the point d to the straight line AQ and also the perpendicular dn to the straight line RQ. It follows that

$$dn = b \sin\frac{z}{a} \quad \text{and} \quad cn = -b \cos\frac{z}{a}. \quad \text{Then} \quad Qn = dp = a + b \cos\frac{z}{a}.$$

We extend dn until it intersects the straight line AD at the point P. If we choose the coordinates $DP = x$ and $Pd = y$, then $x = b + cn$ or

$$x = b - b \cos\frac{z}{a} \qquad \text{and} \qquad y = AQ + dn = z + b \sin\frac{z}{a}. \qquad \text{Since}$$

$b \cos\dfrac{z}{a} = b - x$, we have $b \sin\dfrac{z}{a} = \sqrt{2bx - x^2}$ and

$$z = a \arccos(1 - x/b) = a \arcsin\frac{\sqrt{2bx - x^2}}{b}.$$

When we substitute these values we obtain

$$y = \sqrt{2bx - x^2} + a \arcsin\frac{\sqrt{2bx - x^2}}{b}.$$

If we choose the abscissa on the axis AD with the center C as origin and let $b - x = t$, then $\sqrt{2bx - x^2} = \sqrt{b^2 - t^2}$ and we have the following equation in t and y:

$$y = \sqrt{b^2 - t^2} + a \; \arccos\frac{t}{b},$$

which gives the equation of the *ordinary* cycloid if $b = a$. If $b > a$ or $b < a$, then the curve is called a *curtate cycloid* or *prolate cycloid* respectively. In any case y is an infinite valued function of x or of t, that is, any line parallel to the line AQ intersects the curve in an infinite number of points unless the distance x or t is so large that $\sqrt{2bx - x^2}$ or $\sqrt{b^2 - t^2}$ becomes a complex quantity.

523. Among the curves of this genus, there are other well known curves known as *epicycloids* and *hypocycloids* which arise, as in *figure 106*, when the circle ACB rolls on the circumference of another circle OAQ and a point D taken either outside or inside the moving circle traces a curve Dd. Suppose that the stationary circle has radius $OA = c$, that the moving circle has radius $CA = CB = a$, and that the distance $CD = b$. We take the straight line OD as the axis for the desired curve Dd. From the initial position where O, C, and D lie on a straight line the moving circle goes to the position QcR. We let the arc $AQ = z$, so that the angle $AOQ = \frac{z}{c}$. Then the arc $Qa = AQ = z$, so that the angle $acQ = \frac{z}{a} = Rcd$, the distance $cd = CD = b$, and the point d is on the curve Dd. We drop the perpendicular dP to the axis. Likewise we drop cm and cn perpendicular and parallel to the axis OD respectively. Since the angle $Rcn = AOQ = \frac{z}{c}$, we have the angle

$dcn = \frac{z}{c} + \frac{z}{a} = \frac{(a + c)z}{ac}$.   From   this   we   obtain

$dn = b \sin\frac{(a + c)z}{ac}$,   and $cn = b \cos\frac{(a + c)z}{ac}$.   Then,   since

$OC = Oc = a + c$,   we   have $cm = (a + c)\sin\frac{z}{c}$ and

$Om = (a + c)\cos\frac{z}{c}$. We choose the coordinates $OP = x$ and $Pd = y$

so that

$$x = (a + c)\cos\frac{z}{c} + b \cos\frac{(a + c)z}{ac},$$

and

$$y = (a + c)\sin\frac{z}{c} + b \sin\frac{(a + c)z}{ac}.$$

It should be clear that if $\dfrac{a + c}{a}$ is a rational number, then because of

the commensurability of the angle $\dfrac{z}{c}$ and $\dfrac{(a + c)z}{ac}$, the unknown quan-

tity z can be eliminated so that it is possible to find an algebraic equa-

tion in x and y. In the other cases the curve described in this way will

be transcendental.

We here note that if we take a to be negative, then the result will

be a hypocycloid, since the moving circle lies inside the stationary circle.

Commonly the distance b is taken equal to the radius a; in this case, we

obtain epicycloids and hypocycloids properly speaking. Here we have

described more general curves, and since the equations are no more

difficult, we have seen fit to add them. If the squares x^2 and y^2 are

added, we have $x^2 + y^2 = (a + c)^2 + b^2 + 2b(a + c)\cos\dfrac{z}{a}$. With the

aid of this equation the elimination of z becomes easier when a and c

are commensurable.

524. Besides the case when the radii a and c of both circles are

commensurable so that the curves are algebraic, there is another case

which merits notice, namely, when $b = -a - c$, that is, when the

point D on the curve falls on O, the center of the stationary circle.

Since $b = -a - c$, we have

$$x^2 + y^2 = 2(a + c)^2 \left(1 - \cos\frac{z}{a}\right) = 4(a + c)^2 \left(\cos\frac{z}{2a}\right)^2.$$

It follows that $\cos\dfrac{z}{2a} = \dfrac{\sqrt{x^2 + y^2}}{2(a + c)}$. Since

$$x = (a + c)\left(\cos\frac{z}{c} - \cos\frac{(a + c)z}{ac}\right)$$

and

$$y = (a + c)\left(\sin\frac{z}{c} - \sin\frac{(a + c)z}{ac}\right),$$

we have

$$\frac{x}{y} = - \tan\frac{(2a + c)z}{2ac}$$

and

$$\sin\frac{(2a + c)z}{2ac} = \frac{x}{(x^2 + y^2)^{\frac{1}{2}}}.$$

Furthermore,

$$\cos\frac{(2a + c)z}{2ac} = \frac{-y}{(x^2 + y^2)^{\frac{1}{2}}}.$$

Since

$$\sqrt{x^2 + y^2} = 2(a + c)\cos\frac{z}{2a},$$

we have

$$x = 2(a + c)\cos\frac{z}{2a}\sin\frac{(2a + c)z}{2ac}$$

and

$$y = -2(a + c)\cos\frac{z}{2a}\cos\frac{(2a + c)z}{2ac}.$$

For   example,   suppose $c = 2a$,   then $x = 6a\cos\dfrac{z}{2a}\sin\dfrac{z}{a}$,

$y = -6a\cos\dfrac{z}{2a}\cos\dfrac{z}{a}$, and

$$\sqrt{x^2 + y^2} = 6a\cos\frac{z}{2a}.$$

We let $\cos\dfrac{z}{2a} = q$, so that $\sin\dfrac{z}{2a} = \sqrt{1 - q^2}$, $\sin\dfrac{z}{a} = 2q\sqrt{1 - q^2}$,

and $\cos\dfrac{z}{a} = 2q^2 - 1$. It follows that $q = \dfrac{\sqrt{x^2 + y^2}}{6a}$,

$$y = -6aq(2q^{2-1}) = (1 - 2q^2)\sqrt{x^2 + y^2}$$

$$= \left(1 - \frac{x^2 - y^2}{18a^2}\right)\sqrt{x^2 + y^2},$$

that is,

$$18a^2 y = (18a^2 - x^2 - y^2)\sqrt{x^2 + y^2}.$$

If we let $18a^2 = f^2$ and square both sides of the equation we obtain the sixth degree polynomial

$$(x^2 + y^2) - 2f^2(x^2 + y^2)^2 + f^4 x^2 = 0.$$

Since we are now concerned with transcendental rather than algebraic curves, we will leave this topic and move on to curves whose construction requires both logarithms and circular arcs.

525. We have previously found the curve from the equation $2y = x^i + x^{-i}$, which we transformed into the equation $y = \cos\log x$. We can also express this as arccos $y = \log x$ or $x = e^{\text{arccos } y}$. In *figure 107* we take the straight line AP as the axis with A at the origin. First it should be clear that there is no continuous part of the curve when the

abscissa is negative and that the axis AP intersects the curve in an infinite number of points D, whose distances from A form a geometric progression. That is, $AD = e^{\frac{\pi}{2}}$, $AD^1 = e^{\frac{3\pi}{2}}$, $AD^2 = e^{\frac{5\pi}{2}}$, $AD^3 = e^{\frac{7\pi}{2}}$, etc. There are also an infinite number of intersections between D and A which converge to A, namely, $AD^{-1} = e^{\frac{-\pi}{2}}$, $AD^{-2} = e^{\frac{-3\pi}{2}}$, $AD^{-3} = e^{\frac{-5\pi}{2}}$, etc. The curve oscillates about the axis and is tangent at the points E and F an infinite number of times to the two lines parallel to the axis through A and B, where the distances $AB = AC = 1$. These points of tangency also form a geometric progression. Furthermore, the infinite number of inflections of the curve approach the line BC and finally they move into that line. A particular characteristic of this curve is that the asymptote is not an infinite straight line, but the finite interval BC. This property of the curve keeps it from being an algebraic curve.

526. We should refer also the innumerable species of SPIRALS which can be added to the list of transcendental curves which require either circular arcs alone or a combination of arcs and logarithms. Spirals are associated with some fixed point C, about which the curve turns an infinite number of times, as a kind of center. The nature of these curves is most conveniently expressed by an equation. In *figure 108* we let M be a point on the curve at a distance CM from the center C with angle ACM made by the straight lines CM and CA. Let this angle $ACM = s$, that is, we let s be that arc of a unit circle which measures the angle ACM, and we let the distance $CM = z$. Given any equation in the variables s and z, the resulting curve is a spiral. The angle ACM can be expressed not only by s, but in an infinite number of other ways; $2\pi + s$, $4\pi + s$, $6\pi + s$, etc. as well as $-2\pi + s$, $-4\pi + s$, etc. all

express the same position of the line CM. When these infinite number of values are substituted for s in the equation, the distance CM takes an infinite number of different values, so that when the straight line CM is extended it intersects the curve an infinite number of times unless the value of z becomes complex for these values. We begin with the simplest case, in which $z = as$. For the same position of the straight line CM, z has the values $a(2\pi + s)$, $a(4\pi + s)$, $a(6\pi + s)$, etc. besides $a(2\pi - s)$, $a(4\pi - s)$, $a(6\pi - s)$, etc. Indeed, when we substitute for s the value $\pi + s$, the same straight line CM keeps its position, but the values of z should be taken to be negative. Hence, we should add to the values already given, the following: $-a(\pi + s)$, $-a(3\pi + s)$, $-a(5\pi + s)$ and $a(\pi - s)$, $a(3\pi - s)$, $a(5\pi - s)$, etc. This curve is given in *figure 109*. The straight line AC is tangent to the curve at C and from this point the two branches spiral in both directions about the center C and infinite number of times and each time they intersect the lines BC and AC at right angles, as they move out to infinity. The line BCB^{-1} is the curves diameter. This curve is usually called by the name of its discoverer, the *spiral of Archimedes*. Once the curve is exactly described, the intersections with the line in any direction can be found, as is perfectly clear from the equation $z = as$.

527. Just as the equation $z = as$, which would give a straight line if the coordinates were rectangular, gives the spiral of Archimedes, so other algebraic equations in z and s give other infinite spiral curves provided the equation is such that all of the values of s correspond to real values of z. For example, the equation $z = \dfrac{a}{s}$, which is similar to an equation for the hyperbola related to its asymptotes, gives a spiral which *John* BERNOULLI called a hyperbolic spiral. After an infinite number of turns about the center C, at an infinite distance, it approaches the line

AA as an asymptote. If we consider the equation $z = a \sqrt{s}$, for negative angles s there are no corresponding real distances z. For each positive angle s we obtain a pair of values for z, one of which is positive and
the other negative. The curve spirals about c an infinite number of
times. If the equation in z and s has the form $z = a \sqrt{n^2 - s^2}$, then
the variable z has no real value unless the angle s lies between n and
$- n$. In this case the curve will be bounded, as can be seen in *figure
110*. That is, if the two straight lines EF and EF make an angle equal
to n with the axis ACB through the center C then they will be tangent
to the curve at C and the curve has the form of the lemniscate $ACBCA$.
In a similar way we obtain an infinite number of other transcendental
curves which it would take too long to develop.

528. This discussion might be vastly extended if besides algebraic
equations in z and s we admitted transcendental equations. From this
genus there is one curve which merits special notice. It is expressed by
the equation $s = n \log \dfrac{z}{a}$. In this curve the angle s is proportional to
the logarithm of the distance z. For this reason the curve is called a *logarithmic spiral* and because of many important properties it is especially
noteworthy. The principal property of this curve is that any straight
line from the center C intersects the curve in equal angles. In order to
see this from the equation, in *figure 111* we let the angle $ACM = s$ and
the distance $CM = z$, so that $s = n \log \dfrac{z}{a}$ and $z = ae^{\frac{s}{n}}$. Now we take
a larger angle $ACN = s + v$, so that $CN = ae^{\frac{s}{n}} e^{\frac{v}{n}}$. From the center
C we draw a circular arc ML which will have length equal to zv so that
the distance

$$LN = ae^{\frac{s}{n}}\left(e^{\frac{v}{n}} - 1\right) = ae^{\frac{s}{n}}\left(\frac{v}{n} + \frac{v^2}{2n^2} + \frac{v^3}{6n^3} + \cdots\right).$$

It follows that

$$\frac{ML}{LN} = \frac{v}{\dfrac{v}{n} + \dfrac{v^2}{2n^2} + \dfrac{v^3}{6n^3} + \cdots}$$

$$= \frac{n}{1 + \dfrac{v}{2n} + \dfrac{v^2}{6n^2} + \cdots}.$$

As the difference between the angles, $MCN = v$, vanishes, the ratio $\dfrac{ML}{LN}$ becomes the tangent of the angle which the radius CM makes with the curve. It follows that when v vanishes we have the tangent equal to n, so that the angle is constant. If $n = 1$, the angle will always be half of a right angle. For this reason such a curve is called a semi-rectangular logarithmic spiral.

CHAPTER XXII

On the Solution to Several Problems
Pertaining to the Circle.

529. When we let the radius of a circle be equal to 1, we have seen that half of the circumference is equal to π, or that the arc of a 180 degree angle is equal to 3.14159265358979323846264338, whose common logarithm is 0.49714987269413385435126828288. If this is multiplied by 2.30258 $\cdots$ we obtain the natural logarithms which is equal to 1.14472988585494001741434237. Since the length of the arc of a 180° angle is known, we can find the length of an arc with any number of degrees. Suppose the arc has n degrees, we let z be the length of the arc we wish to find. Since $\dfrac{180}{n} = \dfrac{\pi}{z}$, we have $z = \dfrac{\pi n}{180}$. The logarithm of z can be found if we subtract the logarithm 1.75812263240917221545252526413 from the logarithm of n. If the arc is given in minutes n', then the logarithm 3.53627388279281584796129321l should be subtracted from the logarithm of n. If the arc is given in seconds n'', then the logarithm of the arc is found by subtracting the logarithm 5.31442513317645948047006009 from the logarithms of n or by adding 4.68557486682354051952993990 to the logarithm of n and then subtracting 10 from the resulting characteristic.

530. Conversely from the arc length we can find its sine, tangent, and secant. The angle the arc measures can be expressed in the usual way in terms of degrees, minutes, and seconds. Suppose that z is such an arc length of a unit circle expressed as a decimal. We take the logarithm of z and add 10 to its characteristic, as is usually done in tables of logarithms of sines, tangents, and secants. Then we either subtract it from that logarithm 4.6855748668235405195299939990, or we add it to the logarithm 5.3144251331764594804700600009. In either case we obtain a logarithm of the angle expressed in seconds. In the second method we have to subtract 10 from the characteristic. If we want an angle whose arc is equal to the radius, we can accomplish this with the aid of logarithms quite easily by the golden rule. Since π is to 180 degrees as 1 is to the desired angle. In this way we find that the angle expressed in degrees is 57.295779513082320876798. When the angle is expressed in minutes, we find that it is equal to 3437.74677078493925260788, and that it is equal to 206264.8062470963551564728 seconds. This angle is usually expressed as

$$57° \; 17' \; 44'' \; 48''' \; 22'''' \; 29''''' \; 21''''''.$$

The sine of this angle can be computed from the series found previously and is found to be equal to 0.84147098480514 and the cosine is equal to 0.54030230584341. When we take the quotient of these two numbers we obtain the tangent of this same angle.

531. Now that we have discussed a comparison between an arc of a circle and its sine or tangent, we can answer several other questions about the nature of a circle. First we note that any arc is larger than its sine, except when it vanishes. On the other hand, the cosine is equal to 1 when the angle vanishes, so that the cosine is greater than the arc length; the cosine of a right angle vanishes, so that in this case the cosine is less

than the arc length. From this it follows that there is some angle between $0°$ and $90°$ there is some angle such that the arc is equal to its cosine. We investigate this question in the following problem.

PROBLEM I.

Find the arc of a circle which is equal to its cosine.

SOLUTION.

Let s be the desired arc, so that $s = \cos s$. The easiest method for finding the solution is by the rule which is called the *false position*. In order to use this method we have to know a value which is reasonably close to the solution. We can either make a guess, or take three or more values for s and find their cosines. Suppose we let $s = 30°$. We follow the rule given above: the logarithm of 30 is 1.4471213. From this we subtract 1.7581226 to find that the logarithm of the arc is 9.7189987. But the logarithm of the cosine of 30 is 9.9375306, so that it is clear that the cosine of 30 degrees is much larger than the arc itself, so that the desired arc is larger than $30°$. We suppose now that $s = 40°$. Since the logarithm of 40 is 1.6020600, we subtract 1.7581226 to find that the logarithm of the arc is 9.8439374. But $\log \cos 40° = 9.8842540$, so we know that the desired arc is a little larger than $40°$. Now we suppose that $s = 45°$. Since the logarithm of 45 is 1.6532125, we substract 1.7581226 to find that the logarithm of the arc is 9.8950899. But $\log \cos 45° = 9.8494850$. It follows that the desired angle lies between $40°$ and $45°$. From this we can find a good approximation. If we suppose $s = 40°$, then the error is $+ 403166$, while if we suppose that $s = 45°$, then the error is -456049. The difference in the errors is 859215, so that 859215 is to 403166 as the supposed difference of $5°$ is to the excess of the desired arc to $40°$. It follows that the arc we are seeking is more than $42°$. But these bounds are still too far apart for the exactness we desire. Now let us find closer bounds.

	$s = 42°$	$s = 43°$
log s =	1.6232493	1.6334685
subtract	1.7581226	1.7581226
log s =	9.8651267	9.8753459
	also	also
log cos s =	9.8710735	9.8641275
	+.0059468	-.0112184
	.0171652	

$$171652 : 59468 = 1° : 20', 47''.$$

Hence we have found closer bounds, namely, 42°, 20' and 42°, 21' between which the value s is found. We now express these angles in minutes as follows:

	$s = 2140'$	$s = 2541'$
log s	= 3.4048337	3.4050047
subtract	3.5362739	3.5362739
log s =	9.8685598	9.8687308
log cos s =	9.8687851	9.8686700
	+.0002253	-.0000608
	.0000608	
	.002861	

$$2861 : 2253 = 1' : 47'', 14'''$$

From this we conclude that the arc which is equal to its cosine is 42°, 20', 47'', 14'''. The length of the arc which is equal to its cosine is 0.7390847. This is the solution we sought.

532. In *figure 112* we let ACB be a sector of a circle which is divided into two parts by the chord AB, namely, the segment AEB and the triangle ACB. If the angle ACB is small, the triangle is larger than the segment, while if the angle is sufficiently obtuse, then the segment is larger than the triangle. It follows that there is a sector of the circle which the chord AB divides into two equal parts. Hence we have the following.

PROBLEM II.

Find the sector of a circle ACB which the chord AB cuts into two equal parts, that is, the area of the triangle ACB is equal to the area of the segment AEB.

SOLUTION.

We let the radius $AC = 1$ and the desired arc $AEB = 2s$, so that half of the arc $AE = BE = s$. When the radius CE is drawn, we have $AF = \sin s$, $CF = \cos s$,   so   that   the   triangle $ACB = \sin s \cos s = \frac{1}{2}\sin 2s$. Furthermore, the sector ACB has area equal to s. Since this sector has to be equal to twice the area of the triangle, we have $s = \sin 2s$. Our problem now is to find an arc which is equal to the sine of its double. First we note that the angle must be greater than a right angle, so that s is greater than $45°$. Hence we make the following suppositions.

	$s = 50°$	$s = 55°$	$s = 54°$
$\log s =$	1.6989700	1.7403627	1.7323938
subtract	1.7581226	1.7581226	1.7581226
	9.9408474	9.9822401	9.9742712
$\log \sin 2s =$	9.9933515	9.9729858	9.9782063
	+.0525041	-.0092543	+.0039351
	.0092543		
	.0617584		

$$617584 : 525041 = 5 : 4, 15'$$

It follows that s is near $54°$, $15'$ so we add $s = 54°$ to the above suppositions, and from the errors we conclude that $s = 54°$, $17'$, $54''$. This is the correct answer to the nearest minute. We take the following positions, which differ only in the minute:

$s =$	54°, 17'	$s = 54°, 18'$	$s = 54°, 19'$
	or	or	or
$s =$	3257'	$s = 3258'$	$s = 3259'$
	and	and	and
$2s =$	108°, 34'	$2s = 108°, 36'$	$2s = 108°, 38'$
compl =	71°, 26'	compl = 71°, 24'	compl = 71°, 22'
log $s =$	3.5128178	3.5129511	3.5130844
substract	3.5362739	3.5362739	3.5362739
log $s =$	9.9765439	9.9766772	9.9768105
log sin $2s =$	9.9767872	9.9767022	9.9766171
	+.0002433	+.0000250	-.0001934
		.0001934	
		.0002184	

It follows that $2184 : 250 = 1' : 6'', 52'''$. We conclude that $s = 54°, 18', 6'', 52'''$. If we wish to determine this angle more accurately, we have to use larger tables. Now we make the following suppositions, which differ by 10 seconds.

$s =$	$54°, 18', 0''$	$s = 54°, 18', 10''$	
	or	or	
$s =$	$195480''$	$s = 195490''$	
$2s =$	$108°, 36', 0''$	$2s = 108°, 36', 20''$	
compl. $=$	$71°, 24', 0''$	compl. $= 71°, 23', 40''$	
log $s =$	5.2911023304	5.2911245466	
subtract	5.3144251332	5.3144251332	
	9.9766771972	9.9766994134	
log sin $2s =$	9.9767022291	9.9766880552	
	$+.0000250319$	$-.0000113582$	
	$.0000113582$		
	$.0000363901$		

It follows that $s = 54°, 18', 6'', 52''', 43'''', 33'''''$. We conclude that
the angle $ACB = 108°, 36', 13'', 45''', 27'''', 6'''''$, and that its com-
plement   is   equal   to $71°, 23', 46'', 14''', 32'''', 54'''''$,
log sin $2s = 9.9766924791$ and the sine itself is equal to 0.9477470.
Hence sin $s = AF = BF = 0.8121029$. Twice this number is the length
of the chord $AB = 1.6242058$. Furthermore the cosine $CF = 0.5335143$.
Thus we can construct an approximation to the desired sector, and the
problem is solved.

533. In a similar way we can find the angle whose sine divides a
quadrant of the circle into two equal parts, as in *figure 113*.

PROBLEM III.

Find the ordinate DE which divides a quadrant of the circle ACB
into two parts with equal area.

SOLUTION.

Let the arc $AE = s$, so that the arc $BE = \dfrac{\pi}{2} - s$, since $AEB = \dfrac{\pi}{2}$. The area of the quadrant is $\dfrac{\pi}{4}$. The area of the sector $ACE = \dfrac{1}{2}s$, from which we subtract the triangle

$$CDE = \frac{1}{2}\sin s \cos s,$$

which leaves the region

$$ADE = \frac{1}{2}s - \frac{1}{2}\sin s \cos s,$$

whose double is supposed to equal the quadrant. It follows that

$$\frac{1}{4}\pi = s - \frac{1}{2}\sin 2s,$$

so that

$$s - \frac{\pi}{4} = \frac{1}{2}\sin 2s.$$

We let the arc

$$s - \frac{\pi}{4} = s - 45° = u,$$

so that

$$2s = 90 + 2u, \quad 2u = \cos 2u,$$

and $u = \dfrac{1}{2}\cos 2u$. We need an arc which is equal to its cosine, but this was our first problem. We know that $2u = 42°,\ 20',\ 47'',\ 14'''$, and $u = 21°,\ 10',\ 23'',\ 37''''$. The required arc

$$AE = s = 66°,\ 10',\ 23'',\ 37'''$$

and the arc

$$BE = 23°, 49', 36'', 23'''.$$

It follows that $CD = 0.4039718$, $AD = 0.5960281$, the sine of the arc $AE = DE = 0.9147711$. In this way the quadrant is bisected, the whole circle is divided into eight equal parts, and the problem is solved.

534. Just as a circle is bisected by any straight line drawn through the center, so from any point on the circumference we should be able to draw straight lines which divide the circle into three or more equal parts. We now investigate and find a solution in the case of four parts, as in *figure 114*.

PROBLEM IV.

Given a semicircle AEDB, draw a chord AD which cuts the semicircle into two equal parts.

SOLUTION.

Let the required arc $AD = s$, and draw the radius CD, so that the area of the sector $ACD = \frac{1}{2}s$. We remove the triangle

$$ACD = \frac{1}{2}AC \cdot DE = \frac{1}{2}\sin s$$

from the sector. There remains the segment

$$AD = \frac{1}{2}s - \frac{1}{2}\sin s,$$

which should be equal to half of the semicircle ADB. Since the area of a semicircle is $\frac{1}{2}\pi$, we have

$$s - \sin s = \frac{1}{2}\pi = 90°,$$

or

$$s - 90° = \sin s.$$

We let $s - 90° = u$, so that $\sin s = \cos u$, and $u = \cos u$. From the solution to the first problem we know that

$$u = 42°, 20', 47'', 14''',$$

so that the angle

$$ACD = s = 132°, 20', 47'', 14'''.$$

Then the angle

$$BCD = 47°, 39', 12'', 46'''.$$

The length of the chord $AD = 1.8295422$, and the problem is solved.

535. We have found a segment of the circle which contains one quarter of the area of the whole circle. The diameter is a chord which cuts off a segment which has area equal to half of the whole circle. Now we look for a segment whose area is one third of the whole circle. The following refers to *figure 115*.

PROBLEM V.

Draw two chords AB and AC from a point A on the circumference, which divide the circle into three equal parts.

SOLUTION.

Let the radius of the circle be equal to 1, so that the area of the semicircle is equal to π. Let the arcs $AB = AC = s$. so that the areas of the segments $AEB = AFC = \dfrac{1}{2}s - \dfrac{1}{2}\sin s$. Since the area of the circle is equal to π and the area of the segment AEB should be one third of the area of the circle, we have

$$\frac{1}{2}s - \frac{1}{2}\sin s = \frac{\pi}{3} = 60°,$$

or

$$s - \sin s = 120°.$$

It follows that $s - 120° = \sin s$. We let $s - 120° = u$, then $u = \sin(u + 120) = \sin(60 - u)$. We need to find the arc u which is equal to the sine of the angle $60° - u$. Hence, u must be less than $60°$ so we take the following positions

$u =$	$20°$	$u = 30°$	$u = 40°$
$60 - u =$	$40°$	$60 - u = 30°$	$60 - u = 20°$
$\log u =$	1.3010300	1.4771213	1.6020600
subtract	1.7581226	1.7581226	1.7581226
$\log u =$	9.5429074	9.7189987	9.8439374
$\log \sin(60 - u) =$	9.8080675	9.6989700	9.5340517
	$+.2651601$	$-.0200287$	$-.3098857$

It is clear that the angle u is a little less than $30°$. When the following calculation is made, it is clear that the angle is greater than $29°$. Let $u = 29°$.

$$
\begin{aligned}
60 - u &= \quad 31° \\
\log u &= \quad 1.4623980 \\
\text{subtract} &= \quad 1.7581226 \\
\hline
\log u &= \quad 9.7042754 \\
\log \sin(60 - u) &= \quad 9.7118393 \\
\hline
&\quad +.0075639 \\
&\quad -.0200287 \\
\hline
&\quad .0275926
\end{aligned}
$$

$$275926 : 75639 = 1° : 16', 26''.$$

The angle u is approximately equal to $29°, 16', 26''$. In order to find a more accurate approximation, we take the following positions which differ only by one minute.

	$u =$	$29°, 16'$	$u = 29°, 17'$
		or	or
	$u =$	$1756'$	$u = 1757'$
$60 - u =$		$30°, 44'$	$60 - u = 30°, 43'$
$\log u =$		3.2445245	3.2447718
subtract		3.5362739	3.5362739
$\log u =$		9.7082506	9.7084979
$\log \sin(60 - u) =$		9.7084575	9.7082450
		$+.0002069$	$-.0002529$
		$.0002529$	
		$.0004598$	

$$4598 : 2069 = 1' : 27'', 0'''.$$

We conclude that $u = 29°, 16', 27'', 0'''$, so that the arc

$$s = AEB = 149°, 16', 27'', 0''' = AFC.$$

We conclude that the arc

$$BC = 61°, 27', 6'', 0''',$$

the length of the chords $AB = AC = 1.9285340$, and the problem is solved.

536. We have treated problems in which we had to find an arc which was equal to the sine or cosine of some angle. We now discuss a problem which is more difficult, but we use the same methods. The

following refers to *figure 116.*

PROBLEM VI.

In the semicircle AEB find the arc AE whose sine ED is such that the arc AE is equal to the sum of the two straight lines AD + DE.

SOLUTION.

Since it is immediately obvious that the arc is larger than a quarter of the circle, we try to find its supplement BE. We let the arc $BE = s$, so that the arc $AE = 180° - s$. Since $AC = 1$, $CD = \cos s$, and $DE = \sin s$, we have

$$180° - s = 1 + \cos s + \sin s.$$

Since

$$\sin s = 2 \sin\frac{1}{2}s \, \cos\frac{1}{2}s,$$

and

$$1 + \cos s = 2 \cos\frac{1}{2}s \, \cos\frac{1}{2}s,$$

we conclude that

$$180° - s = 2 \cos\frac{1}{2}s \left(\sin\frac{1}{2}s + \cos\frac{1}{2}s\right).$$

But

$$\cos\left(45° - \frac{1}{2}s\right) = \frac{1}{2^{\frac{1}{2}}}\cos\frac{1}{2}s + \frac{1}{2^{\frac{1}{2}}}\sin\frac{1}{2}s,$$

so that

$$\sin\frac{1}{2}s + \cos\frac{1}{2}s = \sqrt{2}\,\cos\left(45° - \frac{1}{2}s\right).$$

We conclude that $180° - s = 2\sqrt{2}\, \cos\frac{1}{2} s\, \cos\left(45° - \frac{1}{2}s\right)$. After having made this reduction, we take the following positions.

½ s =	20°	½ s = 21°	
45° − ½ s =	25°	45° − ½ s = 24°	
180 − s =	140°	180 − s = 138°	
log(180 − s) =	2.1461280	2.1398791	
subtract	1.7581226	1.7581226	
log(180 − s) =	0.3880054	0.3817565	
log cos ½ s =	9.9729858	9.9701517	
log cos(45 − ½)s =	9.9572757	9.9607302	
log $2\sqrt{2}$ =	0.4515450	0.4515450	
	0.3818065	0.3824269	
error	+.0061989	−.0006704	
	.0006704		
	.0068693		

$$68693 : 61989 = 1° : 54'.$$

It follows that $\frac{1}{2}s$ lies between the bounds 20°, 54′, and 20°, 55′, so we make the following suppositions

½ s =	20°, 54′	½ s =	20°, 55′
45 − ½ s =	24°, 6′	45 − ½ s =	24°, 5′
s =	41°, 48′	s =	41°, 50′
180 − s =	138°, 12′	180 − s =	138°, 10′
	or		or
180 − s =	8292′	180 − s =	8290′
log(180 − s) =	3.9186593		3.9185545
subtract	3.5362739		3.5362739
	0.3823854		0.3822806
log cos ½ s =	9.9704419		9.9703937
log cos(45 − ½ s) =	9.9603919		9.9604484
log $2\sqrt{2}$ =	0.4515450		0.4515450
	0.3823788		0.3823871
error	+.0000066		-.0001065
	.0001065		
	.0001131		

$$1131 : 66 = 1′ : 3′′, 30′′′.$$

We now have

$$\frac{1}{2}s = 20°, 54′, 3′′, 30′′′,$$

so that

$$s = 41°, 48′, 7′′, 0′′′ = BE,$$

and the desired arc

$$AE = 138°, 11′, 53′′, 0′′′.$$

Finally, the lines DE = 0.6665578, AD = 1.7454535, and the problem is solved.

537. Now we will compare arcs with their tangents. Since in the first quadrant an arc is less than its tangent, we seek an arc which is equal to half of its tangent in order to solve the following problem, which refers to *figure 117*.

PROBLEM VII.

Find a sector ACD whose area is half that of the triangle ACE which is bounded by the radius AC, the tangent AE, and the secant CE.

SOLUTION.

We let the arc AD = s, so that the sector ACD = $\frac{1}{2}s$ and the triangle ACE = $\frac{1}{2}\tan s$. It follows that $\frac{1}{2}\tan s$ = s, or $2s$ = $\tan s$. We make the following suppositions

	$s = 60°$	$s = 70°$	$s = 66°$	$s = 67°$
$\log 2s$ =	2.0791812	2.1461280	2.1205739	2.1271048
	1.7581226	1.7581226	1.7581226	1.7581226
$\log 2s$ =	0.3210586	0.3880054	0.3624513	0.3689822
$\log \tan s$ =	0.2385606	0.4389341	0.3514169	0.3721481
	+.0824980	-.0509287	+.0110344	-.0031659

From this we see that s is bounded by 66°, 46′, and 66°, 47′. In order to find more accurate bounds, we take the following positions.

$s =$	$66°, 46'$	$s = 66°, 47'$
	or	or
$s =$	$4006'$	$s = 4007'$
$2s =$	$8012'$	$2s = 8014'$
$\log 2s =$	3.9037409	3.9038493
	3.5362739	3.5362739
$\log 2s =$	0.3674670	0.3675754
$\log \tan s =$	0.3672499	0.3675985
error	$+.0002171$	$-.0000231$
	$.0000231$	
	$.0002402$	

$$2402 : 2171 = 1' : 54'', 14'''.$$

It follows that the arc $s = AD = 66°, 46', 54'', 14'''$, so that the tangent $AE = 2.3311220$, and the problem is solved.

538. With respect to *figure 118* we propose the following

PROBLEM VIII.

Given a quadrant of the circle ACB, find the arc AE which is equal to the chord AE when it is extended to the intersection F.

SOLUTION.

Let the arc $AE = s$, so that the chord $AE = 2 \sin\frac{1}{2}s$ and the ver-sine $AD = 1 - \cos s = 2 \sin\frac{1}{2}s \sin\frac{1}{2}s$. Then the similar triangles ADE and ACF give us the proportion

$$2 \sin\frac{1}{2}s \sin\frac{1}{2}s : 2 \sin\frac{1}{2}s = 1 : s.$$

From this it follows that $s \sin\frac{1}{2}s = 1$. We then take the following positions.

	$s = 70°$	$s = 80°$	$s = 84°$	$s = 85°$
log s =	1.8450980	1.9030900	1.9242793	1.9294189
subtract	1.7581226	1.7581226	1.7581226	1.7581226
	0.0869754	0.1449674	0.1661567	0.1712963
log sin ½ s =	9.7585913	9.8080675	9.8255109	9.8296833
	9.8455667	9.9530349	9.9916676	0.0009796
error	+.1544332	0.0469650	+.0083223	-.0009796

Hence we see that s lies between the bounds 84°, 53' and 84°, 54'. Next we let

	$s = 84°, 53'$	$s = 84°, 54'$
	or	or
	$s = 5093'$	$s = 5094'$
½ s =	42°, 26.5'	½ s = 42°, 27'
log s =	3.7069737	3.7070589
subtract	3.5362739	3.5362739
	0.1706998	0.1707850
log sin ½ s =	9.8292003	9.8292694
	0.9999001	0.0000544
error	+.0000998	-.0000544

From this we obtain the arc $s = AE = 84°, 53', 38'', 51'''$, and the arc $BE = 50°, 6', 21'', 9'''$. This solves the problem.

539. Although an arc of the first quadrant is always less than its tangent, still in the subsequent quadrants there are arcs which are equal to their tangents. We investigate this in the next problem, by means of series.

PROBLEM IX.

Find all arcs which are equal to their tangents.

SOLUTION.

The first arc with this property is infinitely small. Then in the second quadrant there is no such arc since the tangents are negative. In the third quadrant there is such an arc which is just a bit less than 270°. Furthermore, there are such arcs in the fifth, the seventh, etc. quadrants. We let q be equal to a quarter of the circumference of a circle, then the required arc has the form $(2n + 1)q - s$, so that

$$(2n + 1)q - s = \cot s = \frac{1}{\tan s}.$$

If we let $\tan s = x$, then $s = x - \frac{1}{3}x^3 + \frac{1}{5}x^5 - \frac{1}{7}x^7 + \cdots$. It is clear that the larger n becomes, the smaller s becomes, and also x becomes very small. It follows that an approximate value for x is $\frac{1}{(2n + 1)q}$. A closer approximation is given by letting

$$\frac{1}{x} = (2n + 1)q - s = (2n + 1)q$$

$$- \frac{1}{(2n + 1)q} - \frac{2}{3(2n + 1)^3 q^3} - \frac{13}{15(2n + 1)^5 q^5}$$

$$- \frac{146}{105(2n + 1)^7 q^7} - \frac{2343}{945(2n + 1)^9 q^9} - \cdots .$$

Since $q = \frac{\pi}{2} = 1.5707963267948$ the required arc is equal to

$$(2n + 1)1.57079632679 - \frac{0.63661977}{(2n + 1)} - \frac{0.17200817}{(2n + 1)^3}$$

$$- \frac{0.09062596}{(2n + 1)^5} - \frac{0.05892834}{(2n + 1)^7} - \frac{0.04258543}{(2n + 1)^9} - \cdots .$$

If the terms, which are given in arc length, are changed to degree measure, then the desired arc in general form is

$$(2n + 1)90° - \frac{131313''}{2n + 1} - \frac{35479''}{(2n + 1)^3}$$

$$- \frac{18692''}{(2n + 1)^5} - \frac{12155''}{(2n + 1)^7} - \frac{8784''}{(2n + 1)^9}.$$

These are the arcs which solve the problem.

I. $1 \cdot 90°$ - $90°$
II. $3 \cdot 90°$ - $12°, 32', 48''$
III. $5 \cdot 90°$ - $7, 22, 32$
IV. $7 \cdot 90°$ - $5, 14, 22$
V. $9 \cdot 90°$ - $4, 3, 59$
VI. $11 \cdot 90°$ - $3, 19, 24$
VII. $13 \cdot 90°$ - $2, 48, 37$
VIII. $15 \cdot 90°$ - $2, 26, 5$
IX. $17 \cdot 90°$ - $2, 8, 51$
X. $19 \cdot 90°$ - $1, 55, 16$

540. I will not pose more questions of this sort, since a method for solving them should be sufficiently clear from the given examples. Besides, these problems were made up in order that the nature of the circle might be penetrated more deeply, since attempts at these quadrature problems have been unsuccessful by all previous methods. If it should happen in the solution of one or another problem, that the arc might be commensurable with the whole circumference, or the sine or tangent of the arc might be constructed from the radius, then indeed quadratures of the circle have been known. For example, if in the solution of problem VI, the sine DE had been equal to $0.6666666 = \frac{2}{3}$, rather than 0.6665578, then a very nice property of the circle could be used, namely, the arc AE could be constructed equal to the straight lines $AD + DE = 1 + \frac{2}{3} + \sqrt{5/9}$. Even if there is no obvious reason which

makes this kind of quadrature of the circle impossible, still, if there should be such a reason, there does not seem to be a better method for solving the problems than that which we have given in this chapter.

END OF THE SECOND BOOK.

APPENDIX

on

SURFACES

CHAPTER I

On the Surfaces of Solids

1. All of the properties of curves, and the means by which they can be expressed in equations, which we have previously discussed, have very wide application and apply to every curve provided the curve lies in a plane. However, if a curve does not lie in a plane, then the treatment given so far is not sufficient for an adequate discussion of such curves. Curves of this kind have two kinds of curvature, which has been beautifully discussed by the brilliant geometer CLAIRAUT. Since this material is closely connected with the nature of surfaces, which we will now discuss, I have decided not to give a separate treatment, but to explain both simultaneously.

2. Just as lines are either straight or curves, so surfaces are either planar or non-planar. Surfaces which are non-planar are either convex , concave, or convex in one part and concave in another. For example, the surfaces of a globe, a cylinder, and a cone, except for the bases, are convex. The interior of a catenoid is concave. Just as a curve is a straight line if any three of its parts lie on a single straight line, so a surface is planar if any four of its points lie on the same plane. It follows that a surface is non-planar, that is, it is either convex or concave, if it has four points which do not lie in the same plane.

3. The nature of a non-planar surface is easily understood insofar as it is everywhere different from a plane. Just as we investigate the nature of a curve by the distance of each point from some straight line axis, so we study the nature of a surface by the distance of each point from some arbitrarily chosen plane. For this reason, if we are given some surface whose nature we would like to investigate, we choose some arbitrary plane so that we can consider the perpendicular distance of each point of the given surface from the fixed chosen plane. Then we try to find an equation which determines these perpendicular distances. We assert that the nature of the surface can be determined from this equation. conversely, from such an equation each point of the surface can be found, and so the surface itself is so determined.

4. Let a table top represent the fixed plane, to which any surface is to be referred. Choose a straight line AB on this plane as axis, with the origin at A. Let M be any point on the given surface which does not lie on the fixed plane, and let MQ be the perpendicular from M to the plane which passes through the line AB and is perpendicular to the fixed plane, as in *figure 119*. Now we measure the distance from M to Q and drop the perpendicular QP, where P lies on the line AB. We now know the position of the point M if we know the lengths of these straight lines AP, PQ, and QM. Likewise any other point M of the given surface is determined by these mutually perpendicular coordinates, just as any point on a planar curve is determined by a pair of perpendicular coordinates.

5. Since we have three coordinates AP, PQ, and QM, we let $AP = x$, $PQ = y$, and $QM = z$. From the nature of the given surface we can determine the third value z if we are given any two values x and y. In this way we can determine each point M on the surface. The nature of any surface is expressed by an equation in which z is expressed

in terms of the two other coordinates x, and y, and some constants. In this way for any given surface the variable z is set equal to a function of the two variables x and y. Conversely, if z is equal to any function of x and y, then this equation gives rise to some surface, whose nature we find from the equation. This is so, since if we substitute for x and y all positive and negative values we obtain all points Q which lie in its plane, then from the equation of z in terms of x and y, we know the perpendicular distance QM for all points M on the given surface. If the value of z is positive then the point M of the surface lies above the plane APQ, while if the value of z is negative, then the point M lies below that plane. If the value of z vanishes, then the corresponding point M of the given surface lies on that plane. If the value of z is complex and not real, then there is no corresponding point M on the surface. If it should happen that z has several real values, then the perpendicular through the point Q intersects the surface in several points.

6. Concerning various properties of surfaces, we can now make a distinction between continuous or regular surfaces and discontinuous or irregular surfaces. A continuous surface is one such that all of its points can be expressed by a single equation between x, y, and z. That is, where z is equal to a single function of x and y for every point of the surface. An irregular surface is one with its various parts given by different functions. For example, if the given surface in one part is a sphere and in another part is either a cone, or a cylinder, or a plane. Here we confine our attention only to regular surfaces, whose nature is given by a single equation. When this is done, it is easy to treat irregular surfaces, since each part is a regular surface.

7. Among regular surfaces, our first division is between algebraic and transcendental. A surface is called algebraic if its nature can be expressed by an algebraic equation in the coordinates x, y, and z, that

is, when z is equal to an algebraic function in x and y. On the other hand, if z is not equal to an algebraic function in x and y, or if in the equation between x, y, and z there enter transcendental functions such as logarithms or those dependent on circular arcs, then the surface whose nature is expressed in this kind of equation is called transcendental. Examples of such surfaces are $z = x \log y$, or $z = y^x$, or $z = y \sin x$. It is clear that we should first consider algebraic surfaces, and then proceed to transcendental surfaces.

8. Next we should turn our attention to the kind of function of x and y which gives the value of z, as to the number of values of z. First we consider the surfaces which are given by a single-valued function of x and y. Suppose that P is such a single-valued function, that is a non-irrational function of x and y. If $z = P$, then each point Q of its plane corresponds to a single point on the surface. That is, each line perpendicular to the plane APQ intersects the surface in a single point. In this case it is not possible that the value of the straight line QM should be a complex number, since each such straight line gives a real point of the surface. Nevertheless this distinction of functions is not essential to the kinds of surfaces, since it depends on the position of the plane APQ which, along with the choice of axis, is arbitrary. It can happen that the same surface may be given by a function z which is single-valued with respect to one plane, and becomes many-valued with respect to a different plane.

9. Let P and Q be any single-valued functions of x and y. If $z^2 - Pz + Q = 0$, then each perpendicular straight line through the point Q will intersect the surface in either two points or no points, since z has two real values or two complex values. In a similar way, if P, Q, and R are single-valued function of x and y, and $z^3 - Pz^2 + Qz - R = 0$, then z is a three-valued function. In this

case each straight line QM will intersect the surface in three points, if all of the roots of the equation are real, or in only one point, if two of the roots are imaginary. In a similar way we can judge the number of intersections from the equation which defines z.

10. Furthermore, just as in equations for curves, the two coordinates can be interchanged, so in equations for any surface, the three coordinates x, y, and z can be permuted. In the first place, if in the plane APQ we take another straight line Ap perpendicular to AP, then $Ap = y$ and $pQ = x$. In this way the two coordinates x and y have been interchanged. All of the other permutations can be understood in terms of the rectangular parallelepiped $ApQM\xi\pi qPA$ in which we see three fixed mutually perpendicular planes $APQp$, $APq\pi$, and $Ap\xi\pi$. With respect to these, any given surface with point M has the same equation in x, y, and z. In each plane we have two axes, each of which has A as origin. It follows that we have the following six different relations between the coordinates.

For the plane $APQp$, either

$$AP = x \quad PQ = y \quad QM = z,$$

or

$$Ap = y \quad pQ = x \quad QM = z.$$

For the plane $APq\pi$, either

$$AP = x \quad Pq = z \quad qM = y,$$

or

$$A\pi = z \quad \pi q = x \quad qM = y.$$

For the plane $Ap\xi\pi$, either

$$Ap = y \quad p\,\xi = z \quad \xi M = x,$$

or

$$A\,\pi = z \quad \pi\xi = y \quad \xi M = x.$$

In any case, from the fixed point A to the point M on the surface, the straight line AM has the distance

$$\sqrt{x^2 + y^2 + z^2}.$$

11. The same equation in the three coordinates x, y, and z give the relationship of the surface to the three planes which are mutually perpendicular and intersect in the point A. Insofar as the variable z gives the distance of some point M of a surface from the plane APQ, so the variable y gives the distance of the same point M from the plane APq, and the variables x gives the distance from the plane $Ap\xi$. If we know the distance of the point M from each of these three planes, then we know its true position. We should pay attention to these three planes to which any surface can be referred by means of the three variables x, y, and z. If one of the planes, for example APQ, is horizontal, then the other two are vertical since they contain the horizontal lines AP or Ap respectively.

12. With the three mutually perpendicular planes fixed, we refer a given surface to the three planes. From any point M on the surface we draw the straight lines perpendicular to the planes: MQ, Mq, and $M\xi$ to APQ, APq, and $A\pi\xi$ respectively. We let the distances $MQ = z$, $Mq = y$, and $M\xi = x$. When we complete the paralellepiped we have three sets of equal sides from the point A, that is $AP = x$, $Ap = y$, and $A\pi = z$, from which the position of the point M is found. It is clear that if the variables x, y and z lie in the region indicated by the figure, then we consider their values to be positive. If they lie on the opposite

side of their plane, then they are to be considered as negative.

13. If in the equation in x, y, and z, the variable z, which measures the perpendicular distance to the plane APQ, has only even exponents, then z always has two equal values, one positive and one negative. It follows that the surface is so situated that on either side of the plane APQ there are parts which are similar and equal, so that the solid which is bounded by the surface, is also divided by the plane APQ into two similar and equal parts. Just as when a plane curve is divided into two similar and equal parts by a straight line, that line is called a diameter, so in the solid case, that plane which divides the solid into two similar parts, is called a *diametral plane*. We see then that if the variable z has only even exponents in the equation, then the plane APQ is diametral.

14. In a similar way we understand that if in the equation of a surface the variable y, which measures the distance to the plane APq, has only even exponents, then the plane APq is diametral. Likewise, if the variable x has only even exponents, then the plane $Ap\xi$ is diametral. Hence we can see immediately from the equation in the three variables x, y, and z for any surface, whether or not any of the three planes APQ, APq, or $Ap\xi$ is a diametral plane. It can happen that two or even all three of these planes are diametral. For example, for a sphere whose center is at A, since the radius

$$AM = \sqrt{x^2 + y^2 + z^2} = a,$$

we have $x^2 + y^2 + z^2 = a^2$. It follows that each of the three planes divides the sphere into two similar and equal parts.

15. In *figure 120* we represent the three planes, with respect to which a given surface is to be expressed by an equation. The three mutually perpendicular planes which intersect in the point A are denoted

by $QQ^1Q^2Q^3$, $TT^1T^2T^3$, and $VV^1V^2V^3$. If we consider these three planes to be extended to infinity in each direction, then all of space is divided into eight regions represented in the figure by AX, AX^1, AX^2, AX^3, AX^4, AX^5, AX^6, and AX^7. If the variables x, y, and z are in the first region, then they all have positive values. In the other regions one, two, or all three of the variables will have negative values. The sign of each of the variables should be clear from the following table.

Region AX	Region AX^1	Region AX^2	Region AX^3
$AP = +x$	$AP' = -x$	$AP = +x$	$AP' = -x$
$AR = +y$	$AR = +y$	$AR = +y$	$AR = +y$
$AS = +z$	$AS = +z$	$AS' = -z$	$AS' = -z$

Region AX^4	Region AX^5	Region AX^6	Region AX^7
$AP = +x$	$AP' = -x$	$AP = +x$	$AP' = -x$
$AR' = -y$	$AR' = -y$	$AR' = -y$	$AR' = -y$
$AS = +z$	$AS = +z$	$AS' = -z$	$AS' = -z$

16. It will be rather convenient to give each of these eight regions a numerical symbol, so that we may refer to them more easily. Each of these eight regions has A as a boundary point and they are divided by the three mutually perpendicular planes. The planes are determined by the three mutually perpendicular lines Pp, Qq, and Rr which also intersect in the point A. The eight regions can be determined by a choice of three of the letters P, Q, R where the choice of a letter is either upper case or lower case. The principal, or first, region PQR is that part of space contained in the parallelepiped formed by the three straight lines AP, AQ, AR, extended indefinitely beyond P, Q, and R. The region Pqr will be that point of space contained in the paralellepiped formed by the extensions of AP, Aq, and Ar. When we let $AP = x$, $AQ = y$, $AR = z$, and also $Ap = -x$ $Aq = -y$, $Ar = -z$, then we

distinguish the regions as follows:

	I. PQR	II. PQr
between	$AP = +x$	$AP = +x$
the	$AQ = +y$	$AQ = +y$
coordinates	$AR = +z$	$Ar = -z$

	III. PqR	IV. pQR
between	$AP = +x$	$Ap = -x$
the	$Aq = -y$	$AQ = +y$
coordinates	$AR = +z$	$AR = +z$

	V. Pqr	VI. pQr
between	$AP = +x$	$Ap = -x$
the	$Aq = -y$	$AQ = +y$
coordinates	$Ar = -z$	$Ar = -z$

	VII. pqR	VIII. pqr
between	$Ap = -x$	$Ap = -x$
the	$Aq = -y$	$Aq = -y$
coordinates	$AR = +z$	$Ar = -z$

17. These regions are more or less closely located one to another. In the first place there are regions which have two common coordinates, so that they have a common tangent plane, and these are called

conjugate regions. Then, if two coordinates differ, the regions have a common tangent line, and the regions are called *disjoint*. Finally, if the signs of all three coordinates are different, then the regions have only the point A in common, and the regions are said to be *opposite*. The following table gives the relationship between each pair of regions.

Region	*Conjugates*			*Disjuncts*			*Opposite*
PQR I	PQr II	PqR III	pQR IV	Pqr V	pQr VI	pqR VII	pqr VIII
PQr II	PQR I	Pqr V	pQr VI	PqR III	pQR IV	pqr VIII	pqR VII
PqR III	Pqr V	PQR I	pqR VII	PQr II	pqR VIII	pQR IV	pQr VI
pQR IV	pQr VI	pqR VII	PQR I	pqr VIII	PQr II	PqR III	Pqr V
Pqr V	PqR III	PQr II	pqr VIII	PQR I	pqR VII	pQr VI	pQR IV
pQr VI	pQR IV	pqr VIII	PQr II	pqR VII	PQR I	Pqr V	PqR III
pqR VII	pqr VIII	pQR IV	PqR III	pQr VI	Pqr V	PQR I	PQr II
pqr VIII	pqR VII	pQr VI	Pqr V	pQR IV	PqR III	PQr II	PQR I

18. It is clear that each region has three conjugates, three disjunctions, and one opposite. It is clear from the table how each region is

related to any other. The order of the numbers designating each region is worthy of notice. In order that this may be seen more clearly, we put only the numbers into the following.

1	2	3	4	5	6	7	8
2	1	5	6	3	4	8	7
3	5	1	7	2	8	4	6
4	6	7	1	8	2	3	5
5	3	2	8	1	7	6	4
6	4	8	2	7	1	5	3
7	8	4	3	6	5	1	2
8	7	6	5	4	3	2	1

This should be useful in the discussion which follows.

19. We have already noted that whenever the equation for a surface has only even exponen ts for z, then the surface has two similar and equal parts. That is, the part in the first region is equal to that in the second region. In a similar way in the third and fifth, and fourth and sixth, and finally, in the seventh and eighth. This information comes from the first two columns of the table. However, if the variable y has only even exponents, then the first and third regions correspond as well as the second with the fifth, the fourth with the seventh, and the sixth with the eighth. Finally, if x has only even exponents in the equation, then the first region corresponds to the fourth, the second with the sixth, the third with the seventh, and the fifth with the eighth. That is,

if in the equation there are only even exponents for

variable

z	y	x
belongs to regions	belongs to regions	belongs to regions
1, 2, 3, 4, 5, 6, 7, 8	1, 2, 3, 4, 5, 6, 7, 8	1, 2, 3, 4, 5, 6, 7, 8
2, 1, 5, 6, 3, 4, 8, 7	3, 5, 1, 7, 2, 8, 4, 6	4, 6, 7, 1, 8, 2, 3, 5

20. In order that the parts of a surface which lie in the disjoint regions 1 and 5 should be equal, it is necessary that the equation remain the same even if y and z take negative values. This will be the case if in each term of the equation the total degree of both y and z taken together is either always even or always odd. In this case the first region corresponds with the fifth, the second with the third, the fourth with the eighth, and the sixth with the seventh. In a similar way, if in the equation for the surface both variables x and z together have a total degree which is in every term even or in every term odd, then the first region corresponds with the sixth, the second with the fourth, the third with the eighth, and the fifth with the seventh. That is

If in each term of the equation for the surface the total degree
of the two variables is always odd or is always even

variables

y and z	x and z	x and y
corresponding regions	corresponding regions	corresponding regions
1, 2, 3, 4, 5, 6, 7, 8	1, 2, 3, 4, 5, 6, 7, 8	1, 2, 3, 4, 5, 6, 7, 8
5, 3, 2, 8, 1, 7, 6, 4	6, 4, 8, 2, 7, 1, 5, 3	7, 8, 4, 3, 6, 5, 1, 2

If in each term of the equation for the surface
the total degree of the three variables
is always odd or is always even,
then the opposite regions correspond

1, 2, 3, 4, 5, 6, 7, 8

8, 7, 6, 5, 4, 3, 2, 1.

21. If two or even all three of these conditions hold simultaneously, then four or all eight of the regions contain congruent parts of the surface. To be specific,

If both x and y taken individually

have even exponents in each term

then the following four regions have congruent parts

1, 2, 3, 4, 5, 6, 7, 8

3, 5, 1, 7, 2, 8, 4, 6

4, 6, 7, 1, 8, 2, 3, 5

7, 8, 4, 3, 6, 5, 1, 2.

If both x and z taken individually

have even exponents in each term

then the following four regions have congruent parts

1, 2, 3, 4, 5, 6, 7, 8

2, 1, 5, 6, 3, 4, 8, 7

4, 6, 7, 1, 8, 2, 3, 5

6, 4, 8, 2, 7, 1, 5, 3.

If both y and z taken individually

have even exponents in each term

then the following four regions have congruent parts

1, 2, 3, 4, 5, 6, 7, 8

2, 1, 5, 6, 3, 4, 8, 7

3, 5, 1, 7, 2, 8, 4, 6

5, 3, 2, 8, 1, 7, 6, 4.

22. If one of the variables has even exponents everywhere and the other two taken together have joint degree which is everywhere even or is everywhere odd, then four of the regions are congruent according to the following schemes.

If z has even exponents everywhere

while x and y taken together have joint degree

which is everywhere odd or everywhere even,

then the following four regions have congruent parts

1, 2, 3, 4, 5, 6, 7, 8

2, 1, 5, 6, 3, 4, 8, 7

7, 8, 4, 3, 6, 5, 1, 2

8, 7, 6, 5, 4, 3, 2, 1.

If y everywhere has even exponents

while x and z taken together have joint degree

which is everywhere odd or everywhere even

then the following four regions are congruent

1, 2, 3, 4, 5, 6, 7, 8

3, 5, 1, 7, 2, 8, 4, 6

6, 4, 8, 2, 7, 1, 5, 3

8, 7, 6, 5, 4, 3, 2, 1.

If x has even exponents everywhere
while y and z taken together
have joint degree which is everywhere
odd or everywhere even,
then the following four regions have congruent parts

1, 2, 3, 4, 5, 6, 7 ,8

4, 6, 7, 1, 8, 2, 3, 5

5, 3, 2, 8, 1, 7, 6, 4

8, 7, 6, 5, 4, 3, 2, 1.

All three of these cases occur simultaneously if all three variables x, y, and z taken together have joint degree which is everywhere even or everywhere odd

23.   There remain the following cases where four regions have congruent parts.

If x and y taken together,
and y and z taken together
have joint degrees which are
everywhere even or everywhere odd,
then the following four regions have congruent parts

1, 2, 3, 4, 5, 6, 7, 8

5, 3, 2, 8, 1, 7, 6, 4

7, 8, 4, 3, 6, 5, 1, 2

6, 4, 8, 2, 7, 1, 5, 3.

The same congruences are produced if also the two variables x and z have joint degree which is everywhere even or everywhere odd. Those parts of the surface which are contained in disjoint regions will be congruent, if in the equation each pair of variables taken together have joint degree which is everywhere even or everywhere odd. Since there are three such pair, we should note that if two pair have this property, then the third pair will also have the property.

24. We remark that whenever there are conditions sufficient to produce four congruent parts, and there is another new condition, not contained in those already considered, which implies congruence in two regions, then each part will be congruent to each other part, so that all eight parts are congruent to each other. The equation for a surface of this kind has to satisfy all of the properties we have considered simultaneously. That is, each variable taken by itself must have an even exponent; from this it follows that any pair taken together has even joint degree, and also all three taken together will have a joint even degree.

25. As to whether any given equation in three variables has one, two, or all three of these properties, we note that it is easy to see whether each variable has even exponents. It is no more difficult to decide whether all three variables taken together have a joint degree which is everywhere even or everywhere odd. It is more difficult to decide whether only two of the variables have this property. If we make the substitutions in the equation: $x = nz$, or $y = mz$, or $x = ny$, and

notice whether in the first two cases the variable z has an even exponent everywhere, or in the last case the variable y has an even exponent everywhere. If this is the result, then the two variables taken together have joint degree which is everywhere even or everywhere odd. In this case there are at least two parts of the surface which are congruent.

CHAPTER II

On the Intersection of a
Surface and an Arbitrary Plane.

26. Just as the intersection of two lines, either curved or straight, consists of points, so the intersection of surfaces are straight lines or curves. The intersection of two planes is a straight line, which is seen in elementary geometry. The intersection of a sphere with a plane is a circle. It is a great help in understanding a surface, if we know the curves formed by the intersection of the surface and certain given planes. In this way we also know an infinite number of points on the surface, since by the methods of the previous chapter, individual values of the variable z correspond to individual points on the surface.

27. Since we have related a surface to three mutually perpendicular planes, we should first investigate the intersection of the surface with these planes. If we take the first plane APQ, as in *figure 121*, we have the variables $AP = x$, $AQ = y$, and since the third variable z gives the distance of a point on the surface from this plane, we see that when $z = 0$, we have points of the surface which also lie on the plane APQ. For this reason the remaining equation in x and y gives the curve in which the surface intersects the plane APQ. In a similar way, if we let $y = 0$, the equation in x and z expresses the intersection of the surface and the plane APR. Likewise, when $x = 0$ the equation in y and z gives the intersection of the surface and the plane AQR.

28. We have already noted that if the surface is a sphere with center at the point A and radius equal to a, then the equation of the surface is $x^2 + y^2 + z^2 = a^2$. We will use this as an example of these intersections. When $z = 0$, the equation $x^2 + y^2 = a^2$ gives the intersection of the sphere and the plane APQ, which is clearly a circle with center at A and radius equal to a. In a similar way, if we let $y = 0$, the intersection of the sphere with the plane APR gives the circle with equation $x^2 + z^2 = a^2$. In the same way, if we let $x = 0$, the equation $y^2 + z^2 = a^2$ gives the equation of the circle which is the intersection of the sphere and the plane AQR. This is sufficiently clear, since sections through the center of the sphere give great circles, that is circles with the same radius as that of the sphere.

29. It is hardly more difficult to find the intersection of a surface with a plane which is parallel to one of these three planes. Consider a plane which is parallel to the plane APQ at a distance equal to h. All of the points on the surface which have this same distance from the plane APQ, which is given by the variable z, lie on the intersection with this parallel plane. We let $z = h$ to obtain the equation of this intersection. In this case we have an equation in the two rectangular coordinates x and y which expresses the intersection. In the same way the intersections with planes which are parallel to the planes APR or AQR can be obtained, so we will not repeat the details.

30. Hence, if for one of the variables z we let z be equal to the constant h in the equation for the surface, we obtain the equation of the intersection of the surface and the plane parallel to the plane APQ at a distance equal to h. When we successively let h take all possible values,

both positive and negative. We obtain all of the intersections of the surface and the planes parallel to the plane APQ. In this way we are able to divide the surface into an infinite number of sections. With a knowledge of all of the sections, we know the whole surface. Indeed, all of these sections are expressed by a single equation in the coordinates, x and y and the indeterminate constant h. From this one equation we obtain the equations of all of the sections which are curves which are similar, or at least nearly so.

31. All of the sections by planes parallel to the plane APQ will be congruent (and the same is true if the plane is APR or AQR) if the equation in x and y is the same no matter what the value of h might be. This will happen only if the variable z, for which h is substituted, does not occur in the equation for the surface. If the third variable z does not enter into the equation for the surface, then all of the planer sections parallel to the plane APQ will be congruent. This property is expressed by the equation of the surface, insofar as the equation involves only the variables x and y. In a similar way, if the variable x or the variable y is lacking in the equation for the surface, then all of the planar sections parallel to the plane AQR or the plane APR respectively, will be congruent.

32. Not only is a surface of this kind easily conceived by the mind, but it can also be constructed in a material way. Suppose that in the given equation there is no variable z, so that the equation is only in the coordinates $AP = x$ and $AQ = PM = y$, as in *figure 122*. From this we obtain the curve BMD in the plane APQ. We now think of the straight line which is perpendicular to the plane and extended indefinitely, moving along the curve BMD. This motion will trace out the whole surface given by the equation. It should be clear that if the curve BMD is a circle, then the surface which arises will be a right

cylinder. If the curve BMD is an ellipse, then the surface which is generated will be a scalene cylinder. If the curve BMD is made up of straight line segments, then the resulting surface will be that of a prism.

33. Surfaces of this kind are made up of cylinders and prisms, it is convenient to refer to this genus of surfaces as cylindrical or prismatic. The individual species of this genus are determined by the plane figure BMD from which it arises in the way previously described. That figure BMD is called the *base*. Whenever one of the three variables x, y, and z is missing from the equation, then the surface given by the equation is cylindrical or prismatic. If the two variables y and z are missing from the equation, since x is equal to a constant, the curve BMD becomes a straight line perpendicular to the axis AD so that the surface is a plane perpendicular to the plane APQ.

34. After this genus of surfaces, the next most worthy of notice is that which arises from a homogeneous equation in the three variables x, y, and z. That is, in the equation the joint degree in each term is the same. An example of such an equation is $z^2 = mxz + x^2 + y^2$. From the equation we see that all sections by planes parallel to one of the three principal planes will be similar. Indeed, if we give the constant value h to the variable z, then it is clear from the equation $h^2 = mhx + x^2 + y^2$ that we obtain an infinite number of similar figures as the parameter proportional to h varies. Since the sections are not only similar, but increase proportionately to the distance from the plane APQ, it follows that there is a straight line from the point A through corresponding points in each section.

35. Suppose that we have such a homogeneous equation in the three variables x, y, and z, and that we give to the variable z the value $AR = h$, as in *figure 123*. We obtain the figure $TSsMm$ in the plane

parallel to plane APQ which passes through the point R. We obtain an equation in x and y so that $RV = x$ and $VM = y$. With the section $TSsMm$ given, we think of an indefinitely extended straight line through the point A and a point on the section. As this line moves along the section the whole surface is generated. It is clear that if the figure $TSsMm$ is a circle with center at R then the surface is a right circular cone. If R is not the center of the circle, then the surface is a scalene cone. If the figure is composed of straight line segments, then the surface will be of a pyramid. For this reason, the surfaces corresponding to equations of this genus will be called *conical* or *pyramidal*.

36. From what we have seen it should be clear that if an equation in three variables x, y, and z is homogeneous, so that the surface is conical or pyramidal, then not only are all sections by planes parallel to the principal plane APQ similar and proportional to their distance from the vertex A, but for the same reason, we know also that all sections by planes parallel to the plane APR or the plane AQR have the same property, so that corresponding sides preserve the ratio to their distance from A. Later we will show that absolutely all sections of this kind of solid by parallel planes, that is, by all planes parallel to any plane through the point A, will be similar to each other and be proportional to the distance of the plane from the vertex A.

37. We now progress to a genus of surfaces which has a wider range. Let Z be any function of z. Suppose we have a homogeneous function in the three variables x, y, and z. We let $Z = H$ when $z = h$, so that the equation becomes homogeneous in x, y, and H. Now sections by planes parallel to the plane APQ will be similar, and proportional not to the distance h but proportional rather to the parameter H. From this it follows that the lines through corresponding points of the sections are not straight lines, but curves depending on the function Z. We cannot

now say that sections parallel to any plane will be similar to each other.

38. This genus includes the two genera we first discussed. Indeed, if $Z = z$ or $Z = \alpha z$, since we have a homogeneous equation in x, y, and z, the surface will be conical. Likewise, if $Z = \alpha + \beta z$, we have the same result, with this difference: the vertex of the cone is not at the point A. Indeed, if $Z = \dfrac{b - z}{b}$, then the vertex of the cone lies at a distance equal to b from A. If we let $b = \infty$, then the surface is no longer conical but cylindrical and $Z = 1$. It follows that the equation for cylindrical surfaces is homogeneous in the variables x, y, and the constant 1. No matter how the equation is composed from the variables x and y, if z is lacking, then it can always be made homogeneous in x, y, and 1 by powers of 1. Hence, as we have already seen, every equation which lacks one variable gives a cylindrical surface.

39. Among the solids whose sections by planes parallel to one of the principal planes are all similar, those most worthy of notice are those in which these sections are all circles whose centers lie on the same straight line AR, perpendicular to the plane APQ. These solids are turned on a lathe, and so they are called *turned*. For solids of this kind the general equation for the surface is $Z^2 = x^2 + y^2$. Whatever value is given to z so that $Z = H$, we obtain for the section parallel to the plane APQ, the equation $H^2 = x^2 + y^2$, which is a circle with radius equal to H and center on the straight line AR. If $Z^2 = z^2$, then we have the right circular cone; if $Z^2 = a^2$, we have a cylinder, and if $Z^2 = a^2 - z^2$. the result is a sphere. These are all species of turned solids.

40. Now we consider solids, all of whose sections by planes perpendicular to the axis AP are triangles with vertex T on the straight line DT parallel to the axis AP, as in *figure 124*. Let AVB be the base of this solid, that is, the section by the plane APQ, which is some curve.

Let the straight line DT lie at a distance $AD = c$ from the axis AB. We then have the three variables $AP = x$, $PQ = y$, and $QM = z$. Let the distance PV be any function of x which we call P. Then from the similar triangles VQM and VPT we have $\dfrac{P}{c} = \dfrac{P - y}{z}$, that is $z = c - \dfrac{cy}{P}$, for any solid of this kind, we equate $\dfrac{cy}{c - z}$ to any function of x. This solid differs from a cone in that it is defined on the straight edge DT, while the cone is defined on a single point. If the base AVB happens to be a circle, then the resulting solid is that which WALLIS discussed and called a *wedge-cone*

41. Suppose, as in the previous section, that all sections by planes perpendicular to the axis AB are triangles, PTV with a right angle at P, as in *figure 125,* and with vertex T on any curve AT. The base will be the curve AVB. We choose the three variables $AP = x$, $PQ = y$, and $QM = z$. On the curve AVB the straight line PV is given by some function of x which we call P. Then the line PT also becomes a function of x which we call Q. From the similar triangles we have $\dfrac{P}{Q} = \dfrac{P - y}{z}$, so that $z = Q - \dfrac{Qy}{P}$, or $Pz + Qy = PQ$. We also have $\dfrac{z}{Q} + \dfrac{y}{P} = 1$, which is a constant. If the variables y and z never have an exponent greater than 1, then the solid belongs to the genus we have described here.

42. Since we have been considering solids such that all sections by planes parallel to a principal plane are similar, now we consider those in which these sections are at least affine, that is, corresponding abscissas have ordinates which are proportional to each other, consider *figure 126,* in which the three principal sections are ABC, ACD, and ABD. All of the sections parallel to ACD must be affine figures. We suppose that the

base $AC = a$ and the altitude $AD = b$. We take as coordinates $Aq = p$ and $qm = q$ where q is some function of p. Now consider some parallel section PTV, where we let $AP = x$. In this section we let the base PV be some function of x which we call P and the altitude PT is some function of x which we call Q. Now we let $PQ = y$ and $QM = z$. From the affine property, we have $\dfrac{a}{p} = \dfrac{P}{y}$ and $\dfrac{b}{q} = \dfrac{Q}{z}$. It follows that $y = \dfrac{Pp}{a}$ and $z = \dfrac{Qq}{b}$.

43. Hence, if we have for some solid all three of the principal sections ABC, ACD, and ABD, then the nature of the solid is determined by the fact that all sections parallel to ACD are affine to that section. In the first place, we have P and Q which are function of x, then q is a function of p. From the two variables x and p, the two coordinates y and z are determined. If we want an equation in x, y, and z, then because q is a function p, that is, we have an equation in p and q, we can substitute in this equation $p = \dfrac{ay}{P}$ and $q = \dfrac{bz}{Q}$. But P and Q are both functions of x, so that we now have an equation in the three coordinates x, y, and z, by which we have shown that the solid belongs to this genus. It is clear that if we let $x = 0$, then we must have $P = a$ and $Q = b$.

44. If in the equation for a surface the two variables y and z have the same exponent in each term, then all sections perpendicular to the axis AP will have straight line figures. This is the case since when any constant value is substituted for x, we obtain a homogeneous equation in y and z which gives one or more straight lines. Since the exponents of the two coordinates y and z are the same everywhere, whether they be even or odd, this kind of surface has two congruent parts in the first and fifth regions, as well as in the second and third, and likewise in the other

regions as indicated in the table given in section 20.

45. We have seen many bodies in which there are an infinite number of sections which are straight lines, namely, those just treated as well as cylinders and cones. If we consider solids such that sections which include the axis AP are straight lines, we have a wider genus as in *figure 127*, we let $AKMP$ be a section of the solid through the axis AP, with an angle $MPV = \phi$. We let $AP = x$, $PQ = y$, and $QM = z$. Then $\dfrac{z}{y} = \tan\phi$ and the straight line $PM = \dfrac{z}{\sin\phi}$. Suppose now that KM is a straight line, so that $\dfrac{z}{\sin\phi} = \alpha x + \beta$, where α and β are constants depending only on the angle ϕ, that is, they do not depend on y or z. There are functions R and S which do not depend on y or z such that $x = Rz + S$, or $x = Ry + S$. If we let T be a function which depends on y or z only to the first power, and let S be independent of y and z, then all solids of this kind have the general equation $x = T + S$.

46. It is easy to determine any section through the axis AP for any surface whose nature is given by an equation in x, y, and z. We let the angle $VPM = \phi$ be the angle made by the two planes $AKMP$ and $ACVP$. We let the straight line $PM = v$, which will be the desired ordinate. Then we have $QM = z = v \sin\phi$ and $PQ = y = v \cos\phi$. Now if we substitute these values for y and z into the equation for the surface we obtain an equation in the two variables x and v. From this equation we know the nature of the section $AKMP$. In a similar way all of the sections which pass through one of the other principal axis AQ or AR can be found, as in *figure 121*. These three axes, AP, AQ, and AR upon which the three variables x, y, and z depend can be permuted, so that anything we learn about one axis will be true of the other two.

47. We take the plane APQ as the standard plane to which we

refer all sections of the surface. Any section will be parallel to this plane, or it will make an angle with it. In the latter case the extension of the plane will intersect the plane APQ in a straight line. In the former case, when the plane is parallel to the plane APQ, the nature of the section is found by substituting a constant for the variable z. In the latter case, when the section makes an angle with the plane APQ, we have just given the nature of the section provided we let the straight line AP or the straight line AQ be the intersection of the plane with the plane APQ. In order to obtain all possible sections, we consider the other two standard planes.

48. Let the straight line ES, which is parallel to the axis AP, as in *figure 128,* be the intersection of the section plane with the standard plane APQ. Let the angle made by the section plane ESM with the plane APQ be $QSM = \phi$. We let the distance $AE = f$. Since $AP = x$, $PQ = y$, and $QM = z$, we have $ES = x$ and $QS = y + f$. Now if we take the straight line ES as the axis for the section we have the abscissa $ES = x$, and the ordinate $SM = v$. Since the angle $QSM = \phi$, we have $QM = z = v \sin \phi,$ $\qquad SQ = y + f = v \cos \phi,$ $\qquad$ so $\qquad$ that $y = v \cos \phi - f$. If in the equation in x, y, and z for the surface we substitute $y = v \cos \phi - f$ and $z = v \sin \phi$, we obtain the equation in x and v for the desired section ESM. If the intersection ES is perpendicular to the axis AP, then it will be parallel to the other principal axis in the plane APQ. When we permute the variables x and y the section is found in the same way.

49. Suppose now that the intersection ES with the plane APQ has any position, as in *figure 129*. We construct AE perpendicular to the axis AP, then we draw ETX parallel to the axis AP. We let $AE = f$ and the angle $TES = \theta$. After taking the three variables $AP = x$, $PQ = y$, and $QM = z$, we drop the perpendicular QS from Q to the

line ES and then join Q to M by the straight line MS. Then the angle $QSM = \phi$ is the angle between the plane of the section and the plane APQ. The desired coordinates for the plane of the section are $ES = t$ and $SM = v$. We draw the perpendiculars from S to the extensions of EX and QP to give ST and SV. Then we have

$$QM = z = v \sin \phi, \ QS = v \cos \phi, \ SV = v \cos \phi \sin \theta,$$

and $QV = v \cos \phi \cos \theta$. Next we have

$$ST = VX = t \sin \theta$$

and

$$ET = t \cos \theta.$$

Finally we have

$$AP = x = t \cos \theta + v \cos \phi \sin \theta,$$

and

$$PQ = y = v \cos \phi \cos \theta - t \sin \theta - f.$$

When these values are substituted for x, y, and z, we obtain the equation of the desired section.

50. If we have the equation for any solid, we can easily find the equation of the section by any plane. In the first place, it should be clear that if the equation for the solid is an algebraic equation in the three coordinates x, y, and z, then all of its sections will also we algebraic. Furthermore, since the equation for the section in the coordinates t and v is obtained by substituting in the equation for the solid,

$$z = v \sin \phi, \ x = t \cos \theta + v \cos \phi \sin \theta,$$

and

$$y = v \cos \phi \cos \theta - t \sin \theta - f,$$

it is clear that the variables t and v cannot appear with a total degree higher than the total degree of the original equation in x, y, and x. It can happen, however, that the equation for a section may have a smaller total degree due to cancellations after the substitutions.

51. If the equation for a surface has only total degree 1 for each term so that it has the form $\alpha x + \beta y + \gamma z = a$, then each section will be a straight line. In this case the surface is a plane, as should be clear after some consideration, and it will be demonstrated more clearly later. From Euclid we know that the intersection of two planes must be a straight line. In a similar way, we understand that any surface whose nature is contained in the general equation

$$\alpha x^2 + \beta y^2 + \gamma z^2 + \delta xy + \epsilon xz + \zeta yz + ax + by + cz + e^2 = 0$$

will have for every section an equation which is either a straight line or at most a second order curve. There is absolutely no section whose nature cannot be expressed by a second degree equation.

CHAPTER III

On Sections of Cylinders, Cones, and Spheres.

52. Since these solids are usually considered in elementary solid geometry, it may be useful to consider these sections before we proceed to less familiar solids. In the first place, Euclid considers two species of cylinder, the *right* and the *scalene*. A right cylinder is one such that all sections perpendicular to the axis are congruent circles, whose centers lie on one straight line. A scalene cylinder is such that all sections by planes, not perpendicular to the axis, but with some other angle of inclination, are congruent circles. We can express this more conveniently by saying that an oblique or scalene cylinder has for all of its sections, by planes perpendicular to the axis, congruent ellipses whose centers lie on a straight line which is called the axis of the cylinder.

53. Suppose that a cylinder, which is either right or scalene, has its axis CD perpendicular to a table top, as in *figure 130*. The base is $AEBF$, which is the section by the plane of the table top. The base may be a circle or an ellipse, but we have assumed it to be some ellipse with center at C and conjugate axes AB and EF. Whatever is discussed concerning a scalene cylinder is easily adapted to a right cylinder. We let one semi-axis $AC = BC = a$ and the other $CE = CF = c$. Now we choose the coordinates $CP = x$, $PQ = y$, and $QM = z$. From the

nature of an ellipse we have $a^2c^2 = a^2y^2 + c^2x^2$. This is also an equation for the cylinder, since the third variable z does not enter into the equation, because all sections parallel to the plane CPQ are congruent.

54. All of the sections of this cylinder by planes parallel to the base are congruent: circles if it is right and ellipses if it is scalene. Then all sections by planes which are perpendicular to the plane APQ will be a pair of parallel straight lines which coincide if the plane is tangent to the cylinder. The two lines become complex if the plane has no real intersection with the cylinder. This is immediate from the equation, since if either x, or y, or $x \pm \alpha y$ is set equal to a constant to represent a plane perpendicular to the base, then the equation has two simple roots. In this way we have determined all sections by planes parallel to one of the three principal planes.

55. In order to investigate the other sections, we suppose that the straight line GT is the intersection of the plane of the section and the plane of the base, which we first consider to be parallel to one of the conjugate axes EF and perpendicular to the extension of the other, AB, at G. Given this, we let the distance $CG = f$ and the angle of inclination of the plane of the section, GTM, and the plane of the base be measured by the angle ϕ. Let D be the intersection of the plane GTM and the axis of the cylinder. We draw the straight line DG, so that $DGC = \phi$, $DG = \dfrac{f}{\cos \phi}$, and $CD = \dfrac{f \sin \phi}{\cos \phi}$. From any point M on the desired section we draw MT parallel to DG. Since $TQ = f - x$ and the angle $QTM = \phi$, we have $TM = \dfrac{f - x}{\cos \phi}$, and $QM = \dfrac{(f - x)\sin \phi}{\cos \phi} = z$. We draw MS parallel to TG so that it is perpendicular to DG, $MS = TG = PQ = y$, and $DS = \dfrac{x}{\cos \phi}$.

56. Now we take the straight lines DS and SM as coordinate axes for the desired section. We let $DS = t$ and $SM = u$. It follows that $y = u$, $x = t \cos \phi$ and since $z = \dfrac{(f - x)\sin \phi}{\cos \phi}$, we have $z = f \tan \phi - t \sin \phi$. When we substitute these values in the equation for the cylinder $a^2c^2 = a^2y^2 + c^2x^2$, we obtain the equation for the desired section, $a^2c^2 = a^2u^2 + c^2t^2(\cos \phi)^2$. This indicates that the section will be an ellipse with its center at the point D, one of its principal axes lies on the straight line DG and the other is perpendicular to this. The semiaxis falling on the straight line DG (when $u = 0$) is equal to $\dfrac{a}{\cos \phi}$. If we draw the straight line BH parallel to GD, then $BH = \dfrac{a}{\cos \phi}$ is one of the semiaxes of the desired section. The conjugate semiaxis is equal to $c = CE$.

57. Now let the section of a cylinder obtained in this way be an ellipse with conjugate semiaxis $\dfrac{a}{\cos \phi}$ and c. If in the base $AEBF$, $AC = a$ is the major semiaxis, then since $\dfrac{a}{\cos \phi}$ is greater than a, the ellipses of the sections will be more elongated than in the base. On the other hand, if $c > a$, that is, if the intersection GT were parallel to the major axis in the base, then in the section both axes could become equal and the section would become a circle. This would be the case when $\dfrac{a}{\cos \phi} = c$ or $\cos \phi = \dfrac{a}{c}$. Since in the triangle BCH the angle at C is a right angle and the angle $CBH = \phi$, we have $\cos \phi = \dfrac{BC}{BH} = \dfrac{a}{BH}$. It follows that when we take $BH = CE$, the section becomes a circle. This can happen in two ways, namely, when the straight line $BH = CE$, is either above or below the plane of the base. The reason why this cylinder is called scalene, is that these two circles lie in planes which are

oblique to the axis CD.

58. Suppose now that the straight line GT, which is the intersection of the plane of the section with the plane of the base, is situated obliquely, as in *figure 131*. Let $GC = f$ be the perpendicular dropped from C, the center of the base, let the angle $BCG = \theta$, and let $CGD = \phi$ be the angle of inclination, which is equal to the angle QTM, where QT is drawn perpendicular to GT. Then $DG = \dfrac{f}{\cos \phi}$, and $CD = \dfrac{f \sin \phi}{\cos \phi}$. Let M be a point on the desired section, with MQ perpendicular to the base and QP perpendicular to the axis. Then we let $CP = x$, $PQ = y$, and $QM = z$, so that $a^2 c^2 = a^2 y^2 + c^2 x^2$. We drop the two perpendiculars PV and QT to the intersection line GT. Then $GV = x \sin \theta$, and $PV = f - x \cos \theta$. Since the angle $QPW = \theta$, we have $QW = y \sin \theta$, $PW = VT = y \cos \theta$, and $QT = f - x \cos \theta + y \sin \theta$. When we draw the line MT, since the angle $MTQ = \phi$ we have $TM = \dfrac{z}{\sin \phi}$ and $QT = \dfrac{z \cos \phi}{\sin \phi}$.

59. When we complete the rectangle $GSMT$ and let $DS = t$, $SM = GT = u$, we obtain

$$u = GV + VT = x \sin \theta + y \cos \theta.$$

Since

$$QT = f - x \cos \theta + y \sin \theta,$$

we have

$$QT - CG = y \sin \theta - x \cos \theta,$$

so that

$$DS = TM - DG = \frac{y \sin \theta - x \cos \theta}{\cos \phi} = t.$$

Since

$$x \sin \theta + y \cos \theta = u,$$

and

$$y \sin \theta - x \cos \theta = t \cos \phi,$$

we have

$$y = u \cos \theta + t \sin \theta \cos \phi,$$

and

$$x = u \sin \theta - t \cos \theta \cos \phi.$$

When these values are substituted in the equation $a^2c^2 = a^2y^2 + c^2x^2$ for x and y, we obtain

$$a^2c^2 = a^2u^2(\cos \theta)^2 + 2a^2ut \sin \theta \cos \theta \cos \phi + a^2t^2(\sin \theta)^2(\cos \phi)^2$$

$$+ c^2u^2(\sin \theta)^2 - 2c^2ut \sin \theta \cos \theta \cos \phi + c^2t^2(\cos \theta)^2(\cos \phi)^2.$$

Clearly this is the equation for an ellipse with center at D, but the coordinates DS and SM are not perpendicular to the principal axes unless $a = c$, that is, unless the cylinder is right.

60. In order to study this section more closely, let $aMebf$ be the curve in *figure 132,* whose equation in the coordinates $DS = t$, and $MS = u$ we have found in the preceding section. For the sake of brevity we let this equation take the form $a^2c^2 = \alpha u^2 + 2\beta tu + \gamma t^2$, where

$$\alpha = a^2(\cos \theta)^2 + c^2(\sin \theta)^2$$

$$\beta = (a^2 - c^2)\sin \theta \cos \theta \cos \phi$$

$$\gamma = a^2(\sin \theta)^2(\cos \phi)^2 + c^2(\cos \theta)^2(\cos \phi)^2.$$

Let ab and ef be the principal conjugate axes of the section. We choose new coordinates by calling Mp the new ordinate and letting $Dp = p$ and $Mp = q$. We let the angle $aDH = \zeta$, then $u = p \sin \zeta + q \cos \zeta$, and $t = p \cos \zeta - q \sin \zeta$. When these values are substituted into the equation we have

$$a^2c^2 = (\alpha(\sin \zeta)^2 + 2\beta \sin \zeta \cos \zeta + \gamma(\cos \zeta)^2)p^2$$

$$+ (2\alpha \sin \zeta \cos \zeta + 2\beta(\cos \zeta)^2 - 2\beta(\sin \zeta)^2 - 2\gamma \sin \zeta \cos \zeta)pq$$

$$+ (\alpha(\cos \zeta)^2 - 2\beta \sin \zeta \cos \zeta + \gamma(\sin \zeta)^2)q^2.$$

61. Since the coordinates are referred to the principal axes, the coefficient of pq should be equal to zero. Since $2 \sin \zeta \cos \zeta = \sin 2\zeta$ and $(\cos \zeta)^2 - (\sin \zeta)^2 = \cos 2\zeta$, we have $(\alpha - \gamma)\sin 2\zeta + 2\beta \cos 2\zeta = 0$. It follows that $\tan 2 \zeta = \dfrac{2\beta}{\gamma - \alpha}$, so that both the angle aDH and the positions of the principal diameters are known. Furthermore these semiaxes are given as follows:

$$aD = \frac{ac}{(\alpha(\sin \zeta)^2 + 2\beta \sin \zeta \cos \zeta + \gamma(\cos \zeta)^2)^{\frac{1}{2}}}$$

$$eD = \frac{ac}{(\alpha(\cos \zeta)^2 - 2\beta \sin \zeta \cos \zeta + \gamma(\sin \zeta)^2)^{\frac{1}{2}}}.$$

62. Since

$$2\beta = 2\frac{(\gamma - \alpha)\sin \zeta \cos \zeta}{(\cos \zeta)^2 - (\sin \zeta)^2},$$

when this is substituted in the above expressions, we obtain

$$aD = \frac{ac\sqrt{(\cos\zeta)^2 - (\sin\zeta)^2}}{(\gamma(\cos\zeta)^2 - \alpha(\sin\zeta)^2)^{\frac{1}{2}}} = \frac{ac\sqrt{2\cos 2\zeta}}{((\alpha + \gamma)\cos 2\zeta - \alpha + \gamma)^{\frac{1}{2}}}$$

$$eD = \frac{ac\sqrt{(\cos\zeta)^2 - (\sin\zeta)^2}}{(\alpha(\cos\zeta)^2 - \gamma(\sin\zeta)^2)^{\frac{1}{2}}} = \frac{ac\sqrt{2\cos 2\zeta}}{((\alpha + \gamma)\cos 2\zeta + \alpha - \gamma)^{\frac{1}{2}}}.$$

The product of these two semiaxes is

$$aD \cdot cD = \frac{2a^2c^2\cos 2\zeta}{(2\alpha\gamma(1 + (\cos 2\zeta)^2) - (\alpha^2 + \gamma^2)(\sin 2\zeta)^2)^{\frac{1}{2}}}.$$

However, since $(\gamma - \alpha)\sin 2\zeta = 2\beta\cos 2\zeta$, we have

$$(\alpha^2 + \gamma^2)(\sin 2\zeta)^2 = 4\beta^2(\cos 2\zeta)^2 + 2\alpha\gamma(\sin 2\zeta)^2.$$

It follows that

$$aD \cdot eD = \frac{2a^2c^2\cos 2\zeta}{(4\alpha\gamma(\cos 2\zeta)^2 - 4\beta^2(\cos 2\zeta)^2)^{\frac{1}{2}}}$$

$$= \frac{a^2c^2}{(\alpha\gamma - \beta^2)^{\frac{1}{2}}} = \frac{ac}{\cos\phi}.$$

63. In a similar way, since

$$aD^2 = \frac{2a^2c^2\cos 2\zeta}{(\alpha + \gamma)\cos 2\zeta - \alpha - \gamma}$$

and

$$eD^2 = \frac{2a^2c^2\cos 2\zeta}{(\alpha + \gamma)\cos 2\zeta + \alpha - \gamma},$$

we have

$$aD^2 + eD^2 = \frac{4a^2c^2(\alpha + \gamma)(\cos 2\zeta)^2}{4\alpha\gamma(\cos 2\zeta)^2 - 4\beta^2(\cos 2\zeta)^2} = \frac{(\alpha + \gamma)a^2c^2}{\alpha\gamma - \beta^2}$$

From this we conclude that

$$aD + eD = \frac{ac\left(\alpha + \gamma + 2\sqrt{\alpha\gamma - \beta^2}\right)^{\frac{1}{2}}}{(\alpha\gamma - \beta^2)^{\frac{1}{2}}}$$

and

$$aD - eD = \frac{ac\left(\alpha + \gamma - 2\sqrt{\alpha\gamma - \beta^2}\right)^{\frac{1}{2}}}{(\alpha\gamma - \beta^2)^{\frac{1}{2}}}.$$

It follows that the semiaxes aD and eD are roots of the following equation,

$$(\alpha\gamma - \beta^2)x^4 - (\alpha + \gamma)a^2c^2x^2 + a^4c^4 = 0.$$

But we have

$$\sqrt{\alpha\gamma - \beta^2} = ac \cos \phi.$$

64. Since $aD \cdot eD = \dfrac{ac}{\cos \phi}$, and ϕ is the angle which the plane of the section makes with the plane of the base, we conclude to the following beautiful theorem.

THEOREM

If any cylinder is cut by any plane, then the product of the principal axes of the section is to the product of the principle axes of the base of the cylinder as 1 is to the cosine of the angle made by the plane of the section with the plane of the base.

Since all parallelograms formed by the conjugate axes have the same area as the rectangle formed by the principal axes, it follows that such parallelograms for any section have the same ratio to such a parallelogram for the base.

65. We can find a better way to define the nature of the intersection of a scalene cylinder by a plane. In *figure 133* we let the ellipse *AEBF* be the base of the cylinder, with the semiaxes $AC = BC = a$, and $EC = CF = c$. Let the straight line CD perpendicular to the base be the axis of the cylinder. Suppose that the cylinder is intersected by a plane whose intersection with the plane of the base is the straight line *TH*, which is oblique with respect to the axis AB. We drop the perpendicular CH from the center C to this line, and let the angle $GCH = \theta$. We suppose that the axis of the cylinder intersects the plane of the section in the point D. When we draw the straight line DH, the angle $CHD = \phi$ is the angle of inclination made by the plane of the section and the plane of the base. We let $CG = f$, then $GH = f \sin \theta$, $CH = f \cos \theta$, $DH = \dfrac{f \cos \theta}{\cos \phi}$, and $CD = \dfrac{f \cos \theta \sin \phi}{\cos \phi}$. Since the triangle DCG has a right angle at C, we have $DG = \dfrac{f \sqrt{1 + (\sin \theta \sin \phi)^2}}{\cos \phi}$. The sine of the angle DGH is equal to $\dfrac{\cos \theta}{(1 + (\sin \theta \sin \phi)^2)^{\frac{1}{2}}}$, the cosine of the same angle is equal to $\dfrac{\sin \theta \cos \phi}{(1 + (\sin \theta \sin \phi)^2)^{\frac{1}{2}}}$, and the tangent of that angle is equal to $\dfrac{\cos \theta}{\sin \theta \cos \phi}$.

66. Now from any point on the section M drop the perpendicular MQ to the plane of the base. We draw the ordinate QP and let $CP = x$, $PQ = y$, so that $a^2 c^2 = a^2 y^2 + c^2 x^2$. We draw QT parallel to CG and we let GR be perpendicular to QT. Now we have $GR = y$ and $QR = f - x$. Since the angle $TGR = GCH = \theta$, we have $GT = \dfrac{y}{\cos \theta}$ and $TR = \dfrac{y \sin \theta}{\cos \theta}$. It follows that

$$QT = f - x + \frac{y \sin \theta}{\cos \theta}.$$

Since the triangles CDG and QMT are similar, we have $\dfrac{CG}{DG} = \dfrac{QT}{TM}$ and $\dfrac{CG}{CG - QT} = \dfrac{DG}{DS}$, where MS is drawn parallel to GT. It follows that

$$DS = \frac{(x \cos \theta - y \sin \theta)\sqrt{1 + (\sin \theta \sin \phi)^2}}{\cos \theta \cos \phi}.$$

When we let $DS = t$ and $Ms = u$, we have

$$x \cos \theta - y \sin \theta = \frac{t \cos \theta \cos \phi}{\left(1 + (\sin \theta \sin \phi)^2\right)^{\frac{1}{2}}}$$

and $y = u \cos \theta$. These two equations can be solved for t and u, but the solutions will involve some complication.

67. Now suppose we take the axis EF to be the diameter of the base which is parallel to TH, rather than the principal axis, and the conjugate axis AB, which when extended intersects TH at G. We keep the other variables the same as before: $CG = f$, $GCH = \theta$, $CHD = \phi$, $CA = CB = m$, $CE = CF = n$. The straight line QP is drawn parallel to the diameter EF, and we let $CP = x$, $PQ = y$, so that $m^2 n^2 = m^2 y^2 + n^2 x^2$, $GT = MS = y$ and

$$DS = \frac{DG \cdot x}{CG} = \frac{x\sqrt{1 + (\sin \theta \sin \phi)^2}}{\cos \phi}.$$

Now when we let $DS = t$, and $MS = u$ we have

$$x = \frac{t \cos \phi}{\left(1 + (\sin \theta \cos \theta)^2\right)^{\frac{1}{2}}}$$

and $y = u$. If we let the angle $CGD = \eta$, since $\cos \eta = \dfrac{CG}{DG}$, we have

$x = t \cos \eta$. Then the desired section has the equation $m^2 n^2 = m^2 u^2 + n^2 t^2 (\cos \eta)^2$ with respect to the conjugate diameters. The semidiameter in the direction of $DS = \dfrac{m}{\cos \phi}$ and the other is equal to n. The angle GSM made by the two diameters has a tangent equal to $\dfrac{\cos \theta}{\sin \theta \cos \phi}$. The cosine of this angle is equal to $\dfrac{\sin \theta \cos \phi}{(1 + (\sin \theta \sin \phi)^2)^{\frac{1}{2}}} = \sin \theta \cos \eta$. In this way the nature of the section can easily be known.

68. Now that we have considered sections of a cylinder, we move on to a consideration of sections of a cone, both right and scalene. These two differ only in that sections perpendicular to the axis are ellipses if the cone is scalene, while they are circles if the cone is right. In *figure 134* let $OaebfO$ be any cone with vertex at O and axis Oc. Let the plane of a table top be perpendicular to the axis of the cone and let the table top represent the plane which passes through the vertex of the cone. Let AB and EF be straight lines in the plane of the table which are parallel to the principal axes ab and ef of any section of the cone by a plane perpendicular to the axis of the cone. From any point M on the section $aebf$ we drop the perpendicular MQ to the plane of the table and from Q we draw PQ perpendicular to AB. If we let $OP = x$, $PQ = y$, and $QM = z$, then also the abscissa of the section $cp = x$, and the ordinate $pM = y$. Since the axes ab and ef have a constant ratio to $Oc = QM = z$, when we let $ac = bc = mz$ and $ec = fc = nz$, we have $m^2 n^2 z^2 = m^2 y^2 + n^2 x^2$. This is the natural equation in three variables for the surface of a cone.

69. Since all of the sections perpendicular to the axis Oc are ellipses, which is clear when we let z be a constant in the equation

$m^2n^2z^2 = m^2y^2 + n^2x^2$. In a similar way we easily know the sections which are perpendicular to the straight line AB or the straight line EF. Suppose the cone is intersected by a plane perpendicular to AB at the point P. We let $OP = a$, and the equation for that section is $m^2n^2z^2 = m^2y^2 + n^2a^2$. This is the equation in the coordinates $Pp = z$ and $pM = y$ for a hyperbola with center at P, transverse semiaxis equal to $\dfrac{a}{m}$ and conjugate semiaxis equal to $\dfrac{na}{m}$. In the same way, if we let y be constant, the section perpendicular to the straight line EF is seen to be a hyperbola with center on the straight line EF.

70. If the plane which intersects the cone is perpendicular to the plane $AEBF$ but is perpendicular to neither AB nor to EF, it is till easy to find the equation of the section. Suppose that this plane intersects the plane of the base $AEBF$ in the straight line BE as in *figure 135*, and let $OB = a$, $OE = b$. From any point M of the section we drop the perpendicular MQ and from Q we draw the ordinate QP, so that $OP = x$, $PQ = y$, and $QM = z$. From the nature of the cone we have $m^2n^2z^2 = m^2y^2 + n^2x^2$. Since $\dfrac{a}{b} = \dfrac{a-x}{y}$, we have $y = b - \dfrac{bx}{a}$. In the section we choose the coordinate $BQ = t$, and $QM = z$, then

$$\frac{b}{(a^2+b^2)^{\frac{1}{2}}} = \frac{y}{t}, \quad \text{so} \quad \text{that} \quad y = \frac{bt}{(a^2+b^2)^{\frac{1}{2}}} \quad \text{and}$$

$$a - x = \frac{at}{(a^2+b^2)^{\frac{1}{2}}}. \quad \text{Let} \ \sqrt{a^2+b^2} = c, \quad \text{so that} \quad y = \frac{bt}{c} \quad \text{and}$$

$x = a - \dfrac{at}{c}$. After making these substitutions we obtain the following equation in t and a:

$$m^2n^2c^2z^2 = m^2b^2t^2 + n^2a^2c^2 - 2n^2a^2ct + n^2a^2t^2.$$

Let $t - \dfrac{n^2a^2c}{m^2b^2 + n^2a^2} = GQ = u$, since $BG = \dfrac{n^2a^2c}{m^2b^2 + n^2a^2}$, we have $m^2 n^2 c^2 z^2 = (m^2 b^2 + n^2 a^2) u^2 + \dfrac{m^2 n^2 a^2 b^2 c^2}{m^2 b^2 + n^2 a^2}.$

71. This section of the cone is a hyperbola having center at the point G, with transverse semiaxis $Ga = \dfrac{ab}{(m^2 b^2 + n^2 a^2)^{\frac{1}{2}}}$, and conjugate semiaxis equal to $\dfrac{mnabc}{m^2 b^2 + n^2 a^2}$. The asymptotes of this hyperbola which meet the axis Ga at the point G make an angle with the axis Ga whose tangent is equal to $\dfrac{mnc}{(n^2 b^2 + n^2 a^2)^{\frac{1}{2}}}$. In order that the hyperbola be equilateral it is necessary that $m^2 n^2 a^2 + m^2 n^2 b^2 = m^2 b^2 + n^2 a^2$ or $\dfrac{b}{a} = \tan OBE = \dfrac{n\sqrt{m^2 - 1}}{m(1 - n^2)^{\frac{1}{2}}}$. Unless $\dfrac{m^2 - 1}{1 - n^2}$ is positive, an equilateral hyperbola cannot be obtained in this way. In a right cone, where $m = n$, the angle which the asymptote makes with the axis of the section has its tangent equal to m and the angle is aOc, as in *figure 134*.

72. Now suppose that the section is oblique, but its intersection BT with the plane $AEBF$ is perpendicular to the straight line AB, as in *figure 136*. Let $OB = f$ and the angle of inclination of the section plane to the plane of the base be $OBC = \phi$. Suppose that the cutting plane intersects the axis of the cone OC in the point C. Then $BC = \dfrac{f}{\cos \phi}$ and $OC = \dfrac{f \sin \phi}{\cos \phi}$. Let M be any point of the desired section, and let MT be perpendicular to BT. Then drop the perpendicular MQ to the plane of the base, and let QP be perpendicular to OB. When we let $OP = x$, $PQ = y$, and $QM = z$, we have $m^2 n^2 z^2 = m^2 y^2 + n^2 x^2$. We choose coordinates for the section by letting $BT = t$, and $TM = u$. Since the angle $QTM = \phi$, $QM = z = u \sin \phi$,

$$TQ = u \cos \phi = f - x,$$

we have $y = t$, $z = u \sin \phi$, and $x = f - u \cos \phi$. It follows that

$$m^2 n^2 u^2 (\sin \phi)^2 = m^2 t^2 + n^2 (f - u \cos \phi)^2.$$

73.   We let $BC = \dfrac{f}{\cos \phi} = g$, so that $f = g \cos \phi$, and

$x = (g - u) \cos \phi$. Now the equation for the section is

$$m^2 n^2 (\sin \phi)^2 = m^2 t^2 + n^2 g^2 (\cos \phi)^2 - 2n^2 gu (\cos \phi)^2 + n^2 u^2 (\cos \phi)^2.$$

If we let $u - \dfrac{g (\cos \phi)^2}{(\cos \phi)^2 - m^2 (\sin \phi)^2} = SG = s$, and with MS parallel

to BT, we take

$$BG = \frac{g (\cos \phi)^2}{(\cos \phi)^2 - m^2 (\sin \phi)^2} = \frac{f \cos \phi}{(\cos \phi)^2 - m^2 (\sin \phi)^2}$$

$$= \frac{f \cos \phi}{1 - (1 + m^2)(\sin \phi)^2}.$$

If we take the coordinates $GS = s$ and $SM = t$, we have the equation

$$m^2 t^2 + n^2 ((\cos \phi)^2 - m^2 (\sin \phi)^2) s^2 - \frac{m^2 n^2 f^2 (\sin \phi)^2}{(\cos \phi)^2 - m^2 (\sin \phi)^2} = 0.$$

Hence the curve is a conic section having its center at G. It will be a parabola if the center G goes to infinity, and this happens if $\tan \phi = \dfrac{1}{m}$, that is if the straight line BC is parallel to the side of the cone Oa in *figure 134*. In this case we have

$$m^2 t^2 + n^2 f^2 - 2n^2 fu \cos \phi = 0.$$

In *figure 136* the vertex of the parabola is at G when we take

$BG = \dfrac{f}{2 \cos \phi}$ and the latus rectum is equal to $\dfrac{2n^2 f \cos \phi}{m^2}$.

74. Since the section is a parabola if $(\cos \phi)^2 - m^2(\sin \phi)^2 = 0$, it is clear that it will be an ellipse if $(\cos \phi)^2 > m^2(\sin \phi)^2$, that is, if $\tan \phi > \dfrac{1}{m}$. In this case the straight line BC again intersects Oa on the opposite side of the cone. Since $BG = \dfrac{g}{1 - m^2(\tan \phi)^2}$, we have $BG > BC$ where G is the center of the section. It follows that the semiaxis of the desired section in the direction of BC is equal to $\dfrac{mf \sin \phi}{(\cos \phi)^2 - m^2(\sin \phi)^2}$, the conjugate semiaxis is equal to

$$\frac{nf \sin \phi}{\left((\cos \phi)^2 - m^2(\sin \phi)^2\right)^{\frac{1}{2}}},$$

and the semilatus rectum is equal to $\dfrac{n^2}{m} f \sin \phi$. It follows that the section will be a circle if $m = n\sqrt{(\cos \phi)^2 - m^2(\sin \phi)^2}$, that is if

$$m^2 = n^2 - n^2(1 + m^2)(\sin \phi)^2$$

or

$$\sin \phi = \frac{\sqrt{n^2 - m^2}}{n(1 + m^2)^{\frac{1}{2}}} = \sin OBC$$

and

$$\cos \phi = \frac{m\sqrt{1 + n^2}}{n(1 + m^2)^{\frac{1}{2}}}.$$

It follows that no section of this kind can be a circle unless $n > m$.

75. If $m^2(\sin \phi)^2 > (\cos \phi)^2$, that is, if $\tan \phi > \dfrac{1}{m}$, so that the straight line BC diverges from Oa on the opposite side of the cone, then the section will be a hyperbola. Its transverse semiaxis is equal to

$$\frac{mf \sin \phi}{-(\cos \phi)^2 + m^2(\sin \phi)^2},$$ the   conjugate   semiaxis   is   equal   to

$$\frac{nf \sin \phi}{(m^2(\sin \phi)^2 - (\cos \phi)^2))^{\frac{1}{2}}},$$ the semilatus rectum is equal to $\dfrac{n^2}{m} f \sin \phi$,

and the angle made by the asymptotes and the axis at the center G has

its tangent equal to $\dfrac{n}{m} \sqrt{m^2(\sin \phi)^2 - (\cos \phi)^2}$. Hence the hyperbola

will be equilateral if

$$m^2 n^2(\sin \phi)^2 - n^2(\cos \phi)^2 = m^2 = (m^2 + 1)n^2(\sin \phi)^2 - n^2$$

that is if

$$\sin \phi = \frac{\sqrt{m^2 + n^2}}{n(1 + m^2)^{\frac{1}{2}}}$$

and

$$\cos \phi = \frac{m\sqrt{n^2 - 1}}{n(1 + m^2)^{\frac{1}{2}}}.$$

Unless $n > 1$, the equilateral hyperbola cannot be obtained in this way.

76. If the cone is right that is, if $m = n$, then all sections have
been considered, since the position of the straight line AB can be chosen
arbitrarily. However, if the cone is scalene, there remains the investigation of all sections by planes whose intersection with the base plane is
oblique with respect to the straight line AB. Let BR be the intersection
of the cutting plane and the plane of the base $AEBF$ in *figure 137*. Let
$OB = f$, the angle $OBR = \theta$, and the angle of inclination of the cutting
plane to the base plane be ϕ. Then we drop the perpendicular OR from
O to BR, so that $OR = f \sin \theta$ and $BR = f \cos \theta$. Next we draw the
straight line RC in the cutting plane, so that $RC = \dfrac{f \sin \theta}{\cos \phi}$ and

$OC = \dfrac{f \; \sin \theta \; \sin \phi}{\cos \phi}$, since the angle $ORC = \phi$. Now if we take a section which is perpendicular to the axis of the cone OC and project it orthogonally onto the plane of the base, then its principal axes will lie in the direction of the straight lines AB and EF, one as m and the other as n.

77. In this projected section we draw the diameter EF parallel to BR, so that the angle $BOe = \theta$. Let aOb be the conjugate diameter. We let the semidiameters $Oa = \mu$ and $Oe = \nu$, then

$$\mu = \frac{\sqrt{m^4(\sin \theta)^2 + n^4(\cos \theta)^2}}{\left(m^2(\sin \theta)^2 + n^2(\cos \theta)^2\right)^{\frac{1}{2}}}$$

and

$$\nu = \frac{mn}{\left(m^2(\sin \theta)^2 + n^2(\cos \theta)^2\right)^{\frac{1}{2}}}.$$

We also have

$$\tan BOb = \frac{n^2\cos \theta}{m^2\sin \theta}$$

so that

$$\sin BOb = \frac{n^2\cos \theta}{\left(m^4(\sin \theta)^2 + n^4(\cos \theta)^2)\right)^{\frac{1}{2}}}$$

and

$$\cos Bob = \frac{m^2\sin \theta}{\left(m^4(\sin \theta)^2 + n^4(\cos \theta)^2\right)^{\frac{1}{2}}}.$$

Since the angle $ObR = \theta + BOb$, we have

$$\sin \, ObR \; = \; \frac{m^2(\sin \, \theta)^2 \; + \; n^2(\cos \, \theta)^2}{\left(m^4(\sin \, \theta)^2 \; + \; n^4(\cos \, \theta)^2\right)^{\frac{1}{2}}},$$

$$\cos \, ObR \; = \; \frac{(m^2 \; - \; n^2)\sin \, \theta \, \cos \, \theta}{\left(m^4(\sin \, \theta)^2 \; + \; n^4(\cos \, \theta)^2\right)^{\frac{1}{2}}}.$$

But we also have

$$\mu\nu \; = \; \frac{mn \, \sqrt{m^4(\sin \, \theta)^2 \; + \; n^4(\cos \, \theta)^2}}{m^2(\sin \, \theta)^2 \; + \; n^2(\cos \, \theta)^2}.$$

78. Since $OR \; = \; f \, \sin \, \theta$ we have

$$Ob \; = \; \frac{OR}{\sin \, ObR} \; = \; \frac{f \, \sin \, \theta \, \sqrt{m^4(\sin \, \theta)^2 \; + \; n^4(\cos \, \theta)^2}}{m^2(\sin \, \theta)^2 \; + \; n^2(\cos \, \theta)^2}$$

and

$$Rb \; = \; \frac{(m^2 \; - \; n^2)f \, \sin \, \theta \, \cos \, \theta}{m^2(\sin \, \theta)^2 \; + \; n^2(\cos \, \theta)^2}.$$

since the angle at R in the triangle RbC is a right angle, the angle CbR has its tangent equal to

$$\frac{m^2(\sin \, \theta)^2 \; + \; n^2(\cos \, \theta)^2}{(m^2 \; - \; n^2)\sin \, \theta \, \cos \, \theta \, \cos \, \phi}.$$

Hence, the angle CbR is known. Now from any point M on the section we draw the line MT, to the straight line RT, parallel to Cb and from M we draw MS to Cb parallel to RT. We let $bT \; = \; MS \; = \; t$ and $bS \; = \; TM \; = \; u$, which should be considered as oblique coordinates for the desired section, where $\tan \, bSM \; = \; \dfrac{m^2(\sin \, \theta)^2 \; + \; n^2(\cos \, \theta)^2}{(m^2 \; - \; n^2)\sin \, \theta \, \cos \, \theta \, \cos \, \phi}$. It is clear that these coordinates will be rectangular when the cone is right since in that case $m \; = \; n$.

79. From the point M on the section we drop the perpendicular MQ to the plane $AEBF$. Next we draw TQ parallel to the diameter ab, and from Q we draw the ordinate QP parallel to the other diameter ef. When we let $OP = x$, $PQ = y$, and $QM = z$, from the nature of a cone we have

$$\mu^2 \nu^2 z^2 = \mu^2 y^2 + \nu^2 x^2,$$

since, when we consider the conic section through the point M parallel to the base, its semidiameters parallel to ab and ef are μz and νz. Since the triangle COb is a right triangle with sides OC and Ob, we have the hypotenuse Cb equal to the quotient whose numerator is equal to

$$f \sin \theta \sqrt{m^4(\sin \theta)^2 + n^4(\cos \theta)^2 - (m^2 - n^2)^2(\sin \theta)^2(\cos \theta)^2(\sin \phi)^2}$$

and whose demoninator is

$$(m^2(\sin \theta)^2 + n^2(\cos \theta)^2)\cos \phi.$$

Since the triangles TMQ and bCO are similar we have

$$TM(u) : TQ(ob - x) : QM(z) = bC : Ob : OC$$

so that $x = Ob - \dfrac{Ob \cdot u}{Cb}$, $z = \dfrac{OC \cdot u}{Cb}$, and $y = t$. Hence

$$\mu^2 \nu^2 \cdot OC^2 \cdot u^2 = \mu^2 \cdot Cb^2 \cdot t^2 + \nu^2 \cdot Ob^2 (Cb - u)^2.$$

80. When this equation is expanded we obtain

$$0 = \mu^2 \cdot Cb^2 \cdot t^2 + \nu^2 (Ob^2 - \mu^2 \cdot Oc^2)u^2$$

$$- 2\nu^2 \cdot Ob^2 \cdot Cb \cdot u + \nu^2 Ob^2 \cdot Cb^2.$$

We substitute $u - \dfrac{Ob^2 \cdot Cb}{Ob^2 - \mu^2 \cdot Oc^2} = s$. That is, if we take

$$bG = \frac{Ob^2 \cdot Cb}{Ob^2 - \mu^2 \cdot Oc^2}$$

$$= \frac{Cb}{1 - (m^2(\sin\theta)^2 + n^2(\cos\theta)^2)(\tan\phi)^2},$$

and call $GS = s$, then G will be the center of the conic section whose equation in the coordinate t and s is

$$\mu^2 \cdot Cb^2 \cdot t^2 + \nu^2(Ob^2 - \mu^2 \cdot Oc^2)s^2 = \frac{\mu^2 \nu^2 \cdot Ob^2 \cdot Oc^2 \cdot Cb^2}{Ob^2 - \mu^2 \cdot Oc^2}.$$

The transverse semidiameter is equal to $\dfrac{\mu \cdot Ob \cdot Oc \cdot Cb}{(Ob^2 - \mu^2 \cdot Oc^2)^{\frac{1}{2}}}$, the conjugate semidiameter is equal to $\dfrac{\nu \cdot Ob \cdot Oc}{(Ob^2 - \mu^2 \cdot Oc^2)^{\frac{1}{2}}}$, and the semilatus rectum is equal to $\dfrac{\nu^2 \cdot Ob \cdot Oc}{\mu \cdot Cb}$. Furthermore it is clear that if

$$\tan\phi < \frac{1}{(m^2(\sin\theta)^2 + n^2(\cos\theta)^2)^{\frac{1}{2}}},$$ or if $\tan\phi < \dfrac{\nu}{mn}$, then the

curve will be an ellipse. If $\tan\phi = \dfrac{\nu}{mn}$, then the section will be a parabola, while if $\tan\phi > \dfrac{\nu}{mn}$, it will be a hyperbola.

81. The third solid whose plane sections we will investigate is the sphere. It is clear from elementary geometry that all of these plane sections are circles. In order that the method we have been using may become clearer, we will obtain analytically the information which is ordinarily treated synthetically. That is, from the equation for a solid, we should be able to obtain any of its sections. Let C be the center of the sphere in *figure 138,* where the plane of a table top is considered to contain the center, so that this section is a great circle with radius $CA = CB = a$. This radius is also the radius of the sphere. Let the

straight line DT be the intersection of the cutting plane and the plane of the table, to which we drop the perpendicular CD from C. We let the length of CD be f and the angle of inclination be ϕ.

82. Let M be any point on the section. From M we drop MQ perpendicular to the plane of the table. We take CD as the axis and construct the perpendicular QP. When we let $CP = x$, $PQ = y$, and $QM = z$, from the nature of a sphere we have $x^2 + y^2 + z^2 = a^2$. From M we also draw MT perpendicular to DT, and we draw QT also perpendicular to DT. Since both QT and MT are perpendicular to DT, and the angle $MTQ = \phi$, the angle of inclination of the cutting plane to the plane of the base, if we take DT and MT as coordinates for the desired section and let $DT = t$, $TM = u$, then $MQ = u \sin \phi$, and $TQ = u \cos \phi$. Then $CP = x = f - u \sin \phi$, $PQ = y = t$, and $QM = z = u \sin \phi$. When these values are substituted into the equation for the sphere, we obtain this equation for the section

$$f^2 - 2fu \cos \phi + u^2 + t^2 = a^2.$$

83. It should be clear that this is the equation for a circle. Indeed, if we let $u - f \cos \phi = s$, then

$$f^2(\sin \phi)^2 + s^2 + t^2 = a^2,$$

so that the radius of the section is equal to $\sqrt{a^2 - f^2(\sin \phi)^2}$. If we draw the straight line Dc from D parallel to TM, with cC perpendicular to the plane of the table, we have $Dc = f \cos \phi$ and $Cc = f \sin \phi$, since $CD = f$ and the angle $CDc = \phi$. Hence, when we consider the coordinates s and t, we see that c is the center of the section and that $\sqrt{CB^2 - Cc^2}$ is the radius of this circle, as is clear from Euclid's Elements. In a similar way we could investigate the section by any plane of any solid, provided we have the equation in three variables of the solid.

84. In order that this whole operation might become perfectly clear, we suppose that we have any solid whatsoever, whose nature is expressed by an equation in the three variables $AP = x$, $PQ = y$, and $QM = z$ as in *figure 139*. The first two of these coordinates lie in the plane of a table, and z is perpendicular to this plane. Suppose that the solid is cut by some plane whose intersection with the plane of the table is the straight line DT and the angle of inclination is equal to ϕ. Let the straight line $AD = f$ and the angle $ADE = \theta$. When we drop the perpendicular AE from A to DE, we have $AE = f \sin \theta$, and $DE = f \cos \theta$. Next we drop a perpendicular MT from any point M on the section to the line DT. When we join Q to T, the angle $MTQ = \phi$, the angle of inclination. Now if we take $DT = t$ and $TM = u$ as coordinates for the section, we have $QM = u \sin \phi$ and $TQ = u \cos \phi$.

85. We drop the perpendicular TV from T to the axis AD. Since the angle $TDV = \theta$, we have $TV = t \sin \theta$ and $DV = t \cos \theta$. Furthermore the angle $TQP = \theta$, so that $PV = u \sin \theta \cos \phi$ and $PQ - TV = u \cos \theta \cos \phi$. From this information we can express the coordinates x, y, and z in terms of t and u as follows.

$$AP = x = f + t \cos \theta - u \sin \theta \cos \phi$$

$$PQ = y = t \sin \theta + u \cos \theta \cos \phi$$

$$QM = z = u \sin \phi.$$

Finally, if these values are substituted in the equation in x, y and z for the solid, we obtain an equation in t and u, that is, the coordinates of the section, which should indicate the nature of the section. This method is almost the same as that which we used before in section 50.

CHAPTER IV

On the Change of Coordinates.

86. Just as there are innumerably different ways in which an equation for a curve in a plane can be transformed by changing the position of the origin or changing the direction of the axis, or both of these changes, so in the present business there are many more varieties of change. First of all, in the same plane where two coordinates are located, these coordinates can be varied infinitely. Then this very plane which contains the two coordinates can be changed, and in this way the first variety is increased infinitely. For instance, given an equation in three rectangular coordinates, we can always find another equation in three other coordinates which are also rectangular and this can be done in infinitely more ways than could be done with two coordinates in the case of equations of curves.

87. Suppose first that we change only the origin on the axis, so that the coordinates y and z remain the same and the new ordinate differs from x only by a constant. Suppose that the new ordinate is t, so that $x = t \pm a$. When we substitute this in the equation for a surface, we obtain an equation in the coordinates t, y, and z which is *a priori* different, but the surface is the same. In a similar way the other coordinates y and z can be increased or decreased by constants. If we let

$x = t \pm a$, $y = u \pm b$, and $z = v \pm c$, there arises an equation in the variables t, u, and v for the same surface, and the new coordinates are parallel to the previous ones. Although the equation is more general than before, still it hasn't changed very much .

88. Since the three rectangular coordinate which express the equation for the surface are referred to three mutually perpendicular planes, we suppose that the plane in which the two coordinates x and y are located remains fixed. In this first plane we take any other straight line CT, as in *figure 140*, other than AP, and let this be the axis. Since the old coordinates for the axis AP were $AP = x$, $PQ = y$, and $QM = z$, for the new axis CT, the coordinate $QM = z$ will remain the same, but the two others will be $CT = t$ and $TQ = u$, where QT is drawn perpendicular to the new axis CT. In order to find the equation in the new coordinates t, u, and z, we draw CR parallel to the original axis AP and from C we drop the perpendicular CB to AP. We let $AB = a$, $BC = b$, and the angle $RCT = \zeta$. Finally, we draw TR perpendicular to CR and from T we drop the perpendicular TS to QP.

89. Then in the triangle TCR we have $TR = t \sin \zeta$ and $CR = t \cos \zeta$. In the triangle QTS, the angle at Q is also equal to ζ so that $TS = u \sin \zeta$, and $QS = u \cos \zeta$. Now we have

$$AP = x = CR + TS - AB = t \cos \zeta + u \sin \zeta - a$$

and

$$QP = QS - TR - BC = y = u \cos \zeta - t \sin \zeta - b.$$

When we substitute these values for x and y in the equation for the given surface, we obtain an equation in the new coordinates t, u and z which expresses the nature of the same surface. This equation is much more general than the original one, since it contains three new arbitrary constants a, b, and ζ which were not in the original equation. This is the general equation when we keep fixed the plane in which the coordinates x and y lie.

90. Now we allow the plane with the first two coordinates x and y to change. First the change will be such that the new plane intersects the original plane APQ in the straight line AP, as in *figure 141*, where AP will remain as the axis for the new coordinates. Let APT be the new plane with the angle of inclination $QPT = \eta$. From any point M we drop the perpendicular MT to the new plane and this will be the new third coordinate. The three new coordinates will be $AP = x$, $PT = u$, and $TM = v$. We draw TR perpendicular to PQ and TS perpendicular to QM, so that $TR = u \sin \eta$, $PR = u \cos \eta$, $TS = v \sin \eta$, and $MS = v \cos \eta$. It follows that $PQ = y = u \cos \eta - v \sin \eta$ and $QM = z = v \cos \eta + u \sin \eta$. When these values are substituted in the given equation for y and z, we obtain a new equation in the coordinates x, u, and v which expresses the nature of the same surface.

91. Now suppose that the new plane intersects the plane APQ in any line CT with angle of inclination equal to η, as in *figure 140*. We take this straight line CT as the axis in the new plane. First we need the equations for the coordinates in the plane APQ with respect to the axis CT, which we have already found. That is, if we let $AB = a$, $BC = b$, the angle $TCR = \zeta$, and the coordinates $CT = p$, $TQ = q$, and $QM = r$, then

$$x = p \cos \zeta + q \sin \zeta - a,$$

$$y = q \cos \zeta - p \sin \zeta - b,$$

and $z = r$. Now from the preceding section, using the new coordinates t, u and v, we have $p = t$, $q = u \cos \eta - v \sin \eta$, and

$r = v \cos \eta + u \sin \eta$. When we make these substitutions we have

$$x = t \cos \zeta + u \sin \zeta \cos \eta - v \sin \zeta \sin \eta - a$$

$$y = - t \sin \zeta + u \cos \zeta \cos \eta - v \cos \zeta \sin \eta - b$$

$$z = u \sin \eta + v \cos \eta.$$

92. We have taken the coordinates t and u in the new plane as well as some arbitrary line as an axis. In this way we have found the most general equation for a given surface. As in *figure 140*, we have taken AP, PQ, QM as the coordinates t, u, v, where AP represents the intersection of the new plane with the plane where the principal coordinates x and y lie. We let CT be the new axis to which we refer the most general coordinates, which we call $CT = p$, $TQ = q$, and $QM = r$. Furthermore we let the lines AB and BC be fixed lines with the angle $CTR = \theta$. Then from section 89 we have

$$t = p \cos \theta + q \sin \theta - AB$$

$$u = - p \sin \theta + q \cos \theta - BC$$

$$v = r.$$

When these values are substituted into the expression given in the preceding section, we obtain

$$x = p \left(\cos \zeta \cos \theta - \sin \zeta \cos \eta \sin \theta \right)$$

$$+ q \left(\cos \zeta \sin \theta + \sin \zeta \cos \eta \cos \theta \right) - r \sin \zeta \sin \eta + f$$

$$y = - p \left(\sin \zeta \cos \theta + \cos \zeta \cos \eta \sin \theta \right)$$

$$- q \left(\sin \zeta \sin \theta - \cos \zeta \cos \eta \cos \theta \right) - r \cos \zeta \sin \eta + g$$

$$z = -p \sin \eta \sin \theta + q \sin \eta \cos \theta + r \cos \eta + h,$$

where f, g, and h come from calculations with the fixed straight lines.

93. It is clear that the most general equation for any surface depends on six arbitrary constants which can be determined in any way while the nature of the surface remains the same. No matter how simple and succinct the equation for a surface might be in the coordinates x, y, and z, when the most general equation is expressed in terms of p, q, and r, due to the large number of arbitrary constants, will necessarily be very complicated, especially if x, y, and z have higher exponents. It is hardly possible to give an example in which it would be convenient to find the most general equation. The usefulness of the most general equation might be see from the possibility of choosing the arbitrary constants in such a way that the equation would take its simplest form. However, because of the lengthy calculations involved, this labor would generally become overwhelming. Still, in what is to follow, this method of the most general equation will not be without usefulness. From it we will obtain and demonstrate some very important properties.

94. Although the general equation is usually very complicated, still if we consider the total degree of the coordinates, that total degree is always equal to the total degree of the original coordinates x, y and z. Hence, since the equation for a sphere $x^2 + y^2 + z^2 = a^2$ is second degree, the most general equation cannot be of degree higher than two in the coordinates p, q, and r. This degree of the equation for any surface will give an essential property of the nature of that surface. No matter how the position of the coordinates may change, this degree always remains the same. This property is similar to that which we previously observed with regard to curves. For this reason we divided the curves according to their order. In the same way it will be convenient to assign

the order to a surface according to the degree of its equation. For us a first order surface will be one whose equation is first degree. We will assign a surface to the second order if its equation is second degree. Hence the following orders are obtained from the degree of their equations.

95. If we compare these ideas with the results discovered previously concerning sections of surfaces by planes, we found that the order of the section was always equal to the order of the surface. Suppose that the equation for a surface in the coordinates x, y, and z assigned the surface to the nth order, than any section would be given in the rectangular coordinates t and u. But we saw in section 85 that the equation in t and u can be found when we substitute the following values in the equation for the surface:

$$x = f + t \cos \theta - u \sin \theta \cos \phi$$

$$y = t \sin \theta + u \cos \theta \cos \phi$$

$$z = u \sin \phi.$$

It is clear that the equation for the section can have no higher degree than that of the equation in x, y, and z. Indeed the total degree will remain the same.

96. It follows that a surface of the first order can have sections by planes only of the first order, that is straight lines. Then from the sections of second order surfaces we obtain only second order curves, that is conic sections. A conical surface is also of the second order, since its equation has the form

$$z^2 = \alpha x^2 + \beta y^2.$$

In a similar way, from a third order surface the sections by a plane will be third order curves, and so forth. It can happen that the equation for some section admits a factorization. In this case the section will consist of two or more curves of lower order. For example, a conic section through the vertex consists of two straight lines, but taken together they are a curve of the second order, as we noted previously.

97. Now that we have defined the order of a surface, let us first consider those surfaces which belong to the first order. The equation which expressed their nature is $\alpha x + \beta y + \gamma z = a$. Since all sections by a plane will be straight lines, it is clear that these surfaces cannot be any thing other than planes. If there were any convexity or concavity in the surface, then there would be sections which were not straight lines. Although in the higher orders there are surfaces which have some sections which are straight lines (as for example, cylinders, cones, and others which we will see) still there are other sections which are curves. Just as in the case of curves, we have seen that a curve whose intersection with a straight line can have no more than a single point must be a straight line, so a surface whose intersection with a plane is always a straight line must be a plane.

98. From the most general equation this property is very clearly demonstrated. When we form the most general equation from the equation $\alpha x + \beta y + \gamma z = a$, we obtain an equation in p, q, and r according to section 92. By a judicious choice of the six arbitrary constants, nothing prevents the vanishing of the coordinates p and q, so that the equation is reduced to the form $r = f$. This equation expresses the nature of the same surface, but the equations show that the given surface must be a plane parallel to the plane containing the two coordinates p and q. It can also be arranged that $r = 0$, so that it is clear that the plane in which p and q are the coordinates is the desired surface.

99.   Since   it   is   clear   that   the   surface   with   equation
$\alpha x + \beta y + \gamma z = a$ is a plane, we should find its position with respect
to the plane in which the coordinates x and y lie. Let M be any point
on this surface in *figure 142*. Let the three coordinates $AP = x$,
$PQ = y$, and $QM = z$. First we let $z = 0$, to obtain the equation
$\alpha x + \beta y = a$. This expresses the equation of the intersection of the
desired surface with the plane APQ. This is clearly the straight line
BCR, whose position with respect to the axis AP is such that the
straight line AB which is perpendicular to the axis AP has length equal
to $\dfrac{a}{\beta}$ and $AC = \dfrac{a}{\alpha}$. Hence the tangent of the angle ACB will be equal
to $\dfrac{\alpha}{\beta}$. Then the sine of this angle is equal to $\dfrac{\alpha}{(\alpha^2 + \beta^2)^{\frac{1}{2}}}$, and its

cosine is equal to $\dfrac{\beta}{(\alpha^2 + \beta^2)^{\frac{1}{2}}}$. Next we extend QP until it intersects

the line BC at R. Since $CP = x - \dfrac{a}{\alpha}$, we have

$$CR = \frac{x\sqrt{\alpha^2 + \beta^2}}{\beta} - \frac{a\sqrt{\alpha^2 + \beta^2}}{\alpha\beta}$$

and $PR = \dfrac{\alpha x}{\beta} - \dfrac{a}{\beta}$.

100.   From Q we drop the perpendicular QS to BC. When we
join M to S it is clear that the angle MSQ measures the angle of inclination of the given surface to the plane APQ. Since $PR = \dfrac{\alpha x - a}{\beta}$, we
have

$$QR = \frac{\alpha s + \beta y - a}{\beta} = \frac{-\gamma z}{\beta}.$$

Since the angle $RQS = ACB$ we have $QS = \dfrac{-\gamma z}{(\alpha^2 + \beta^2)^{\frac{1}{2}}}$. It follows that

$$\tan QSM = \frac{-\sqrt{\alpha^2 + \beta^2}}{\gamma},$$

so that

$$\cos QSM = \frac{\gamma}{(\alpha^2 + \beta^2 + \gamma^2)^{\frac{1}{2}}}.$$

Hence the desired surface has an angle of inclination with respect to the plane with coordinates x and y whose tangent is equal to $\dfrac{-\sqrt{\alpha^2 + \beta^2}}{\gamma}$. Likewise the angle of inclination of the same surface with respect to the plane with coordinates x and z has its tangent equal to $\dfrac{-\sqrt{\alpha^2 + \gamma^2}}{\beta}$, and the angle of inclination with respect to the plane with coordinates y and z has its tangent equal to $\dfrac{-\sqrt{\beta^2 + \gamma^2}}{\alpha}$.

CHAPTER V

On Second Order Surfaces.

101. We have defined a second order surface to be one whose equation in x, y, and z has degree two, that is, the joint degree of any term is no more than two and some term has joint degree equal to two. If an algebraic equation is given for a surface, then the order can be assigned to the surface immediately. Since every first order surface has been shown to be a plane, in this chapter we will consider second order surfaces. Among these there is a greater diversity than was found among second order curves, as should be clear to anyone who considers the problem attentively. I will endeavor to explain these diverse genera clearly. In the higher orders the multitude of genea increases to such a degree that we have to restrict ourselves to only the second order.

102. Since the nature of a second order surface is expressed by a second degree equation in x, y, and z, cylinders and cones, both right and scalene, as well as the sphere, are contained in this second order. We have already studied the properties of those surfaces. All of the surfaces belonging to this order have equations contained in this general equation:

$$\alpha z^2 + \beta yz + \gamma xz + \delta y^2 + \epsilon xy + \zeta x^2 + \eta z + \theta y + \iota x + \kappa = 0.$$

No matter what three coordinates may be chosen, the equation always has this form. The various genera of surfaces contained in this order are determined by the different relationships among the coefficients. Although the same surface can be expressed by an infinite number of different equations, still there are an infinite number of different surfaces in this order.

103. Just as with plane curves our principal division was between those which went to infinitey and those which were contained in a bounded region, so too, with surfaces belonging to any order we have two classes. The first class is that of those which go to infinity, while the second class contains those which lie in a bounded region. Hence, the cylinder and the cone belong to the first class, while the sphere belongs to the second. There are no surfaces of the second class in any odd order, since any surface with an odd order has plane sections of the same order. But we have seen that every odd order curve goes to infinity, so that every surface with an odd order must also go to infinity.

104. Whenever a surface goes to infinity at least one of the three variables x, y, and z must go to infinity. We shall assume that z goes to infinity whenever the surfaces goes to infinity. In order to investigate that part which goes to infinity we suppose that $z = \infty$, and give special consideration to the term αz^2, in particular, whether this term is present or not. Suppose the first term is present, then in comparison, the terms ηz and κ will vanish, so that the equation for that part going to infinity will have the form

$$\alpha z^2 + \beta yz + \gamma xz + \delta y^2 + \epsilon xy + \zeta x^2 + \theta y + \iota x = 0.$$

In this equation all of the terms which are not infinite, or are at least less infinite than αz^2 will vanish.

105. We suppose that all terms of degree two are present. Whatever the surface might be, the most general equation will have all terms
of the highest degree. Hence, this hypothesis that all terms of degree two
are present takes nothing away from the generality of our solution.
Furthermore, when the terms yz and xz are present, the terms θy and ιx
will vanish in comparison. There remains the equation

$$\alpha z^2 + \beta yz + \gamma xz + \delta y^2 + \epsilon xy + \zeta x^2 = 0.$$

From this equation we obtain

$$z = \frac{-\beta y - \gamma x \pm \sqrt{(\beta^2 - 4\alpha\delta)y^2 + (2\beta\gamma - 4\alpha\epsilon)xy + (\gamma^2 - 4\alpha\zeta)x^2}}{\alpha}.$$

This equation expresses the nature of that part of the surface which goes
to infinity.

106. If a surface has some part which goes to infinity it will be
congruent with some infinite part of the surface with equation

$$\alpha z^2 + \beta yz + \gamma xz + \delta y^2 + \epsilon xy + \zeta x^2 = 0,$$

so that this surface becomes a kind of asymptote for that surface with
the general equation. Since this equation has all three variables in the
second degree, it is the equation for a cone with its vertex at the origin,
where all of the coordinates vanish simultaneously. It follows that any
surface which goes to infinity has a cone which is an asymptote for the
given surface. That is, the infinite part of the given surface is essentially
congruent with that part of the cone in that there is always a bounded
distance between the two surfaces. Just as we distinguished branches of
a curve which go to infinity by straight line asymptotes, so now we can
distinguish parts of a surface which go to infinity by asymptotic cones.

107. Whenever the asymptotic cone is real, the given surface actually goes to infinity. Indeed, the parts of the two surfaces which go to

infinity are congruent. It follows that from the nature of the asymptotic cone we discover the nature of the given surface. If the asymptotic cone is complex, then the given surface has no part which goes to infinity, so that it is contained in a bounded region. It follows that in order to determine whether a given surface belongs to the second class, which lies in a bounded region, we have only to determine whether the asymptotic cone is complex. This occurs when the whole surface is reduced to a single point. Indeed, if the cone has any point other than the vertex, then it necessarily goes to infinity. We have already seen that a straight line which passes through the vertex of a cone and any other point of the cone must lie in the cone.

108. When the asymptotic cone, expressed by the equation

$$\alpha z^2 + \beta yz + \gamma xz + \delta y^2 + \epsilon xy + \zeta x^2 = 0,$$

reduces to a single point, then all of the sections through the vertex likewise must reduce to a single point. In the first place, if $z = 0$, then the equation $\delta y^2 + \epsilon xy + \zeta x^2 = 0$ should have no solution except $x = 0$ and $y = 0$. This requires that $4\delta\zeta > \epsilon^2$. The same should be true if $x = 0$ or $y = 0$, so that $4\alpha\delta > \beta^2$ and $4\alpha\zeta > \gamma^2$. We conclude that unless $4\delta\zeta > \epsilon^2$, $4\alpha\delta > \beta^2$, and $4\alpha\zeta > \gamma^2$, the surface with the equation

$$\alpha z^2 + \beta yz + \gamma xz + \delta y^2 + \epsilon xy + \zeta x^2 + \eta z + \theta y + \iota x + \kappa = 0$$

must have a part which goes to infinity.

109. These three conditions are still not sufficient for a surface to be confined to a bounded region. A further condion that is required is that the value of z, is the asymptotic equation found above, be complex. That is, the expression

$$(\beta^2 - 4\alpha\delta)y^2 + 2(\beta\gamma - 2\alpha\epsilon)xy + (\gamma^2 - 4\alpha\zeta)x^2$$

must always be negative when any values other than zero are substituted for the variables x and y. Since $\beta^2 - 4\alpha\delta$ and $\gamma - 4\alpha\zeta$ are supposed to be negative, this condition is equivalent to

$$(\beta\gamma - 2\alpha\epsilon)^2 < (\beta^2 - 4\alpha\delta)(\gamma^2 - 4\alpha\zeta).$$

That is, if

$$\alpha\epsilon^2 + \delta\gamma^2 + \zeta\beta^2 < \beta\gamma + 4\alpha\delta\zeta,$$

provided α is positive, since we divided that equation by α. Now with $\alpha > 0$, $4\alpha\zeta > \gamma^2$, $4\alpha\delta > \beta^2$, and $4\delta\zeta > \epsilon^2$, it follows that both δ and ζ are positive.

110. It follows that a surface of the second order is confined to a bounded region if in its equation the four following conditions are satisfied: $4\alpha\zeta > \gamma^2$, $4\alpha\delta > \beta^2$, $4\delta\zeta > \epsilon^2$, and

$$\alpha\epsilon^2 + \delta\gamma^2 + \zeta\beta^2 < \beta\gamma\epsilon + 4\alpha\delta\zeta.$$

Hence, the first genus of second order surfaces are defined to be all those species which do not go to infinity, but are contained in a bounded region. The sphere belongs to this genus since its equation is

$$z^2 + y^2 + x^2 = a^2.$$

In this case $\alpha = 1$, $\delta = 1$, $\zeta = 1$, $\beta = 0$, $\gamma = 0$, and $\epsilon = 0$, so that the four conditions are satisfied. More generally, this is also true for a surface with equation

$$\alpha z^2 + \delta y^2 + \zeta x^2 = a^2$$

in which α, δ, and ζ are all positive, so that such a surface is always bounded, unless one or two of the coefficients vanishes.

111. In view of these four conditions by which a surface is confined to a bounded region, if a definite second degree equation is given, we can immediately decide whether the surface expressed by the equation has

parts going to infinity or not. If only one of these four conditions fails to be satisfied, then the surface certainly goes to infinity. In this case, several subdivisions must be made according to the kinds of parts which go to infinity. The first subdivision consists of those surfaces such that

$$\alpha\epsilon^2 + \delta\gamma^2 + \zeta\beta^2 > \beta\gamma\epsilon + 4\alpha\delta\zeta,$$

in which case the surface goes to infinity and has a cone as an asymptote, as we have already seen. This is the diametrically opposite case to the previous one in which the surface is contained in a bounded region.

112. We also have the intermediate case in which, although the surface goes to infinity, it is intermediate between the two preceeding just as the parabola is intermediate between the ellipse and the hyperbola. This is the case when

$$\alpha\epsilon^2 + \delta\gamma^2 + \zeta\beta^2 = \beta\gamma\epsilon + 4\alpha\delta\zeta.$$

It follows that

$$\alpha z = -\beta y - \gamma x + y\sqrt{\beta^2 - 4\alpha\delta} + x\sqrt{\gamma^2 - 4\alpha\zeta}.$$

Hence the asymptotic equation

$$\alpha z^2 + \beta yz + \gamma xz + \delta y^2 + \epsilon xy + \zeta x^2 = 0$$

has two linear factors, which are either real, or complex, or equal to each other. This threefold diversity gives the three genera of surfaces which go to infinity. Thus, we have all together five genera of second order surfaces which we now will consider in greater detail.

113. Since we can change the position of the three axis to which the coordinates are parallel, the general equation can be put into a simpler form. We will find the simplest form of the equation which still contains all of the species contained in the most general form. Since the geneal equation for a second order surface has the form

$$\alpha z^2 + \beta yz + \gamma xz + \delta y^2 + \epsilon xy + \zeta x^2 + \eta z + \theta y + \iota x + \kappa = 0,$$

we want to find an equation in three other coordinates p, q and r which will express the same point as that given by the previous coordinates. From Section 92 we recall that

$$x = p(\cos k \cos m - \sin k \sin m \cos n)$$
$$+ q(\cos k \sin m \cos n) - r \sin k \sin n$$

$$y = -p(\sin k \cos m + \cos k \sin m \cos n)$$
$$- q(\sin k \sin m - \cos k \cos m \cos n) - r \cos k \sin n$$

$$z = -p \sin m \sin n + q \cos m \sin n + r \cos n.$$

When we make these substitutions we obtain the equation

$$Ap^2 + Bq^2 + Cr^2 + Dpq + Epr + Fqr + Gp + Hq + Ir + K = 0.$$

114. Now the arbitrary angles k, m, and n can be defined in such a way that the three coefficients D, E, and F vanish. Although the calculations are rather long, we can find the actual angles which bring this result. If someone should doubt this result that the coefficients can be eliminated, he should at least concede that at least two of the coefficients, D and E, can be made equal to zero. If this has been effected, the position of the third axis, to which the ordinate r is parallel in the plane perpendicular to the ordinate p, can easily be changed so that the coefficient F also vanishes. Let $q = t \sin j + u \cos j$ and $r = t \cos j - u \sin j$, so that instead of the term qr we have a new term tu whose coefficient can be made equal to zero by a proper choice of the angle j. In this way the general equation for second order surfaces is brought to the form

$$Ap^2 + Bq^2 + Cr^2 + Gp + Hq + Ir + K = 0.$$

115. In addition, the coordinates p, q and r can be increased or decreased in such a way that the coefficients G, H, and I also vanish. This is accomplished simply by changing the position of the origin. In this way every second order surface is contained in the equation

$$Ap^2 + Bq^2 + Cr^2 + K = 0.$$

From this equation it is clear that each of the three principal planes passing through the origin bisects the surface into two congruent parts. It follows that every second order surface has not only a diametral plane but three mutually perpendicular planes of this kind. The point of intersection of these three planes is the center of the surface, although the center in some cases is at an infinite distance. This is similar to the conic sections, all of which have a center, although the center of a prabola lies at an infinite distance from the vertex.

116. With the equation for all second order surfaces in its simplest form, the first genus of these surfaces has the equation

$$Ap^2 + Bq^2 + Cr^2 = a^2,$$

where all three coefficients A, B, and C are positive. Surfaces which belong to this first genus are not only contained in a bounded region, but each has a center in which the three mutually perpendicular diametral planes intersect. In *figure 143* let C be the center, and let CA, CB, and CD be the mutually penpendicular principal axes, to which the coordinates p, q, and r are parallel. Let $ABab$, ADa, and BDb be the three diametral planes which divide the surface into congruent parts.

117. If we let $r = 0$, then the equation $Ap^2 + Bq^2 = a^2$ gives the nature of the principal section $ABab$, which is an ellipse with center C and semiaxes $CA = Ca = \dfrac{a}{A^{\frac{1}{2}}}$ and $CB = Cb = \dfrac{a}{B^{\frac{1}{2}}}$. If we let

$q = 0$, then the equation $Ap^2 + Cr^2 = a^2$ is for the principal section ADa which is also an ellipse with center at C and semiaxes $CA = Ca = \dfrac{a}{A^{\frac{1}{2}}}$ and $CD = \dfrac{a}{C^{\frac{1}{2}}}$. When we let $p = 0$ we obtain the third principal section BDb with equation $Bq^2 + Cr^2 = a^2$, which is also an ellipse with center C and semiaxes $CB = Cb = \dfrac{a}{B^{\frac{1}{2}}}$ and $CD = \dfrac{a}{C^{\frac{1}{2}}}$. When we know these three principal sections, or even their semiaxes $CA = \dfrac{a}{A^{\frac{1}{2}}}$, $CB = \dfrac{a}{B^{\frac{1}{2}}}$, and $CD = \dfrac{a}{C^{\frac{1}{2}}}$, the nature of this solid is determined and known. This first genus of surfaces of the second order is usually called an *ellipsoid,* since the three principal sections are ellipses.

118. Under this genus there are three species which deserve special notice. The first occurs when all three principal axes CA, CB, and CD are equal. In this case the three principal sections become circles and the solid is a sphere with the equation we have already seen

$$p^2 + q^2 + r^2 = a^2.$$

The second species occurs when only two of the principal axes are equal to each other. For example, let $CD = CB$ or $C = B$ so that the section BDb becomes a circle. From the equation $Ap^2 + B(q^2 + r^2) = a^2$ we understand that all sections parallel to this principal section will also be circles. For this reason the solid is a spheroid, either prolate, if $AC > BC$, or oblate, if $AC < BC$. The third species consists of those solids in which the three coefficients are all unequal, and so these retain the general name of *ellipsoids*

119. The next genus of second order surfaces are respresented by the equation

$$Ap^2 + Bq^2 + Cr^2 = a^2$$

in which none of the coefficients A, B, nor C vanishes but either one or two of these coefficients is negative. If only one is negative, we consider the equation

$$Ap^2 + Bq^2 - Cr^2 = a^2$$

in which A, B, and C are all positive. As for the center and diametral planes, everything is the same as before. In *figure 144* it is clear that the principal section $ABab$ is an ellipse whose semiaxes are $AC = \dfrac{a}{A^{\frac{1}{2}}}$ and $BC = \dfrac{a}{B^{\frac{1}{2}}}$. Both of the other principal sections, Aq and Bs, are hyperbolas with the center at C and conjugate semiaxis equal to $\dfrac{a}{C^{\frac{1}{2}}}$.

120. This species of surface is unbounded since it diverges both up and down along its hyperbolas. Hence this surace has an asymptotic one with equation $Ap^2 + Bq^2 - Cr^2 = 0$, whose vertex is at the center C and whose sides are asymptotes for the hyperbolas. This asymptotic cone lies inside the surface. It will be a right cone if $A = B$, otherwise it will be scalene. The axis of the cone is the straight line CD perpendicular to the plane ABa. Furthermore, all sections perpendicular to the axis CD will be ellipses similar to the ellipse $ABab$. All sections which are perpendicular to the plane $ABab$ will be hyperbolas. Hence these surfaces are usually called *elliptic hyperboloids* circumscribed about their asymtotic cone.

121. We can note three species in the genus also. In the first species, $a = 0$ so that the ellipse shrinks to a point, the hyperbolas become straight lines, and the surface coincides with the asymptotic cone. It follows that the first species containes all cones, both right and scalene. Hence, we have a new subdivision. There is one species if $A = B$ so that the ellipse $ABab$ becomes a circle. In this case the surface is round or turned; indeed, this surface is generated if any of the hyperbolas is rotated about the conjugate axis. The third species is simply the general form of the genus.

122. The third genus occurs when two of the coordinates of the terms p^2, q^2, and r^2 are negative, so that the equation becomes

$$Ap^2 - Bq^2 - Cr^2 = a^2.$$

If we let $r = 0$, the first principal section is the hyperbola $EAFeaf$ in figure 145. The center is at C, the transverse semiaxis is equal to $\dfrac{a}{A^{\frac{1}{2}}}$, and the conjugate semiaxis is equal to $\dfrac{a}{B^{\frac{1}{2}}}$. The second principal section, when $q = 0$, is also a hyperbola AQ, aq, with the same transverse semiaxis, but with conjugate semiaxis equal to $\dfrac{a}{C^{\frac{1}{2}}}$. The third principal section is complex. The whole of this surface lies inside the asymptotic cone. For this reason this genus can be called *hyperbolic hyperboloid* inscribed in the asymptotic cone. If $B = C$ the surface will be round, generated by revolving a hyperbola about its transverse axis. In this case we have a species of its own. If we let $a = 0$, we obtain a cone, which we have already seen as a species of the previous genus.

123. In order to investigate the next genus we allow one of the

coefficients A, B or C to vanish. Suppose that $C = 0$, so that the general equation found in section 114 becomes

$$Ap^2 + Bq^2 + Gp + Hq + Ir + K = 0.$$

By increasing or decreasing the ordinates p and q the terms Gp and Hq can be removed, but Ir remains. By meanes of this term, K can be removed so that the remaining equation is

$$Ap^2 + Bq^2 = ar.$$

There are two cases to be considered. The first case occurs when both coefficients A and B are positive; the second occurs when one of the coefficients is negative. In either case the center of the surface will lie on the axis CD but at an infinite distance.

124. First we suppose that both A and B are positive. This gives the fourth genus with the equation

$$Ap^2 + Bq^2 = ar.$$

The first principal section is given when $r = 0$, which is a single point. The second section is given when $q = 0$, the third when $p = 0$, and both of these will be a parabola. That is, MAm and NAn in *figure 146*. Since every section perpendicular to the axis AD is an ellipse and every section containing this axis is a parabola, the solid of this genus is called an *eliptic paraboloid*. There are two species of this genus. The first species occurs when $A = B$, so that the solid is round, called a *parabolic cone*. The second species occurs when $a = 0$, so that the equation becomes $Ap^2 + Bq^2 = b^2$. This gives cylinders, which are right when $A = B$ and scalene when $A \neq B$.

125. The fifth genus has the equation

$$Ap^2 - Bq^2 = ar.$$

The first principal section is given by $r = 0$, which is two straight lines, Ee and Ff in *figure 147*. These two lines intersect in the point A. Each section parallel to this section will be a hyperbola with center on the axis AD and asymptotes Ee and Ff. The two planes which are perpendicular to the plane ABC and intersect that plane in the lines Ee and Ff approach the surface at an infinite distance, so that the surface has two intersecting planes for an asymptote. The other principal sections, the the planes ACD and ABd, are parabolas. For this reason the surfaces belonging to this genus are called *parabolic hyperboloids* with two planes for an asymptote. If $a = 0$, so that $Ap^2 - Bq^2 = b^2$ we have a species which is a hyperbolic cylinder, all of whose sections perpendicular to the axis AD are congruent hyperbolas. If in addition $b = 0$, then we obtain the two asymptotic planes.

126. Finally, the sixth genus of second order surfaces has the equation

$$Ap^2 = aq,$$

which gives a parabolic cylinder, all of whose sections perpendicular to the axis AD are congruent parabolas, each with its vertex on the straight line AD and with axes parallel to each other. We can reduce all second order surfaces to these six genera, so that there is no recond order surface which does not belong to one of these six genera. Furthermore, in the final genus, if $a = 0$, so that $Ap^2 = b^2$, this equation gives two parallel planes, which constitutes a quasi species of this genus. We have here an analogy to the second order curves, where we saw that two intersecting straight lines are a species of hyperbola, while two parallel lines are a species of parabola.

127. Although we have arrived at these six genera of second order surfaces from the simplest equation, still we can now easily assign the genus for any second order surface from any given second degree equation. Indeed, if the given equation is

$$\alpha z^2 + \beta yz + \gamma xz + \delta y^2 + \epsilon xy + \zeta x^2 + \eta z + \theta y + \iota x + \kappa = 0,$$

we make our decision from the second degree terms. That is, we consider the terms

$$\alpha z^2 + \beta yz + \gamma xz + \delta y^2 + \epsilon xy + \zeta x^2.$$

If $\quad 4\alpha\zeta > \gamma^2, \quad\quad 4\alpha\delta > \beta^2, \quad\quad 4\delta\zeta > \epsilon^2, \quad$ and $\alpha\epsilon^2 + \delta\gamma^2 + \zeta\beta^2 < \beta\gamma\epsilon + 4\alpha\delta\zeta$, then the surface is closed, and it belongs to the first genus, which we have called ellipsoids.

128. If one or more of these conditions fail to hold and $\alpha\epsilon^2 + \delta\gamma^2 + \zeta\beta^2 \neq \beta\gamma\epsilon + 4\alpha\delta\zeta$, then the surface belongs to either the second or third genus and it will be a hyperbolic surface with an asymptotic cone. If it is in the second genus it will circumscribe the cone, while if it is third genus, then it will be inscribed in the cone. If

$$\alpha\epsilon^2 + \delta\gamma^2 + \zeta\beta^2 = \beta\gamma\epsilon + 4\alpha\delta\zeta,$$

then the expression

$$\alpha z^2 + \beta yz + \gamma xz + \delta y^2 + \epsilon xy + \zeta x^2$$

can be written as the product of two linear factors, either complex or real. If they are complex, then the surface belongs to the fourth genus, while if they are real, then the surface belongs to the fifth genus. If the expression has two equal factors, that is, if it is a perfect square, then the surface belongs to the sixth genus. In this way we can easily decide to which genus any given equation belongs. It is more difficult to make a decision about the second and third genera since one can change into the other.

129. In a similar way, surfaces of the third and higher order can be divided into genera. That is, we have to consider the general equation which contains only the terms of highest degree. Consequently, for third order surfaces, we consider the third degree terms, which are

$$\alpha z^3 + \beta y z^2 + \gamma y^2 z + \delta x^2 z + \epsilon x z^2 + \zeta x y z + \cdots .$$

First we determine whether this expression can be written as the product of linear factors. If this not possible, the surface has a third order cone for an asymptote. Since the nature of this cone is expressed by setting the expression equal to zero, there are several different third order cones, and from this diversity we obtain several different genera. Although all second order cones are assigned to the same genus, there are right and scalene cones. Still, in the third order there is a much greater variety of cones.

130. Once these genera have been determined, we consider the case in which the expression can be expressed as the product of linear factors. If the expression has one real linear factor, then the surface has an asymptotic plane. When we set the other factor equal to zero, the equation either has a solution or it does not. If the equation has no nontrivial solution, the surface has only an asymptotic plane. If the equation has a solution, then the surface has two asymptotes, one of which is a plane and the other is a second order cone. If the expression has three linear factors, since one is always real, the other two can both be complex or both be real. This gives rise to two genera. Finally, if all three linear factors are real, depending on whether two or all three are equal to each other, we have two more genera. There are no third order surfaces which lie in a bounded region.

CHAPTER VI

On the Intersection of Two Surfaces.

131. We have given a method for investigating the nature of a section which arises from the intersection of some surface with a plane. Since the curve which constitutes the section lies completely in the cutting plane, we can use two coordinates, which we assume to be in the plane, to express the nature of the curve. In this way, from the knowledge we have of plane curves, we understand the surface. However, if the cutting surface is not a plane, since the section will not lie in a plane, we cannot gain a knowledge of its nature from the relation of two coordinates. It is for this reason that we need a different method for locating the true position of any point on sections of this kind, obtained from equations for the surfaces.

132. The positions of points which do not lie in the same plane can be given by means of three mutually perpendicular planes. For any point, the three distances of that point from each of the planes is assigned to the point. These three variables are required in order to express the nature of the curve which is not in a single plane. Hence, if an arbitrary value is assigned to one of the variables, the other two variables are determined. One equation in the three coordinates is not sufficient for this. Indeed, one equation gives the nature of a whole

surface. For this reason, two equations are required, so that if some value is assigned to one of the variables, the other two will be determined.

133. The nature of any non-planar curve is most conveniently expressed by two equations in three variables, for example x, y, and z which represent mutually perpendicular coordinates. By means of the two equations, two of the variables can be determined by the third. For instance, y and z are equal to some functions of x. We can even choose one of the variables arbitrarily for elimination, so that we can find three equations in only two variables: one in x and y, another in x and z, and a third in y and z. Of these three equations, any one is automatically determined by the other two; given equations in x and y and in x and z, the third can be found by eliminating x from these two.

134. Suppose that some non-planar curve is given. In *figure 148* we let M represent any point on the curve. We arbitrarily choose three mutually perpendicular axes, AB, AC, and AD, by means of which the three mutually perpendicular planes BAC, BAD, and CAD are determined. From the point M on the curve we drop the perpendicular MQ to the plane BAC. From the point Q we draw QP perpendicular to the axis AD. Then AP, PQ, and QM are the three coordinates in which the two equations will determine the nature of the curve. We let $AP = x$, $PQ = y$, and $QM = z$. From the two given equations in x, y, and z we eliminate z to obtain an equation in only the two variables x and y. This will determine the position of the points Q in the plane BAC. All of the points Q corresponding to points M produce a curve EQF, whose nature is given by the equation in x and y.

135. In this way, from the two given equations in the three variables, we easily discover the nature of the curve EQF. This is formed by sending each point M of the curve onto the point Q, where MQ is

perpendicular to the plane *BAC*. The curve *EQF* is called the *projection* of the curve *GMH* onto the plane *BAC*. Just as the projection onto the plane *BAC* is found by eleminating the variable *z*, so the projection of the same curve onto the plane *BAD* or onto the plane *CAD* is obtained by eliminating the variable *y* or the variable *x*. The single projection *EQF* is not sufficient to know the curve *GMH*. However, if we also know, for each point *Q*, the perpendicular *QM* = *z*, then from the projection *EQF* we can easily construct the curve *GMH*. For this information we need, besides the equation in *x* and *y* which expresses the projection, another equation in *z* and *x* or one in *z* and *y*, or even one in the three variables *x*, *y* and *z*, from which we can learn the length of the perpendicular *QM* = *z* for each point *Q*.

136. The equation in *z* and *x* expresses the projection of the curve *GMH* onto the plane *BAD*, the equation in *z* and *y* expresses the projection onto the plane *CAD*, and the equation in the three variables *z*, *y*, and *x* expresses the surface on which the curve *GMH* lies. It is clear first of all that from two projections of the same curve *GMH* onto two planes, that curve *GMH* is known. It is also clear that if a surface on which the curve *GMH* lies is given and also its projection onto any plane, the curve is known. Indeed, the points *M*, from the intersection of the straight line *QM* with the surface, define the desired curve *GMH*, where *QM* is the perpendicular through each point *Q* of the projection.

137. With this understanding of the nature of any non-planar curve, it will not be difficult to define the intersection of any two surfaces. Just as the intersection of two planes is a straight line, so the intersection of two surfaces is either a straight line or a curve, which may lie on a plane or it may not. Whatever kind it may be, each of its points must lie on both surfaces, so that it will satisfy both equation. If both surfaces are expressed by equations in all three coordinates with respect

to the same three principal mutually perpendicular planes, or with respect to the same mutually perpendicular axes AB, AC, and AD, then both of these equations together expresses the nature of the intersection.

138. Suppose we are given two surfaces which intersect each other, each expressed by an equation in three coordinates with respect to the same set of principal axes. Thus we have two equation in the three corrdinates x, y, and z. If we eliminate one of these coordinates, we have an equation in the two other variables which gives the projection of the intersection onto the plane determined by these two coordinates. This method can also be used to investigate the intersection of any surface with a plane. Since the general equation for a plane is $\alpha z + \beta y + \gamma x = f$, we can substitute for z, in the equation for the surface, $z = \dfrac{f - \beta y - \gamma x}{\alpha}$, to obtain the equation for the projection of the intersection onto the plane determined by the coordinates x and y. We also have the equation $z = \dfrac{f - \beta y - \gamma x}{\alpha}$ which gives the length of the perpendicular QM from any point Q on the projection to the point M on the section.

139. If it happens that the equation for a projection has no real solution, as for example $x^2 + y^2 + a^2 = 0$, then this indicates that the two surfaces do not intersect each other. If the equation for the inersection has a single point as a solution, that is, the projection reduces to a single point, then the intersection itself is a single point, so that the two surfaces are tangent at that point. This tangency can be known from the equation. It can happen that the tangency can occur along a line, when there are an infinite number of tangent points. This line may be a straight line or a curve. For example, the line will be straight if a plane is tangent to a cylinder or a cone. A sphere inside a right cone will be

tangent along the circumference of a circle. This tangency can be known from the equation; if for a projection of this kind, we obtain an equation which has two equal roots. The reason for this is that a tangency is nothing but the coincidence of two intersections.

140. In order that this might be clarified, we suppose that a sphere is cut by some plane. We take the equation of the sphere to be $z^2 + y^2 + x^2 = a^2$, and for any plane we have the equation

$$\alpha z + \beta y + \gamma x = f.$$

When we substitute $z = \dfrac{f - \beta y - \gamma x}{\alpha}$ in the equation for the sphere, we obtain the equation in x and y for the projection

$$0 = f^2 - \alpha^2 a^2 - 2\beta f y - 2\gamma f x$$
$$+ (\alpha^2 + \beta^2)y^2 + 2\beta\gamma xy + (\alpha^2 + \gamma^2)x^2.$$

This is clearly an ellipse, provided the equation has real solutions. If it has only complex solutions, then there is no intersection of the sphere by the plane. If the ellipse reduces to a single point, then the plane and sphere are tangent. In order to investigate these cases we consider

$$y = \frac{\beta f - \beta\gamma x \pm \alpha\sqrt{a^2(\alpha^2 + \beta^2) - f^2 + 2\gamma f x - (\alpha^2 + \beta^2 + \gamma^2)x^2}}{\alpha^2 + \beta^2}.$$

If f has such a value that the expression is not real, then there is no point of tangency nor any intersection.

141. Suppose $f = a\sqrt{\alpha^2 + \beta^2 + \gamma^2}$. Then

$$y = \frac{\beta f - \beta\gamma x \pm \alpha x\sqrt{-(\alpha^2 + \beta^2 + \gamma^2)} \pm (-\alpha\gamma a i)}{\alpha^2 + \beta^2}.$$

This equation has no real solution except $x = \dfrac{\gamma a}{(\alpha^2 + \beta^2 + \gamma^2)^{\frac{1}{2}}}$ and

$$y = \frac{\beta a}{(\alpha^2 + \beta^2 + \gamma^2)^{\frac{1}{2}}}.$$ It follows that if $f = a\sqrt{\alpha^2 + \beta^2 + \gamma^2}$, then

the plane with equation $\alpha z + \beta y + \gamma x = f$ is tangent to the sphere at the point with coordinates

$$x = \frac{\gamma a}{(\alpha^2 + \beta^2 + \gamma^2)^{\frac{1}{2}}},$$

$$y = \frac{\beta a}{(\alpha^2 + \beta^2 + \gamma^2)^{\frac{1}{2}}},$$

and $z = \dfrac{\alpha a}{(\alpha^2 + \beta^2 + \gamma^2)^{\frac{1}{2}}}.$ The truth of these values can be proved in

elementary geometry where the tangent plane of a sphere is treated.

142. Now we can derive a general rule for determining whether any surface is tangent to a plane or any other surface. First we use the two equations to eliminate one of the variables. Then we determine whether the resulting equation can be written as the product of two linear factors set equal to zero. If the two linear factors are complex, then there is a point of tangency which can be found by setting either factor equal to zero. If the two linear factors are real and equal to each other, then the surfaces are tangent along a straight line. If this equation has two equal non-linear factors, or if it is divisible by a perfect square, then when the square root is set equal to zero, we obtain the projection of the curve along which the surfaces are tangent. It should also be clear that if that equation has four complex factors, then the surfaces are tangent in two points.

143. In order that this might be explained at greater length, we will investigate the tangency of a cone and a sphere whose center lies on the axis of the cone. The equation for the sphere is $z^2 + y^2 + x^2 = a^2$,

and for the cone $(f - z)^2 = mx^2 + ny^2$, where the vertex of the cone is at a distance f from the center of the sphere. When we eliminate y we obtain

$$(f - z)^2 = na^2 - nz^2 + (m - n)x^2$$

for the projection of the intersection onto the plane of the coordinates x and z. First we suppose that the cone is right, so that $m = n$. Hence we have

$$z = \frac{f \pm \sqrt{n(1 + n)a^2 - nf^2}}{1 + n}.$$

Now if $f = a\sqrt{1 + n}$, then we have a double root, $z = \dfrac{a}{(1 + n)^{\frac{1}{2}}}$, so

that we have tangency along a curve, namely the circle whose projection onto the plane passing through the axis is a straight line perpendicular to the axis.

144. For a scalene cone, where m and n are unequal, the equation seems always to give an intersection, while in fact there is usually no intersection. If $m > n$ then we always have a real equation for the projection of the intersection. It should be noted that the projection's being real does not always indicate a real intersection. In order that the intersection be real, it is not sufficient that the projection be real, but in addition it is required that the length of the perpendicular from the projection to the intersection also be real. Although all real curves have all projections which are real, still we cannot conclude to the reality of the intersection from the reality of the projection. We must always be acutely aware of this precaution, lest we be misled to assume a real intersection from a real projection.

145. We can avoid this inconvenience if we take the projection

onto the plane of the coordinates x and y. In this plane there is no point which does not correspond to a point on the cone, so that if the projection is real, then the intersection also will be real. Since $z = \sqrt{a^2 - x^2 - y^2}$, we have from the other equation

$$f - \sqrt{a^2 - x^2 - y^2} = \sqrt{mx^2 + ny^2}$$

or

$$a^2 + f^2 - (1 + m)x^2 - (1 + n)y^2 = 2f\sqrt{a^2 - x^2 - y^2}.$$

It follows that

$$\left(a^2 - f^2\right)^2 - 2((a^2 - f^2) + (a^2 + f^2)m)x^2$$

$$- 2((a^2 - f^2) + (a^2 + f^2)n)y^2 + (1 + m)^2x^4$$

$$+ 2(1 + m)(1 + n)x^2y^2 + (1 + n)^2y^4 = 0$$

so that

$$\frac{a^2 - f^2 + n(a^2 + f^2) - (1 + m)(1 + n)x^2}{(1 + n)^2}$$

$$\pm \frac{2f}{(1 + n)^2}\sqrt{n(1 + n)a^2 - nf^2 + (m - n)(1 + n)x^2} = y^2$$

and

$$\frac{a^2 - f^2 + m(a^2 + f^2) - (1 + m)(1 + n)y^2}{(1 + m)^2}$$

$$\pm \frac{2f}{(1 + m)^2}\sqrt{m(1 + m)a^2 - mf^2 + (n - m)(1 + m)y^2} = x^2.$$

146. In order that the discovered equation should have factors, it is necessary that either $f^2 = (1 + n)a^2$ or $f^2 = (1 + m)a^2$. In the former case we have

$$y^2 = \frac{na^2 - (1+m)x^2}{1+n} \pm \frac{2fx\sqrt{m-n}}{(1+n)(1+n)^{\frac{1}{2}}},$$

so that if $m < n$ it is necessary that $x = 0$, $y = \pm \dfrac{a\sqrt{n}}{(1+n)^{\frac{1}{2}}}$, and

$z = \dfrac{a}{(1+n)^{\frac{1}{2}}}$. Hence we have two points of tangency in this case, with

each point equidistant from the axis of the cone. However, if $m > n$,
then we need the other equation

$$x^2 = \frac{ma^2 - (1+n)y^2}{1+m} \pm \frac{2fy\sqrt{n-m}}{(1+m)(1+m)^{\frac{1}{2}}},$$

which is not real unless $y = 0$. In this case $x = \pm \dfrac{a\sqrt{m}}{(1+m)^{\frac{1}{2}}}$, and

$z = \dfrac{a}{(1+m)^{\frac{1}{2}}}$. In this case there are also two points of tangency. The

tangent points are located in that part of the cone where it is most nar-
row. In a similar way we have to decide all of the other possible cases of
tangency.

147. There is a much simpler way of finding the tangent plane for
any surface. We use the method of finding tangents to curves, which we
have discussed previously. We suppose that the nature of the surface,
whose tangent plane we are seeking, is given by an equation in the three
coordinates $AP = x$, $PQ = y$, and $QM = z$, as in *figure 149*. From
this equation we want to define the position of the plane tangent to the
surface at the point M. First we consider some plane which cuts the surface at the point M. The tangent of this section at the point M will lie
in the tangent plane. Hence, if we take the tangents at M of two such
sections, the plane defined by these two tangents should be the tangent

plane to the surface at the point M.

148. Hence, we first cut the surface with the plane perpendicular to the plane APQ along the straight line QS parallel to the axis AP. Then we cut the surface at M with a second plane which is perpendicular to the plane APQ but along the straight line QP which is perpendicular to the axis AP. That is, the first cutting plane is perpendicular to the axis AB, while the second one is perpendicular to the axis AP. Let EM be the curve of the first section, whose desired tangent is MS, which meets the straight line QS in the point S. It follows that QS is the subtangent. The second section gives the curve FM whose tangent is the straight line MT and whose subtangent is QT. Having found the tangents, we have the plane SMT which is tangent to the surface at the point M. When we draw the straight line ST, we have the intersection of the tangent plane and the plane APQ. If we draw the perpendicular QR from Q to ST, then QR is to QS as the whole sine is to the tangent of the angle MRQ, the angle of inclination between the tangent plane and the plane APQ.

149. By the method of tangents just discussed we have found the subtangents $QS = s$ and $QT = t$. Then $PT = t - y$ and $PX = s - \dfrac{sy}{t}$ so that $AX = x + \dfrac{sy}{t} - s$. We note this point X, where the straight line ST crosses the axis AP. Since the angle $AXS = TSQ$, $\tan AXS = \dfrac{t}{s}$. From this we know the intersection of the tangent plane and the plane APQ. Then, since $ST = \sqrt{s^2 + t^2}$, we have $QR = \dfrac{st}{(s^2 + t^2)^{\frac{1}{2}}}$. When QM is divided by this quantity QR, we have the tangent of the angle of inclination. That is $\tan MRQ = \dfrac{z\sqrt{s^2 + t^2}}{st}$. Furthermore, we draw MN perpendicular to

MR so that it is perpendicular to both the tangent plane and the surface

at *M*. We know the position of *N* since $QN = \dfrac{z^2\sqrt{s^2 + t^2}}{st}$. We drop

the perpendicular *NV* from *N* to the axis *AP*. Since the angle

$QNV = QST$, we have $PV = \dfrac{z^2}{t} = QW$, and $NW = \dfrac{z^2}{t}$. It follows

that if the position of the point *N* in the plane *APQ* is defined in this

way, then the straight line *NM* will be perpendicular to the surface.

150. We have alrady shown how the intersection of two surfaces
can be investigated by means of projections. How we consider the order
of the projection with respect to the orders of the two surfaces. First of
all, two first order surfaces, that is, two planes, have a projection of their
intersection which is a first order line. We have also seen that the projection cannot have order higher than the second if one of the surfaces is
first order and the other is second order. In a similar way it is clear that
if one surface is third order and the other is first order, the projection
cannot have order greater than the third, and so forth. However, two
second order surfaces can have an intersection whose projections is fourth
order or less. In general, if one surface has order m and the other has
order n, then the projection of the intersection never has order greater
than mn.

151. When neither of the intersecting surfaces is a plane, it is usually the case that the intersection is a non-planar curve. In spite of this,
it can happen that the whole section will lie in a plane. This will be the
case if both equations for the surfaces, taken together, satisfy an equation
$\alpha z + \beta y + \gamma x = f$. One way in which this can happen is that from the
two given equations, the two variables z and y can be defined in terms of
the third x, so that $z = P$ and $y = Q$ where P and Q are functions of
x. Then we see whether there is a number n such that $P + nQ$ removes

all powers of x except the first power and constants. If this is the case, then $P + nQ = mx + k$, so that the intersection will lie in a plane whose equation is given by $z + ny = mx + k$.

152. For example, suppose that the given two second order surfaces are the right cone with equation $z^2 = x^2 + y^2$ and the elliptic hyperboloid with equation $z^2 = x^2 + 2y^2 - 2ax - a^2$. Since in this case, $x^2 + 2y^2 - 2ax - a^2 = x^2 + y^2$, we have $y = \sqrt{2ax + a^2}$ and $z = x + a$. This last equation already shows that the whole intersection lies in a plane, whose position is givenby the eqaution $z = x + a$. With this kind of reasoning many questions concerning the nautre of surfaces can be answered. However, those questions which require more than the method given here, must wait for analysis of the infinite. These two books have prepard the way for that science.

END OF THE APPENDIX

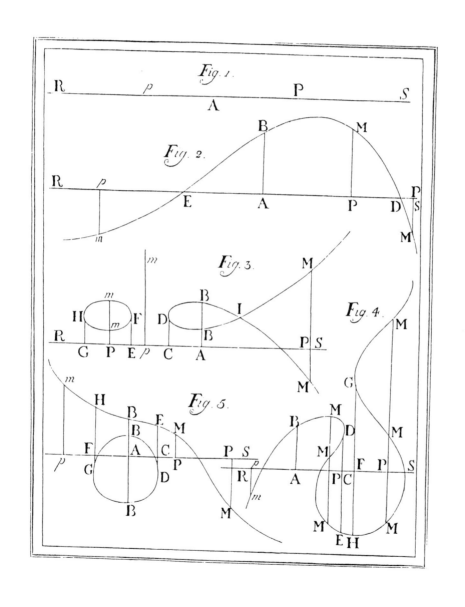

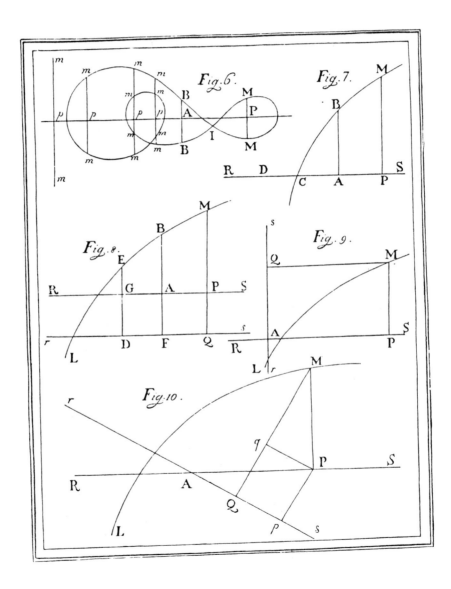

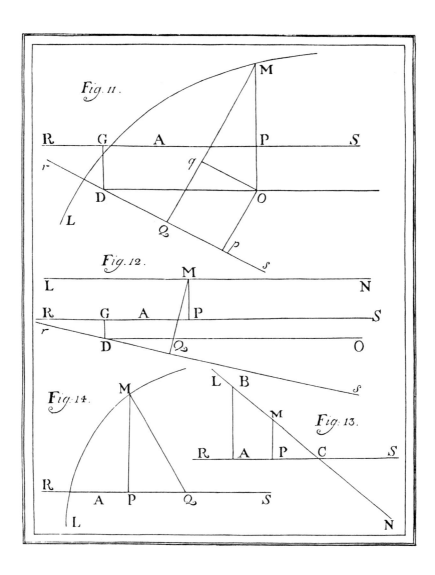

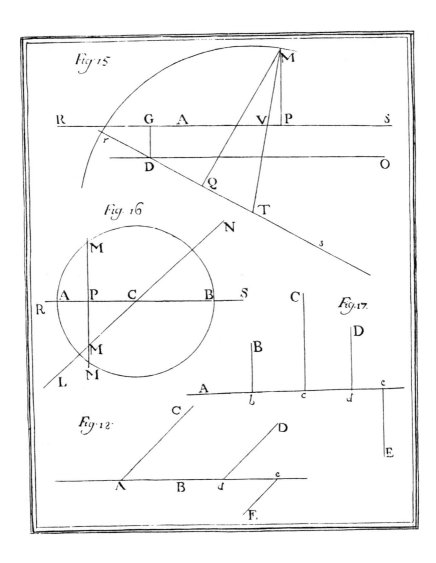

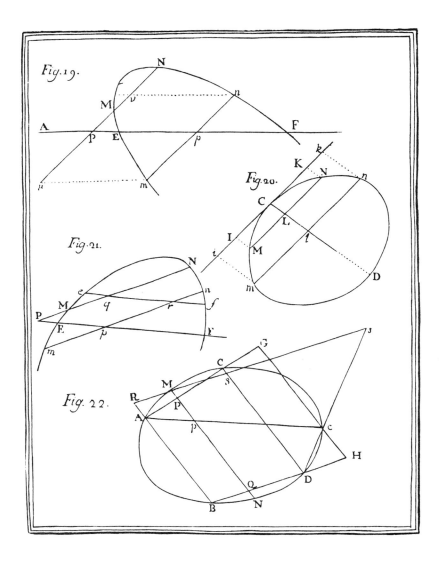

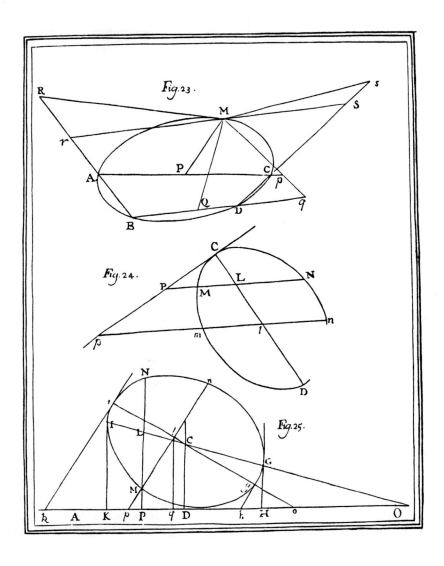

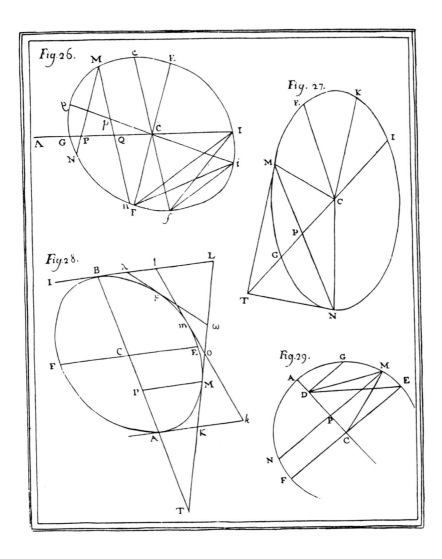

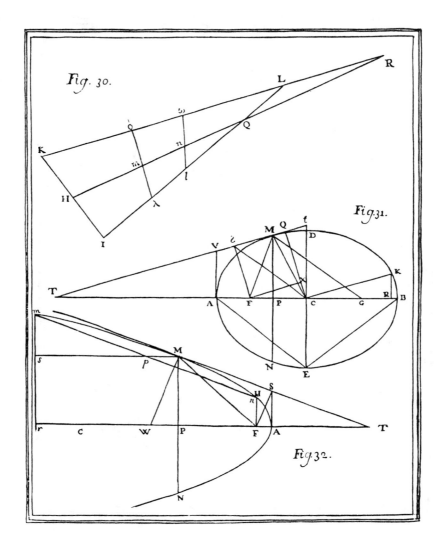

Fig. 30.

Fig. 31.

Fig. 32.

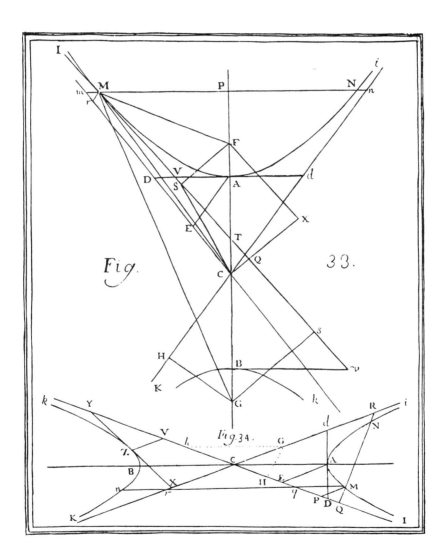

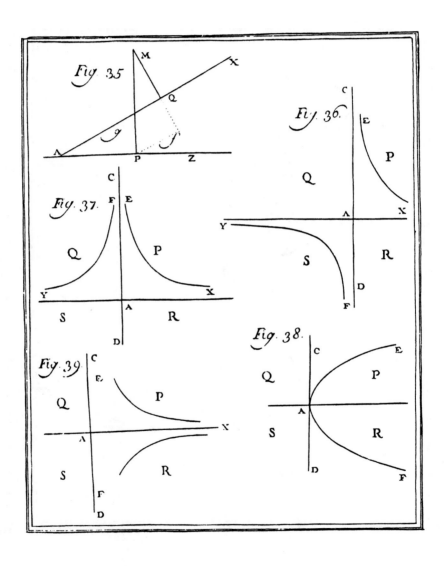

Fig. 35

Fig. 36.

Fig. 37.

Fig. 38.

Fig. 39.

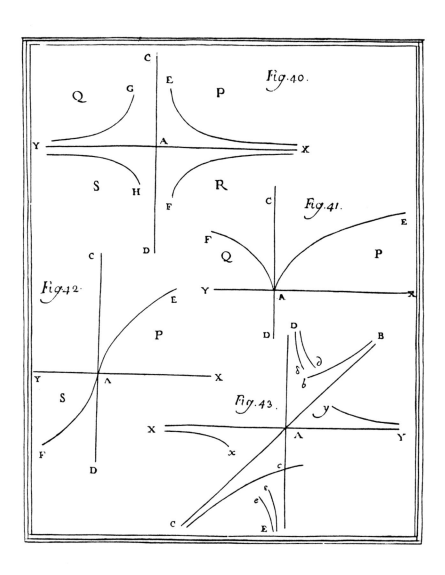

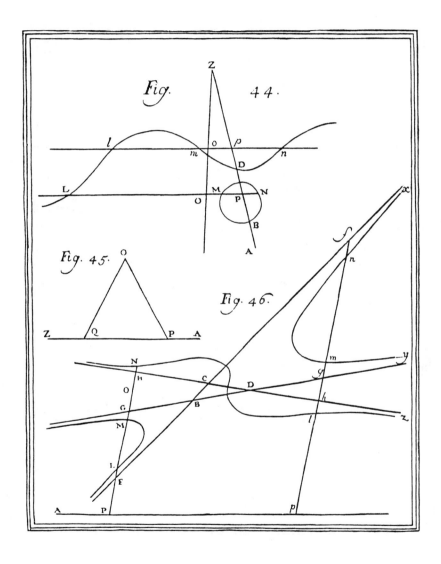

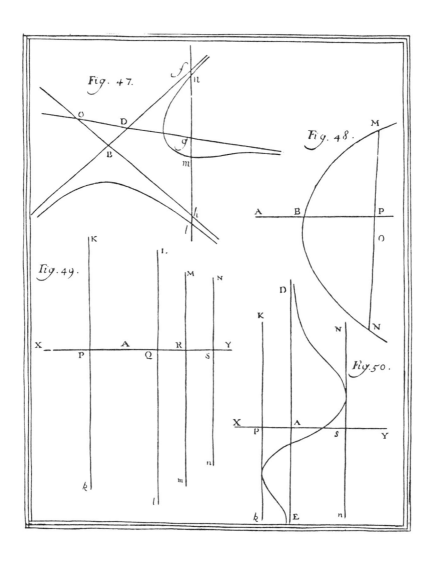

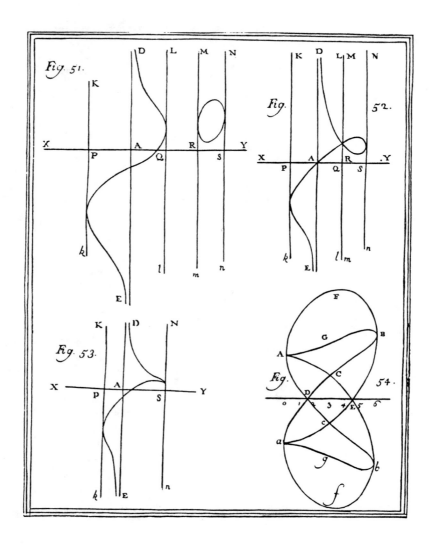

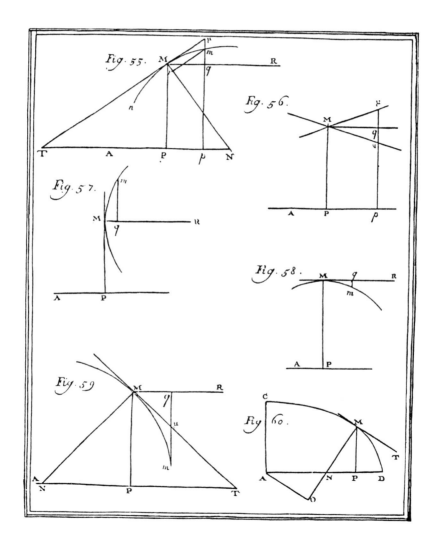

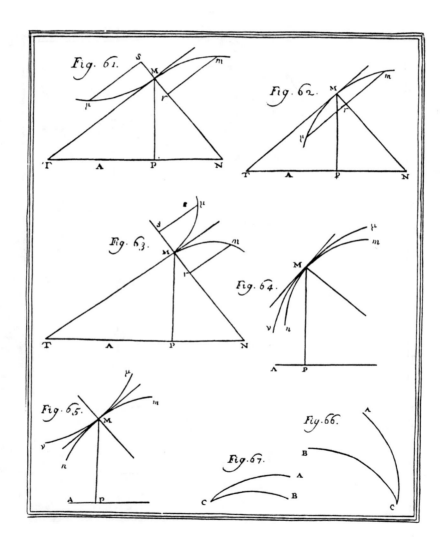

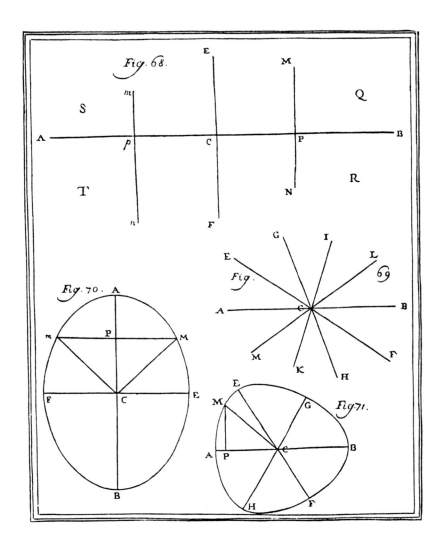

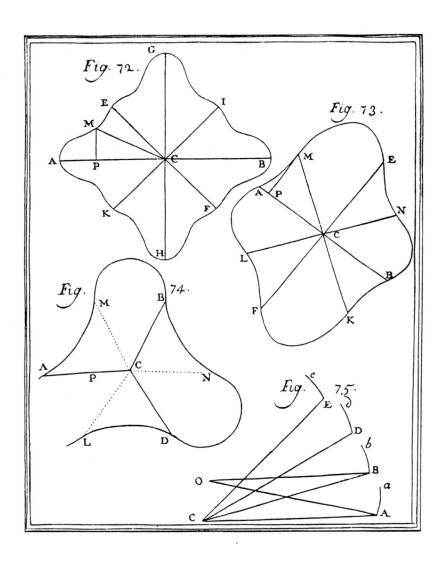

Fig. 72.

Fig. 73.

Fig. 74.

Fig. 75.

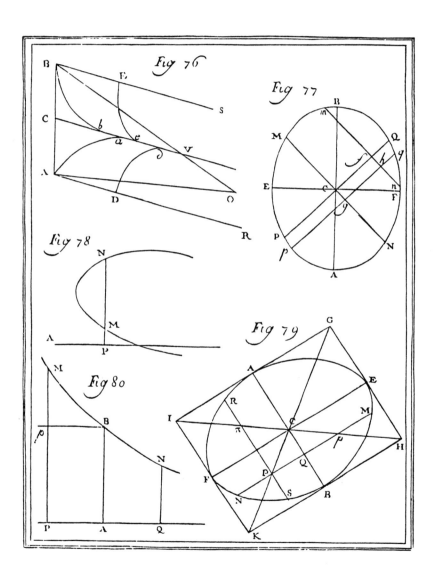

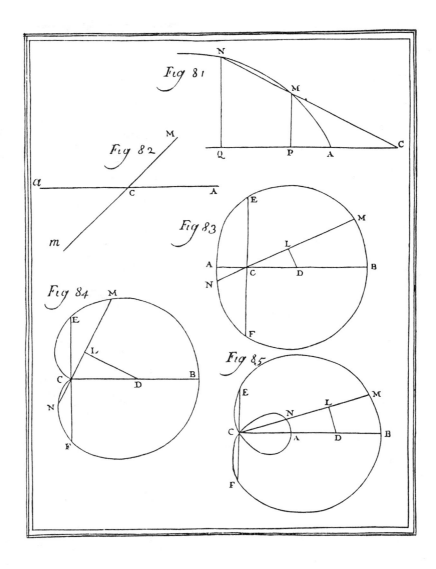

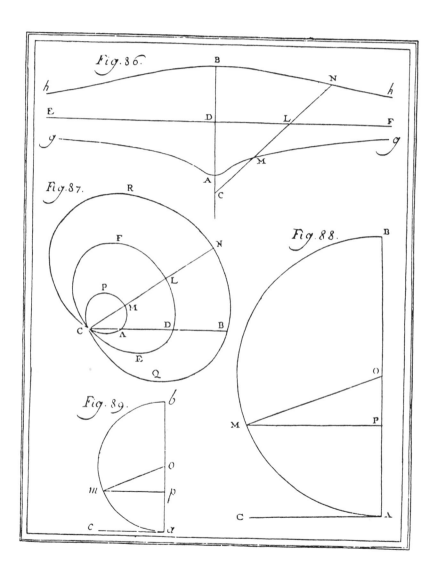

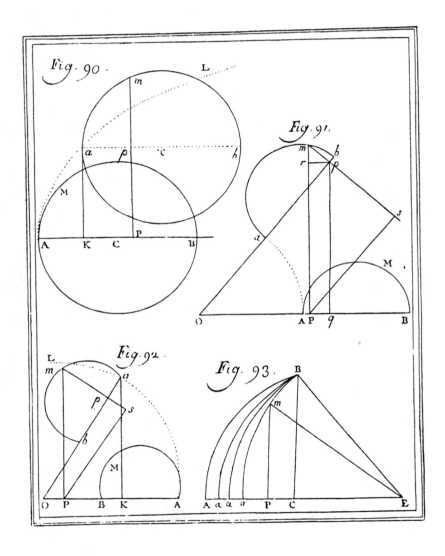

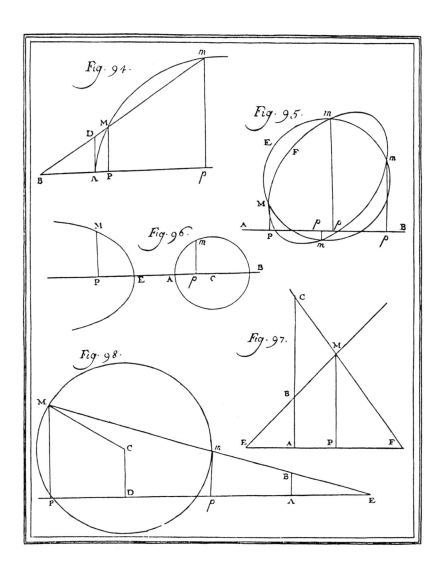

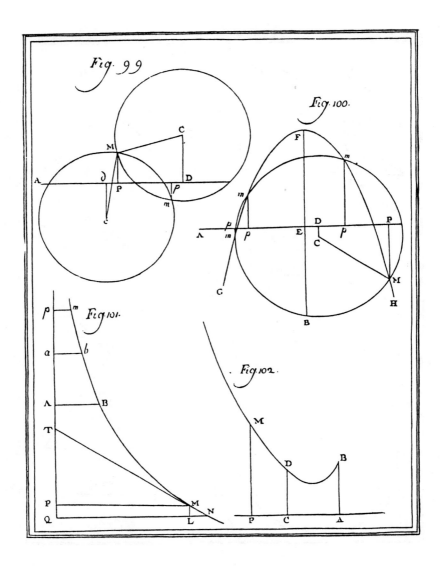

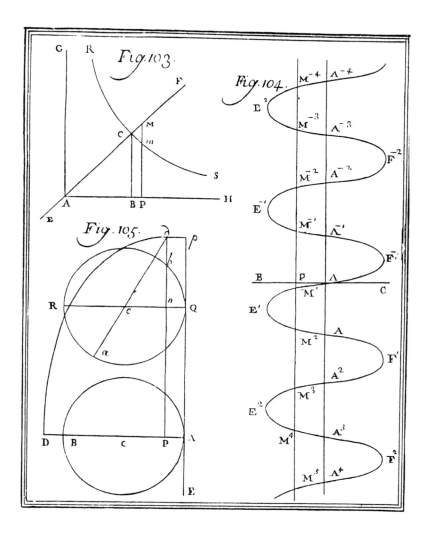

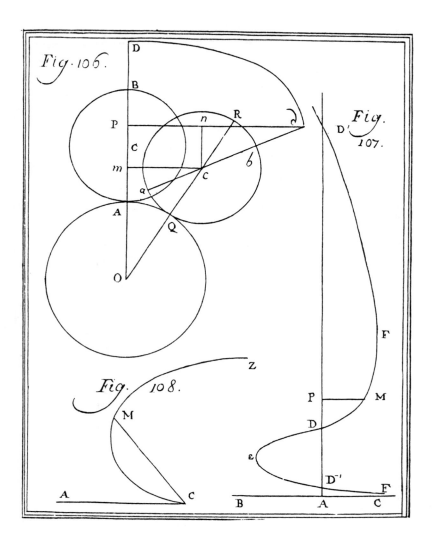

Fig. 106.

Fig. 107.

Fig. 108.

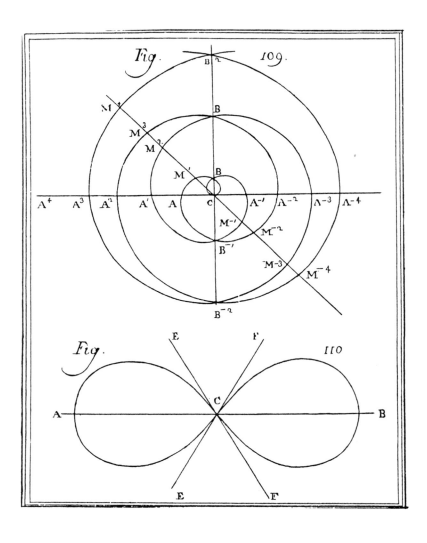

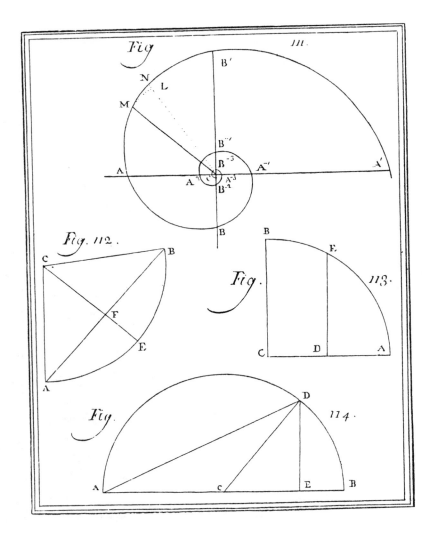

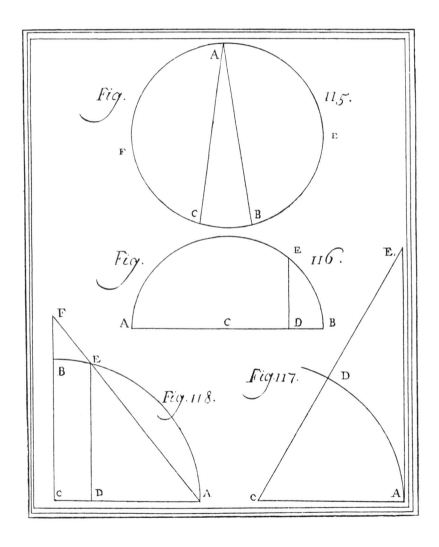

Fig. 115.

Fig. 116.

Fig. 117.

Fig. 118.

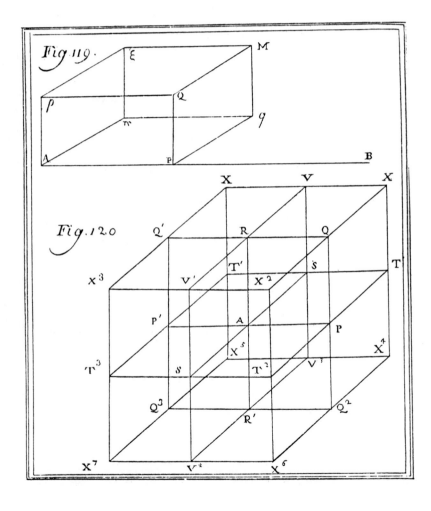

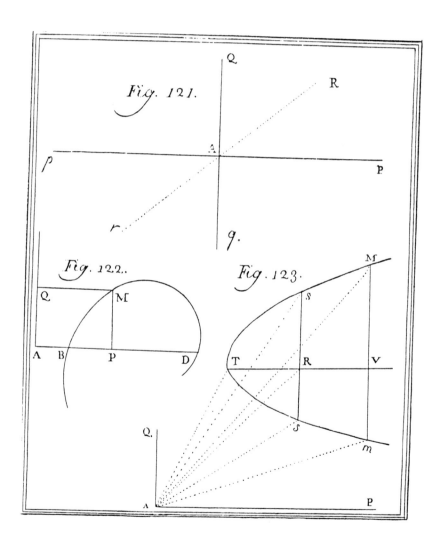

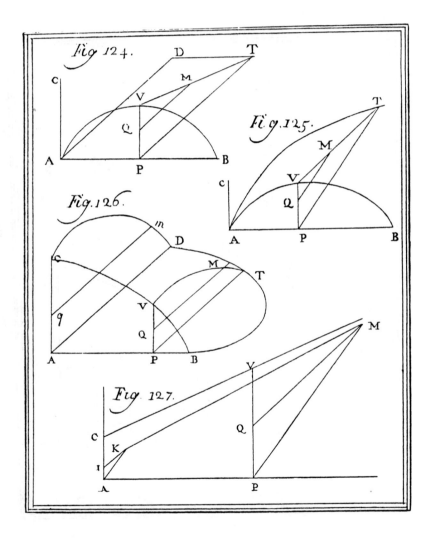

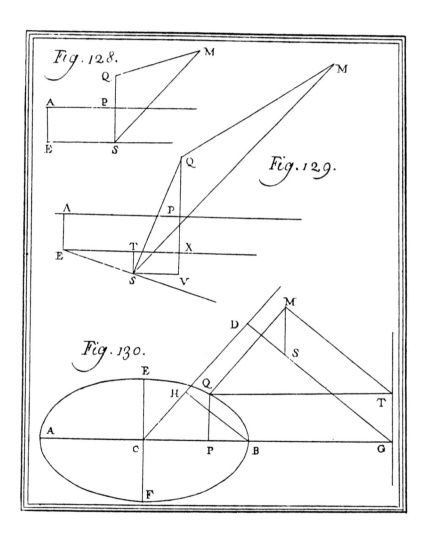

Fig. 128.

Fig. 129.

Fig. 130.

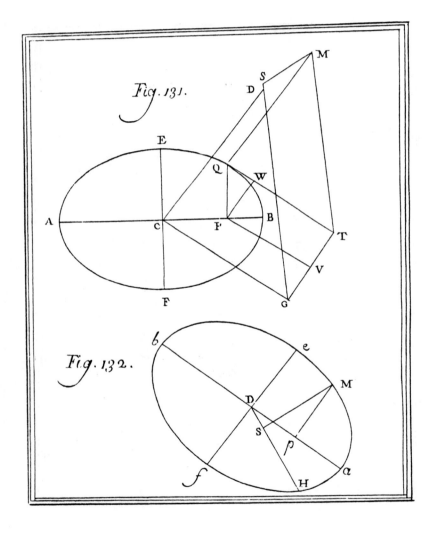

Fig. 131.

Fig. 132.

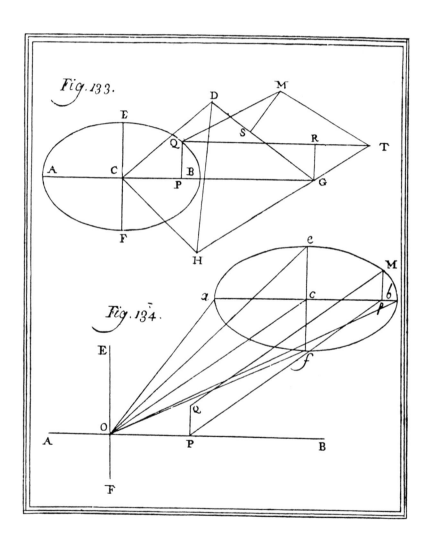

Fig. 133.

Fig. 134.

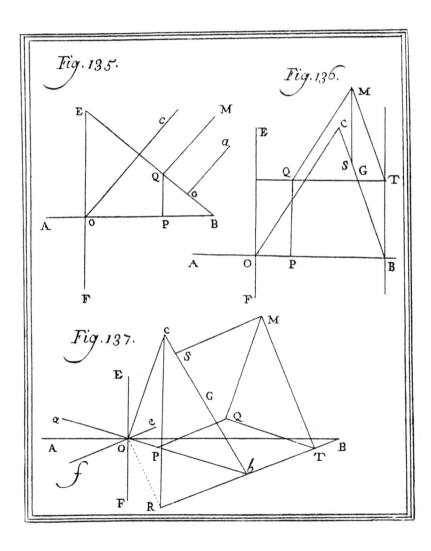

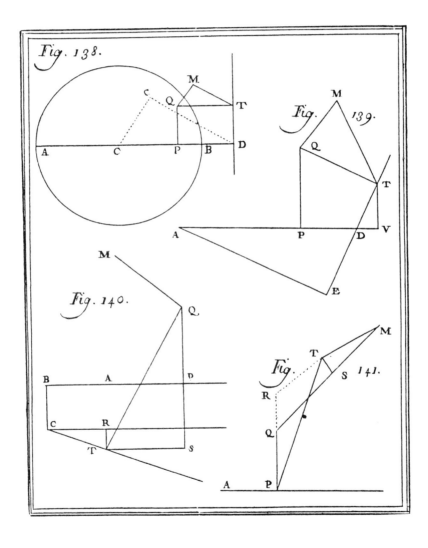

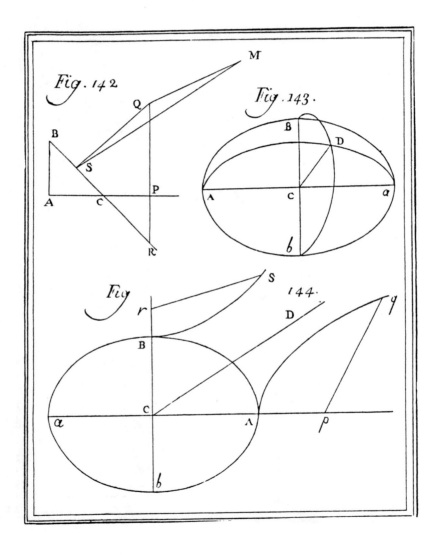

Fig. 142

Fig. 143.

Fig. 144.

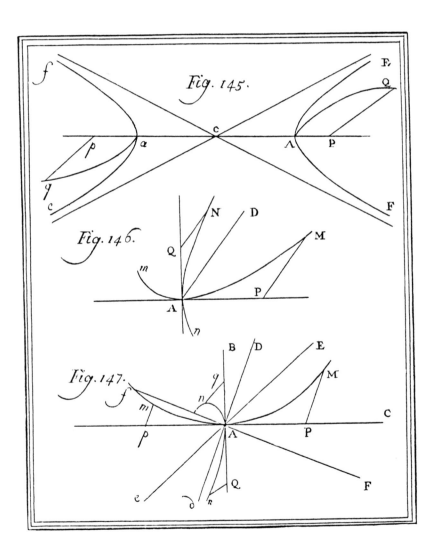

Fig. 145.

Fig. 146.

Fig. 147.

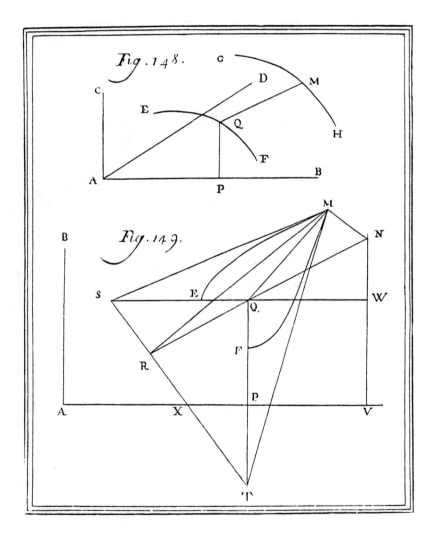

Fig. 148.

Fig. 149.